Dedicated to

My son Subir

Daughter-in-law Sangeeta

Son-in-law Alok

and

Daughter Mita

Nabhi's

Practical Management of Civil Engineering Projects

Incorporating

★ Basic Principles of Management ★ Materials Management
★ Management of Various Types of Construction Works and Practical Problems ★ Management Tools
★ Critical Path Method ★ Resource Management
★ Construction Equipment Management
★ Financial & Accounts Management
★ Total Quality Management
★ Environment Management related to Civil Engineering Works

Useful for

★ Civil Engineers ★ Architects
★ Electrical & Mechanical Engineers
★ Consulting Engineers ★ Contractors
★ Suppliers ★ Finance & Accounts Deptt.
★ Personnel/Administration Deptt. ★ Environment Deptt.
★ Top Management ★ Students of Project Management

by

Er. S.C. Basu Roy

Former Chief Engineer (Civil)
N.D.M.C.

A NABHI PUBLICATION

Nabhi Publications – Why and How

Founded in 1967 by Nabhi Kumar Jain after long many years of service as an understudy of his late father Shri Shanti Prasad Jain, whose sterling contribution to the growth of the book trade in India is universally acknowledged.

The genesis of the publications' success, after a modest beginning, is traced to its founder's yearning for and painstaking efforts to maintain uniformly high standard of quality of all its products with emphasis on simplicity and affordability.

From the stage of birth of an idea to its transformation into a final product the work is overseen by an editorial board manned by highly qualified professionals under the leadership of Shri Nabhi Kumar Jain and its dynamic managing editor, Shri Ajay Kumar Garg.

Nabhi Publications reaffirms its commitment to see the lamp of knowledge illumine every home.

Published by
Ajay Kumar Garg
Managing Editor
NABHI PUBLICATIONS
P.O. Box No. 37
New Delhi -110 001

Editing, Composing & Processing by
NABHI'S BOARD OF EDITORS

Visit us at : www.nabhipublications.com
For complete Bibliography of our Publications and on-line ordering

1st Published, 2009

Price : Rs. 450

ISBN 81-7274-680-6

Caution Note

The authors, editors and publisher have tried their best to incorporate the latest available information in this book. This book is sold with the understanding that authors, editors, publishers and resellers are neither responsible for the result of any action taken on the basis of this work nor for any error or omission, to any person, whether a purchaser of this publication or not. Errors and omissions may kindly be brought to the notice of the Publisher. Any clarification may be supplied, if deemed fit by the Board of Editors.

For binding/manufacturing mistakes and missing pages publisher's responsibility is limited to replacement within one month of purchase by similar edition or the next edition if published and available. All expenses in this connection are to be borne by the purchaser.

All disputes are subject to Delhi Jurisdiction.

Printed by :
Salasar Imaging Systems
Delhi

Rakesh Mishra
Chief Engineer, CPWD

Foreword

I met Shri S.C Basu Roy for the first time in 1992 when I joined NDMC on deputation and was placed under him as S.E.(Projects), and was instantly impressed by his dynamism, intelligence and sincerity of purpose combined with modesty and simplicity. It's a rare combination. I found him a great teacher who was always ready and willing to learn. He had already acquired immense experience of project management of varied nature, yet I found him always ready to learn new techniques of execution as well as management of projects. This is the quality that enabled him to carry along with him his subordinates as well as senior officers with ease. It was a great experience working with him on various projects of NDMC and sharing our experiences even later.

Considering his eagerness to learn and share his knowledge and experiences, it is but natural that he has undertaken the task of putting down in words the gist of his vast experience in the form of books on project management. This is not an easy job, but his expertise in writing is well known. His notes, whether on files or in recording discussions/ minutes of the meeting were always elaborate yet lucid, compact and well structured. He has already brought out two books on practical problems faced by field engineers in project management. This is third in the series. Language used to explain various practical aspects is simple. Use of charts and diagrams makes it easy to comprehend for engineers.

This book covers wide range of topics of project management. It is heartening to see an entire chapter devoted to " Management during pre-construction stage", a subject most important but mostly ignored and neglected in our zeal to meet unrealistic targets fixed generally by people having little knowledge of intricacies of a project. Chapters on project monitoring using CPM/PERT charts for material/resource management and Total Quality Management are surely helpful in understanding/ refreshing the subject. Inclusion of topics of Environmental Management and management of highway projects is an added bonus. A chapter on safety aspects is also of immense help, this being a very sensitive and important subject, often lost sight of for which many engineers have to pay heavily.

In bringing out this book and others preceding it, Shri Basu Roy has done a great service to the engineer's community. I hope they will all benefit by it. This book covering a large spectrum of topics on project management in a compact and concise, scientific manner will go a long way in helping project managers in tackling day to day problems.

Dated : 3rd January 2009 — RAKESH MISRA

New Delhi — Chief Engineer, CPWD

Preface

Besides, technical matters involved in finalizing engineering details of various disciplines (viz, civil, electrical, mechanical, electronics engineering, architectural, horticulture / landscaping etc.), there are large numbers of activities in planning and execution of works. For these reasons there are several complexities in construction, operations and maintenance of projects. For smooth progress / completion of a project / work in time, proper "management" is utmost essential. Effectiveness of management depends on use of appropriate management tools and methods. General management is required for certain routine works. Practical experience of engineers in management for relevant types of works is also very important. As example co-ordination amongst various disciplines (viz. engineering, architectural, horticulture, personnel management, financial & accountancy, administration, materials management, total quality management, safety aspects etc.) is essential.

The basic principles of management are universally applicable in all works, including civil engineering projects. It is important for engineers to have a fair knowledge about application of the management principles, besides practical knowledge in relevant engineering fields. There should be one overall incharge of a project / work, generally known as Project Manager/ Chief Project Manager, who supervises and regulates various components of the project including non-technical activities directly related to the project.

Theoretical knowledge of management or having a management degree may not be enough. In practical, number of problems on management in engineering construction / operation / maintenance are faced. Engineers should have knowledge of these practical / ground management problems. Many projects fail only due to unsatisfactory management.

Realising the importance of management in civil engineering works and in the absence of a book dealing with the practical aspects of management of a construction project, I was prompted to write a book on the subject. It gives immense pleasure to release this first edition of Nabhi's PRACTICAL MANAGEMENT OF CIVIL ENGINEERING PROJECTS, 2009.

Various practical aspects of management in Civil Engineering project / works have been dealt in this book. **Detailed scope** of the book has been stated in Chapter-1. The book has been divided into following three parts:-

Part-I : Basic Principles of Management, Materials Management (Basic) have been given in Chapters 1, 2 and 3.

Part-II : Practical management of various Civil engineering projects (conventional and non-conventional) including related matters like, architectural inputs / design & drawings, total quality management including quality of architectural & structural designs & drawings, financial & accountancy matters, safety matters electrical, mechanical, Air Conditioning, Fire Safety measures, technology selection, use of various management tools at different stages of projects, cost and time control, construction equipment / machinery management etc. have been dealt in Chapters 4 to 18.

Part-III : It deals with environment management related to civil engineering works / projects. The Chapter 19 enumerates certain basic requirements of environment management related to civil engineering projects / works. Chapter 20 relates to water supply management. Chapter 21 relates to management of sewerage and storm water drainage systems and Chapter 22 relates to solid wastes management. Lot of importance has been given by various national and international authorities to environmental matters with specific standards related to various types of projects.

The book is expected to be of immesne practical use for Civil Engineers, Architects, Electrical & Mechanical Engineers, Consulting Engineers, Contractors, Suppliers, Finance & Accounts Deptt., Personnel/Administration Deptt., Environment Deptt., Top Management of the Organisation and Students of various courses on Project Management.

The author is thankful to the publisher, M/s Nabhi Publications for their painstaking editing and composing of the book. The author is also thankful to Shri Bhaskar Basu Roy for his sincere help in bringing out this book smoothly.

Though every care has been taken to incorporate authentic information yet some errors might have crept in inadvertently. Kindly bring such errors to the notice of publisher notwithstanding that the publisher is not legally or morally responsible for any error or omission. Suggestions for improvement of future edition are, however, welcome, which may kindly be sent in writing only.

2nd January, 2009

C - 388, Triveni Apartment
Sheikh Sarai, Phase-I
New Delhi - 110017
Phone: (011) 26014920

Er. S. C. Basu Roy
Author
Former Chief Engineer (Civil)
New Delhi Municipal Council

Contents

PART-III

Source of Further Information

Nabhi Jain has been in the service of book lovers for the last four decades, publishing books on a variety of subjects from the point of view of a common man's requirement, under the banner of Nabhi Publications. Nabhi's books are widely known for their simplicity, authenticity, comprehensiveness and reader-friendliness.

Nabhi has now launched a new enterprise in the service of readers, a grand showroom of books in the heart of Delhi, under the name and style of :

JAIN BOOK DEPOT
C-4, Opp. PVR Plaza, Connaught Place,
New Delhi-110001
Ph. : 011-23416101/02/03, 66307233
Fax : 011-23731117, 23416103
E-mail : sales@jainbookdepot.com
website : www.jainbookdepot.com

Here, readers can access complete range of books on construction, civil engineering, sewerage, road work, electrical, maintenance, horticulture, IS codes, specifications, analysis of rates, schedules of rates, etc. alongwith useful books by Nabhi Publications.

Besides Government publications, the showroom offers a wide variety of books on virtually all subjects viz. Law, Career, Management, Civil Engineering, Career, Competition, Fiction, Fashion, Interior Design, Education, Religion, Children, Travel, Hindi books, Language books, Dictionaries, Directories, Maps & Abstracts. Also there is an exclusive display of Nabhi's books.

Readers can comfortably sit and patiently see the books of their interest, and evaluate themselves which book is better for them. The staff deployed at the showroom is well equipped and offers all necessary help in locating any particular book the reader desires.

The showroom is definitely worth a visit albeit, you will love to come over and again.

NABHI'S OTHER BOOKS

TAXATION

- **Income tax Guidelines & Mini Ready Reckoner** with Tax Planning 2008-09, 2009-10 — 190
- आयकर गाइडलाइन & **Mini Ready Reckoner** साथ में टैक्स प्लानिंग 2008-09, 2009-10 — 190
- **TDS Ready Reckoner**, 2008-09 — 300
- Formation & Management of a **Society**, 2008-09 — 170
- Formation and Management of a **Trust** alongwith Tax Planning, 2008-09 — 350
- How to **Maintain Business & Profession Accounts**, 2009 — 190

COMPANY LAW

- Formation and Management of a **Private Company** alongwith Practical Procedure, 2008 — 310

INDUSTRY, INVESTMENT & FOREIGN EXCHANGE

- Handbook for **NRI**, July 2008 — 490
- **Labour Laws** : One Should Know, 2008 — 110
- **Business Laws** : One Should Know, 2008 — 130
- **Foreign Exchange Manual**, 2006 — 1180

NGOs

- **Handbook for NGOs**, (in 2 Vols.), 2008 (with changes as per Finance Bill, 2008) — 880
- **एन. जी. ओ. हैन्डबुक,** 2008 — 390
- Directory of **Top NGOs in India**, 2008 — 490
- Directory of **Funding and Other Helpful Agencies for NGOs**, 2007 — 290

LAWS FOR DELHI

- **Guide to House Tax in Delhi**, 2003-04, March, 2004 (Reprint 2009) — 350
- **Delhi Shops & Establishments Act & Rules** alongwith Allied Labour Laws, 2008 — 150
- **Delhi Building Bye Laws** 1983 & Building Development Control Regulations as per Master Plan-2021, 2008 — 340
- Formation & Management of a **Co-operative Society** in Delhi, 2007 — 340
- **Delhi VAT** : One Should Know, April, 2005 — 130
- Guide to **New Unit Area System of House** Tax in Delhi 2004-05, October 2004 — 140

GENERAL

- **Cyber Crimes**, Electronic Evidence & Investigation – Legal Issues, 2008 by Vivek Sood — 390
- **Right to Information** : One Should Know, 2008 — 130
- Handbook of **Visa Information**, 2007 — 895
- **Legal Drafting** for Layman, Reprint 2009 — 150
- **How to Make a Will** : Law, Procedure & Enforcement, Reprint 2008 — 90

PART I

★ Scope of the Book

★ Basic Principles of Management

★ Materials Management (Basic)

1

Scope of the Book

1.1 Introduction

Title of the book is "Practical Aspects of Management for Civil Engineering Works".

It is required to know about –

(i) What is Project ?

(ii) What and why Project Management ?

(iii) Practical aspects of Management

(iv) What types of Projects/works will be dealt in this book?

There are wide varieties/types of projects like, -

Projects related to manufacturing of various goods, materials (textiles, steel, copper, cement, wires/cables etc.), agriculture related projects, various research projects, health schemes. Such types of projects will not be dealt in this book.

Management for constructions of buildings, roads, bridges, airports, public health engineering projects, municipal solid waste project, irrigation project, electric power project (Thermal/Hydro-electric), railway project, township project, factory/industrial project etc. will be dealt in this book.

Management skillness is also required for carrying out various Civil Engineering smaller works like, construction and maintenance/operation works in buildings, roads, public health engineering works, irrigation works, small bridges, horticulture works, manufacturing of RCC Hume pipes, pre-cast items, ready-mix concrete, hot mix plants, brick manufacturing plant, crusher plant, large survey works, consultancy services etc. The management for such works will also be dealt in this book.

1.2 Main Factors of Management Process

Materials management (including purchasing) is an important part of management sector. Similarly manpower management, finance management, office management, etc. are also parts of main management. These will be dealt in this book. Following matters of management will also be dealt.

(i) Organised activities

(ii) Objectives

(iii) Relationship among resources

(iv) Working through others

(v) Decision

It is required to know about type of project. Course/methods of management depend on the type of project. Certain basic principle and universality of management are relevant for all types of management. The same is applicable for various management and various types of projects. This will be dealt briefly in next chapter.

1.3 Management of Civil Engineering Project

Detailed management for civil engineering works will be dealt in this book. There are many types of Civil Engineering projects. Basic principles of management are same for various types of Civil Engineering works. However, working out of details for project management are different for different types of projects (viz., building construction project, road construction project, public health engineering works (sewerage, water supply, drainage and solid waste management), railway projects, irrigation project, etc.)

1.4 Components of Management

Management will be dealt in following two parts –

(a) Overall planning and management of various milestones of entire project i.e. in respect of various disciplines and components of the project and dealt from conceptual/ preliminary stage in the organization of owner viz., Government Department/Organisation, private organisations like private/ societies/ associations/Trustees etc. Planning and management of various components/ disciplines (pre-construction, construction and post-construction stages/ period) are done in the organization of the owner (Government/semi-Government/private organizations).

(b) In addition of above stated planning and management process it is also required to carry out detailed management planning/activities for construction stage of different components/disciplines of the project by the contractor/ construction agencies. Detailed activities are to be planned/ processed for effective and efficient management. This part of management (i.e. project construction) is voluminous, difficult and important for smooth progress of construction activities. Monitoring of construction activities is carried out by the owner (Government/semi-Government/private organization) and contractors (construction agencies).

1.5 Basic Principles and Practical Aspects of Management

There are two aspects of management viz., -

(i) Basic principles/universality of management, and

(ii) Practical aspects of management. This aspect is also important particularly for construction projects on ground or in reality. Practical problems are faced on ground for implementing management principles and various management techniques are worked out for construction stage of the project.

1.6 Broadly following are the scope of this book :-

(i) Basic/Universality of management principles

(ii) Project Management for the projects related to Civil Engineering works including works related to environment, material management, etc.

(iii) Scientific management principle

(iv) Management for stages of preliminary/conceptual stage (planning/pre-construction), construction stage and post-construction period. Working out of detailed management techniques in respect of above types of project will be dealt.

(v) As stated above, more stress has been given and details will be dealt in this book about practical aspects of management particularly. Special stress has been given on pre-construction stage of project management.

1.7 Contract Management

It may be stated here that Contract Management is also a part of project management. This part is a vast subject (including arbitration management, certain laws related to building construction works). Hence Contract Management will be dealt in another book.

1.8 Total Quality Management

Total Quality Management of construction project will also be dealt in this book briefly. Labour Management/Personnel Management and construction machinery/ equipment management are also integral parts of project construction management and will be dealt.

It may be stated here that above stated management principles and techniques are also applicable with suitable modifications in respect of non-Civil Engineering projects.

1.9 Environmental Management

Certain parts of Environmental Management will be dealt in this book particularly for projects related to Public Health Engineering works (sewerage, drainage, water supply, Municipal Solid Waste Management).

1.10 Materials Management

Materials Management is an important part of management related to Civil Engineering works/projects. From cost point of view materials are major part (50% to 60% of project cost). This will be dealt in this book. Basic principles of material management has been dealt under Chapter 3. Material Management for certain types of projects have been dealt with respective project management.

It will be possible to co-relate the basic principles of management and material management (dealt under Chapters 2 & 3) with practical aspects of management (dealt under rest of the chapters). The basic principles are to be applied during actual management on ground.

2

Basic Principles of Management

2.1 Introduction

Managerial functions are based on certain principles. Management has been defined as a science. Management principles are universal. Management principles are required for increasing efficiency and to crystallize the nature of management. It takes the inputs from the society and gives the output to the society. Understanding of management principles enables managers to take a more realistic view of organizational problems and their solution. There are different views :-

(i) Management and administration are one.
(ii) Administration is above management.
(iii) Administration is the part of management.

The basic objective of management functions and techniques is to make one an effective manager. Often a confusion is created between two terms :- effectiveness and efficiency. The term "efficiency" is used in engineering way and it refers to the relationship between input and output. Generally, this is also taken as effectiveness.

2.2 Principles of Management

There are different authorities/schools of thought on management principles.

The real father of modern operational management theory is Henry Fayol of France. His first famous book was published in 1916 in French language. However, the book in English language was available for general readers in the world in 1949. He had given the fourteen principles of management as under :-

(1) Division of Work
(2) Authority and Responsibility
(3) Discipline
(4) Unity of Command
(5) Unity of Direction
(6) Subordination of Individual to General Interest
(7) Remuneration of Personnel

(8) Centralisation
(9) Chain of authority and communication ranging the highest to the lowest positions. It can be short-circuited only in special circumstances
(10) Order
(11) Equity
(12) Stability of Tenure
(13) Initiative
(14) Union is strength for establishing team work

Above stated principles are flexible and capable of adoption to every need. It is a matter of knowing how to make use of them, which is a "difficult art" requiring intelligence, experience and "proportion".

2.3 Management Process

The managerial functions may be grouped as planning, organizing, staffing, directing and controlling. Co-ordination is not actually a separate function; it should be treated as essence of management. Above position may be seen as a circular continuous movement as shown in Fig. No. 2.1 :

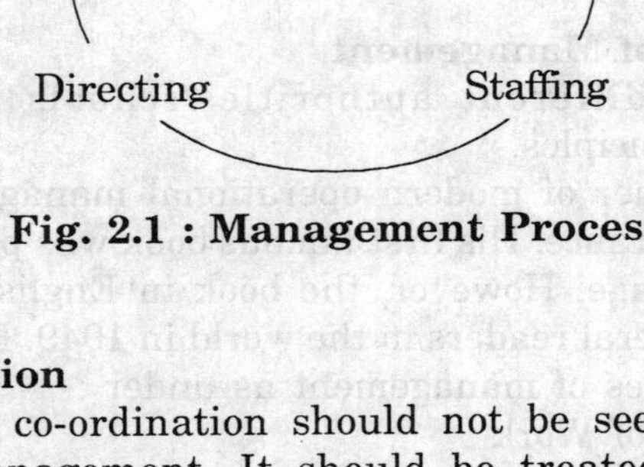

Fig. 2.1 : Management Process

2.4 Co-ordination

Problem of co-ordination should not be seen as a separate function of management. It should be treated as "essence of management".

Co-ordination is not to be confused with co-operation. The co-ordination is related with the synchronization of efforts. For a given situation, it is possible for co-operation to exist without co-ordination.

There are four fundamental principles of co-ordination :

(1) Principles of Direct Contact – Communication in oral form through personal contact is most effective to convey and understanding ideas, suggestions, views, feelings, etc.

(2) Early Beginning - Co-ordination should be started from the stage of planning and policy-making.

(3) Continuity – Co-ordination is a continuous process.

(4) Reciprocal Relation – Integration of all efforts, actions and interests is required.

2.4.1 Techniques of Effective Co-ordination

Following are the "specific techniques" for achieving co-ordination:

(1) Co-ordination by chain of command

(2) Co-ordination by Leadership – Leadership brings individual motivation and persuades the group

(3) Co-ordination by effective communication – Through communication, every person understands his scope and limits of functioning, authority and responsibility and relationship with others. Communication to be effective does not require only a communication network but to keep the network free from any "barrier" which affects flow of messages adversely.

(4) Coordination by Committees

(5) Co-ordination by Control/General Staff

(6) Special Coordinator

2.5 Effective Supervision

(1) By Leadership

(2) Closeness of Supervision – Successful supervisors follow the style of less close supervision. A closer supervision is defined as frequently checking up on subordinates, providing them frequent and detailed instructions; thus limiting their freedom to perform the work in their own way.

(3) Employee-orientation/Human relations

(4) Group Cohesiveness – All members/staff work together for a common goal

(5) Delegation – It becomes important at the supervisory level because supervisory management is the last level for delegation of authority

(6) Other factors

2.6 Planning

Planning is the most important stage of project. Successful completion of project (economically and timely) depends on good and effective planning. Broadly there are three stages of a project:

(i) Pre-construction stage :
   (a) Preliminary/Conceptual Planning
   (b) Detailed Planning (including detailed design) and tendering
(ii) Construction stage
(iii) Post-Construction stage (operation/production/maintenance)

Before going through actual planning, one must be aware that opportunities exist for its performance. At initial stages broad requirements of users/user department should be available. Planning is for future implementation and the same depends on forecast. Good planning also depends on "quality of forecast". In reality it has been found that various authorities (including seniors) do not give necessary importance/priority to "planning process".

2.6.1 Steps in Planning

(1) Perception of Opportunities : It is related with the awareness of the opportunities for deciding whether a programme/ project should at all be carried on. It is a preceding process of actual planning. It is the real starting point of planning.
(2) Establishing Objectives: Next step is the determination of organization objectives for the proposed programme/project. The objectives should be clearly specified and measurable as far as possible. Final product will be compared with this objective.
(3) Establishing Planning premises/factors: Factors are like political factors, ethical standards, government control, fiscal policy, price, demand and availability of various factors of production, etc.
(4) Determining Alternative Course of Action: The management should try to find out these alternatives. Various alternatives are to be examined in the light of planning premises.
(5) Evaluation Alternative Courses: This process presents a difficult problem, because a particular alternative may be best from one point of view but not from other points.
(6) Selecting the Best Course
(7) Formulating Derivative Plans: After formulating the basic plan, various plans are "derived" for various departments, units, activities/disciplines (works) etc. Multi-disciplined large

projects are dealt by various departments/units and derivative plans are prepared for all concerned departments/ units on the basis of basic plan.

(8) Establishing the Sequence of Activities: After formulating basic and derivative plans, the sequence of various activities is determined by adopting suitable techniques.

2.6.2 Making Planning Effective

Following are some factors for making planning effective :

(1) Establishing a Climate for Planning : Every superiors should remove obstacles to planning and arrange facilities/incentives for planning.

(2) *Initiative at Top Level* : Planning to be effective must have the initiative and support of top level management. It is the top level which is responsible for success or failure of any organizational process, and planning is no exception. In reality/on ground it is seen that very less period is given for planning process; proper/effective planning cannot be done in very short period without adequate data/information.

(3) *Participation in Planning* : The best planning is likely to be done when managers/executives are given an opportunity to contribute to plans affecting the areas over which they have authority. User department/users' representatives are required to contribute to the planning process. A method of planning participation may be in the form of "grass roots" budgeting in which budgets/proposals/requirements are prepared by every small units of the organization and then move upward.

(4) *Communication of Planning Elements* : Many planning efforts fail because managers/executives do not adequately emphasise the role of communicating various planning elements. If these are communicated clearly, adequately and timely, the concerned persons are motivated and initiated to take planning process.

(5) *Integration of Long-term and Short-term Plans* : A short-term plan contributes towards the achievement of the long term plan.

2.6.3 Problems faced during Planning Process

The first basic limitation of planning is the lack of accurate information and facts relating to future. Planning concerns future activity and its quality will depend on quality of forecast of future events. Adequate data, information are not available. Problems are faced due to changes made at different stages. The problem of change is more complex in long-range planning. There are failures in planning due to lack of commitment to planning, failure to develop sound strategies, tendency to overlook planning premises, failure to see planning as a rational approach. Sufficient flexibility is not given to planning persons. Rigidity may make managerial work more difficult than it need be. This may result in its delay in work performance, lack of initiative. Planning unit should take into account both time and cost factors. Excessive time spent on securing information and trying to fit all of it into a compact plan may not be suitable in the organisation.

2.7 Objectives

The organizational objectives are achieved through individuals. Each individual must know in advance what he is expected to achieve. The superior can set his subordinates' objectives with their consultation and agreement. When the goals/tasks are carefully set in a network of verifiable measures, it is also required to indicate the resource requirement. People set goals for themselves which create the feeling of commitment and people involved are more likely to meet their targets.

Above stated method of management is known as "Management by Objectives (MBO)". Actionable objectives cannot be established without planning and result-oriented planning is the only kind that makes sense. MBO elicits commitment performance.

2.8 Forecasting

Planning is done for future. Forecasting provide some knowledge for planning. Likely future events are prepared on the basis of forecasting. There is a close relationship between planning and forecasting because forecasting is an essential ingredient of planning. Forecasting merely involves the estimate of future events and provides parameter to planning Forecasts are not always true. Hence, adequate safeguards should be provided in the planning.

Following factors affect the forecasting/planning process :–

Political stability, Government control, Government Fiscal Policy, population trends, employment, productivity, price levels, technological innovations, etc.

It also depends on many factors of resources :– land, labour, capital, management. It depends on natural resources, raw materials and supplies, plant and equipments, financial market. As future cannot be foreseen with accuracy, the managers formulate alternative plans based on alternative premises.

2.9 Decision

Decision taking is one most important part of management. There is process for taking decision on various issues/problems. It has been observed that project suffers at various stages due to late or wrong decision. Sometimes delayed decision is more harmful than taking a decision quickly even with some error.

Following are types of decisions :

(1) Organisational and personal decisions – An executive of organization takes decision.
(2) Routine and Strategic Decision
(3) Policy and Operative Decision
(4) Programmed and Non-programmed decision
(5) Individual and Group Decisions – Group decisions are taken by a group constituted for a specific purpose or by a Standing Committee. Group decisions have certain positive values due to greater participation of individuals and quality in decisions. There are certain negative values such as delay in decision-making process and difficulty in fixing the responsibility of decisions.

2.9.1 Decision making process

The problem raises the necessity of decision and forces an individual to find out alternatives for solving the problem and other decision-making steps. Following are steps in decision making processes :

(1) Problems - The first step is to determine what are the real and correct problems ? Then it is required to know the input data for processing. There are two types-
   (i) Diagnosis
   (ii) Analysis

   A problem is half solved when it is well defined. Analysis of the problem requires to find out who would take decision – individual or group; what information would be required and from where these would be available.

   There are two types of information –
   (i) Primary information (by carrying out survey, opinion poll or certain field works, etc.)
   (ii) Secondary information (which are collected from various sources, existing/old records)

(2) Search for alternatives – The practice of developing alternatives is the best guarantee for ensuring adequate attention on the part of managers. It also avoids bottlenecks in operation as alternatives are available if a decision goes wrong.
(3) Evaluation of alternatives
(4) Comparison of alternatives
(5) Selection of alternatives (considering experience/ experimentation/ research)
(6) Putting decision into action – This requires the communication of decision to subordinates, getting acceptance of the decision and getting their support and cooperation for converting the decision into effective action.
(7) Following up decisions

As safeguard against incorrect/defective decision, the managers are required to institute a system of following up to the decision so as to modify them at the earliest opportune time. Decision making process has been shown on a flow diagram vide figure 2.2.

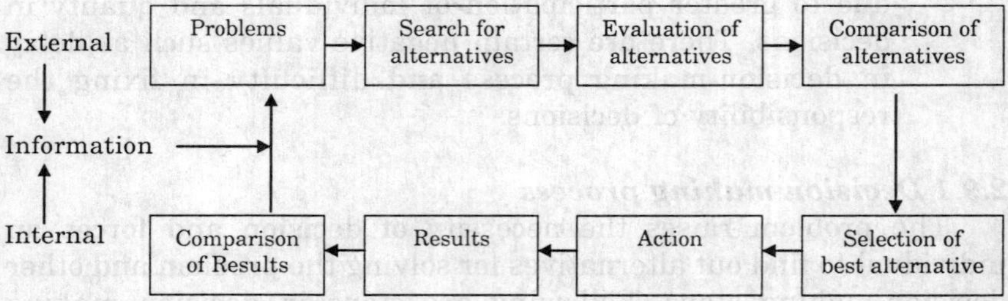

Fig. 2.2 : The Decision Making Process

2.9.2 Human Factor in Decision Making

A major source of limit on rationality is human factors in decision making. The decision-making is influenced directly and indirectly by many human factors like personal value systems, social and economic factors, limitations in human processing.

2.9.3 Free Association Techniques

Brainstorming – A problem is posed and ideas are invited. Quantity rather quality is the primary objective in generating ideas. Ideas are encouraged and criticism of any idea is discouraged. "Chain reactions from idea to idea develop other ideas".

Degree of certainty in the decision making :

(i) Certainty

(ii) Risk

(iii) Uncertainty

Under category of "risk" it has been observed that most of organizational decisions are made under the conditions of risk i.e. some information is available but the same are not sufficient to answer all questions about the outcome of the decision. However, Manager has sufficient information for estimating the probability of outcomes. There are different methods for estimating the probability under condition of risk.

2.10 Organising

"Organising" is important matter related to project management. This covers several aspects like, designing organization structure, concepts of authority and responsibility, delegation of authority, decentralization, line and staff, etc.

Like the Engineer who designs a bridge to meet special needs, the organization planner, in designing the organization structure, applies certain principles in order to meet the various needs. Following are some points/factors :

(1) ***Clear lines of authority*** – Clarity is achieved through delegation by steps or levels from leader to the working level, from the highest manager to the employee who has last responsibility in the organization.

(2) ***Adequate delegation of authority*** – A common problem in the organizational life is that managers often fail to delegate adequately and suffer with various problems like, decision-bottleneck, delay in decision implementation, more pressure on higher level, less motivation to subordinates.

(3) ***Application of ultimate responsibility*** – Responsibility of higher level for the acts of its subordinates is absolute. This is the concept of ultimate responsibility.

(4) ***Minimum Managerial Levels*** – As far as possible, the number of managerial levels should be kept to the minimum. The greater the number of levels, the longer is the chain command.

(5) ***Unity of Direction***

(6) ***Proper emphasis on staff activities*** – A "line activity" (lowest level/ workers) is that which serves the organizational objectives "directly". The contribution of staff (viz., managerial level with authority) activity is "indirect". Here

"Staff" does not mean as general/conventional meaning i.e. subordinate persons worked under senior manager/officer. This will be dealt in detail.

(7) Span of Control

(8) ***Provision of top management*** – There should be proper link between Board of Members, shareholder and the regular management team.

(9) Simplicity

(10) Flexibility

2.10.1 Organisation Charts

There are some advantages for preparing organization charts as under :

(1) Management Tool – Organisation chart serves as a tool by providing the broad picture of authority and its relationships.

(2) Avoiding overlapping and duplication – The chart defines and prescribes responsibility and authority. It indicates functions and their inter-relationships depicting whole result of organizing exercises.

(3) Solving organizational conflicts

(4) Training guides

(5) Easy dealing with outsiders

2.10.2 Span of Management

The grouping of activities to create various departments presents another problem i.e. how many individuals should be placed under one superior. This is the problem of deciding the span of management. This is also called as supervision or control.

Factors for determining Span of Management

(1) ***Capacity of Superior***

(2) ***Capacity of subordinates*** – Efficient and trained subordinates may discharge their functions efficiently without much help of their superiors.

(3) ***Nature of work***

(4) ***Degree of decentralization*** – Higher the degree of decentralization, wider is the span of management. If a Manager clearly delegates his authority, defines it fully, he requires less time to devote to manage his subordinates.

(5) ***Planning*** – If subordinates' jobs are well defined through proper planning process and they know what is expected from them, they are able to perform their task better without consulting superior.

(6) ***Use of staff assistance***

(7) ***Communication Techniques.***

A flow diagram showing levels of Management is given vide figure 2.3.

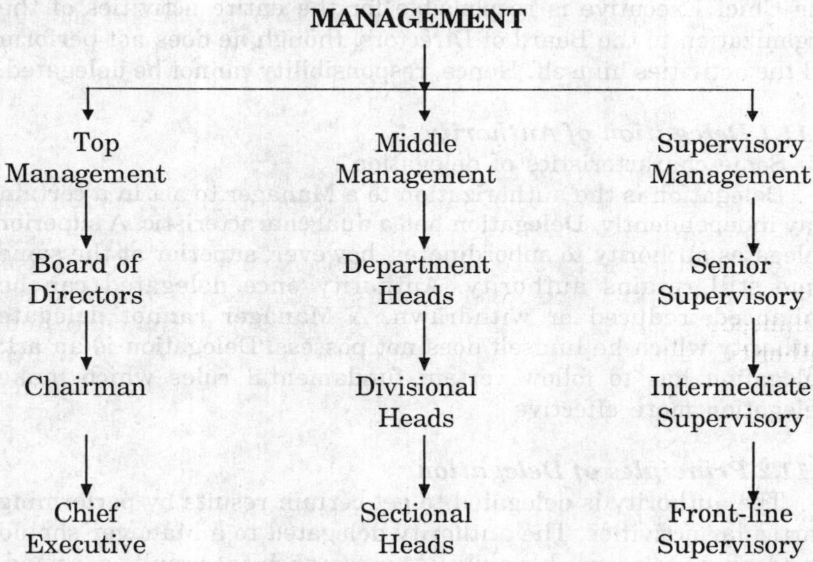

Fig. 2.3 : Levels of Management

2.11 Authority and Responsibility

Authority is the legitimate right to give orders and get orders obeyed. The right is given to a Manager by his superior. The basic objective behind the use of authority is to influence the behaviour of the subordinates in terms of doing right things at right time so that organizational objectives are achieved. Power is used in terms of ability or capacity to do something or to get intended result. A person can have influence on others' behaviour having authority and this is done through power.

Responsibility is an obligation of an individual to perform activities or duties which are assigned to him. Responsibility comes into existence when a person "with authority" accepts the obligation to perform the work. Responsibility is something more than mere duty or activity. In some organizations "accountability" means responsibility.

The authority flows from the superior to a subordinate when duties are assigned and responsibility is the obligation simultaneously extracted from the subordinate for accomplishing the duties. The subordinate Manager, in turn, can delegate authority and duty to his subordinate but "he cannot delegate responsibility". In this way the Chief Executive is responsible for the entire activities of the organization to the Board of Directors, though he does not perform all the activities himself. Hence, responsibility cannot be delegated.

2.11.1 Delegation of Authority

Some characteristics of delegation :

Delegation is the authorization to a Manager to act in a certain way independently. Delegation has a dual characteristic. A superior delegates authority to subordinates; however, superior at the same time still retains authority. Authority once delegated can be enhanced, reduced or withdrawn. A Manager cannot delegate authority which he himself does not possess. Delegation is an art; delegation has to follow certain fundamental rules which make delegation more effective.

2.11.2 Principles of Delegation

The authority is delegated to get certain results by performing particular activities. The authority delegated to a Manager should be adequate to assure his ability to accomplish the results expected.

The authority and responsibility of any Manager should be "co-equal". Authority without responsibility lacks an ultimate purpose. Likewise, responsibility without authority to carry on assigned activities is meaningless.

Unity of command means that a subordinate should be responsible to a single superior and he gets authority from his superior.

Responsibility cannot be delegated. Thus no superior can avoid his responsibility for the total activities assigned to him merely by delegating his authority to his subordinates.

2.11.3 Decentralisation

Decentralisation is both a philosophy of management and a technique. Initially size of organization is smaller and centralization principle is adopted. Subsequently when the organization is expanded/ diversified then decentralization principle should be adopted to achieve success/goal. Centralisation refers to the reservation of authority at the top level of the organization and decentralization refers to "systematic delegation of authority" in the organization.

The term "decentralisation" is used to denote the disposal of physical and functional factors besides the consequence of delegation. In physical decentralization, the physical facilities are distributed over large number of places based on geographical location. The functional decentralization refers to disposal of organizational functions into separate and autonomous units (like, production, marketing, finance, personnel, etc.).

Decentralisation is also done by delegation of authority/power. The need for delegation arises because a Manager can discharge his responsibility only when he has authority to direct and control his subordinates. Delegation and decentralization may be looked in terms of process and result, the former being process and later being the result. Decentralisation is not only a "means" or "tool" for managerial action but it is a "management philosophy" also. Thus decentralization is not used merely to give relief to the work burden of the top management but can be used to denote the role of subordinates.

2.11.4 Benefits of Decentralisation

(1) Reducing burden of top executives
(2) It facilitates diversification of the organization
(3) It helps for market innovations – Marketing innovations/users' requirement ensure better customer satisfaction. In decentralized structure, the higher level people get much time for the creativity and innovations.
(4) It motivates Managers
(5) Encouraging development of Managers
(6) It minimizes levels of organization
(7) Better co-ordination of activities is achieved in decentralization

2.11.5 Appropriate Control Technique

In decentralization various units of the organization work independently and are self-sufficient. They are free to choose their courses of action. There should be appropriate "control techniques" to find out that various units are going in the right direction.

2.12 Line and Staff

The words "staff" and "line" are used in management principles on the basis of their functions. In a general/conventional way (particularly in Government Departments) the public (or in the offices) staff word is used in different way. They mean persons like

Jr Clerk, Sr Clerk, Typist, Stenographer, Jr Engineer, Draftsman, Personal Asstt., Technical Asstt., Foreman, Driver, Peon etc. are the "staff" of the office. "Line" is like Lineman of electricity distribution organization, telephone department, etc. But in the subject of principles of management the unit/division (consisting of various persons – lower and upper rank persons) responsible directly for production (including selling/marketing) / Construction fall under category of "line". The other units (viz., advisors, specialists, finance department/unit, legal advisor etc.) consisting of various persons of upper rank fall under category of "staff". It is required to know the definition, their relationship, authority, responsibility etc. It is better to explain by giving examples. The authority for (i) "planning" and (ii) "doing" is separated under line and staff ; here "planning" is staff function and doing (executing) is "line" function. "Line functions" are those related directly with the attainment of the organizational objectives and "staff functions" help the line functions in attaining the objectives. In manufacturing organization whose objective is to produce and sell goods, production and marketing are line functions; other activities such as finance, personnel, legal, research, etc. are staff functions. There may be staff function within line department. As example, in marketing, which is a line department, the marketing research section exists as a staff function.

2.12.1 Line and Staff Conflict

(a) Line Managers have the complaints about staff personnel on the following lines :

- (i) Line Managers tend to be suspicious of staff personnel who advise a course of action but do not bear the direct responsibility for seeing it implemented.
- (ii) Line Managers feels that staff personnel encroach upon their normal authority.
- (iii) Staff people fail to give fully considered, well-balanced and sound advice. They are more academic and less practical. Since they (staff) are not responsible for end results, many times they give new ideas without much practical applicability.
- (iv) Staff personnel tend to impose their superiority on line managers.

(b) Staff personnel also have complaints against the line managers as under :

- (i) Staff people (viz., Planning/Finance Deptt., Research Wing, Specialists, Legal Deptt., etc.) claim that Managers

(Line) do not make proper use of staff advice. Many Specialists feel that they be consulted during the "planning stage" of a project. This enables them to anticipate problems and to recommend precautionary measures.

(ii) Line Managers generally resist new ideas, while the staff people are more anxious about new thinking and innovations.

(iii) Staff people feel that Line Managers do not give enough authority.

(c) Besides above stated conflicts it may be stated here that the elements of demarcation between line and staff authority are rarely clear. Hence, there is a possibility for overlaps and gaps in authority and responsibility which can aggravate personal relationships. If the staff advice is not sought by the Line Management then they (Staff Management) can feel unneeded and indifferent. This mandatory consultation is often called principle of compulsory staff services.

2.12.2 Line and Staff Organisation Structure

A diagram on this matter is given vide Figure 2.4

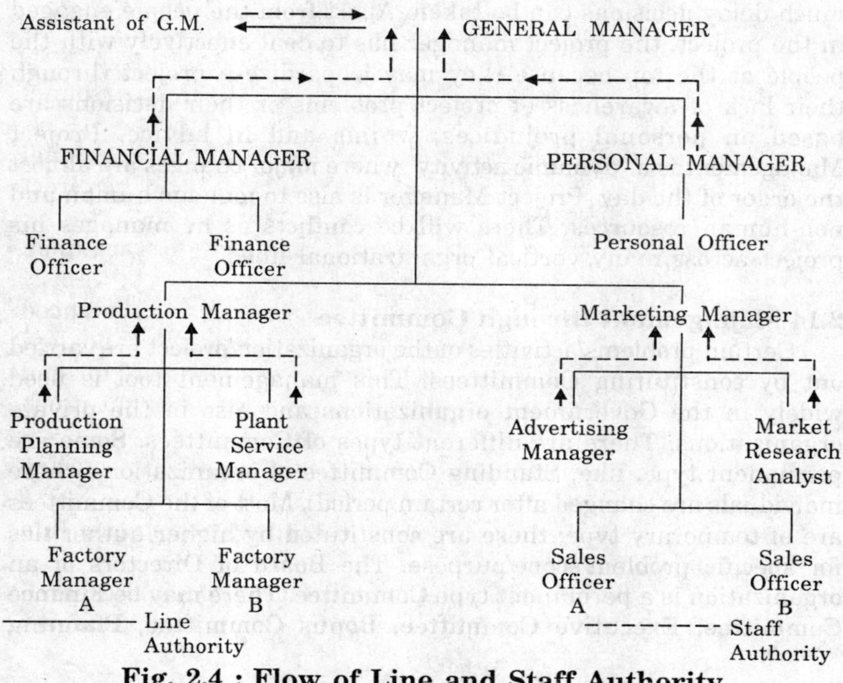

Fig. 2.4 : Flow of Line and Staff Authority

2.13 Project Organisation

The project work can be managed in two forms of organization: (i) pure project organization and (ii) matrix organization. There may be different set up like, project organization, product division, project division. The project organization provides greater flexibility in organization, greater check over the project work, provision of determining exact responsibility and better coordination of organisational resources.

Uncertainties in project organization comes because a Project Manager has to deal with professionals coming in the project from a number of diverse fields; often they have diverse type of approach and interest. Lack of prescribed organizational processes, lack of clearly defined responsibility, lack of clear communication lines and lack of measurement yardsticks for various professionals, – make the job of a project manager often more challenging. There he cannot proceed in traditional way in motivating and controlling them. Apart from managing professionals of different fields, that too without adequate organizationally set procedures, a project manager has to face a very unusual decision pressure that results from the severe penalties to be imposed because of delay in the completion of project. The project manager has to devise a decision process where information can be monitored quickly and without much delay decisions can be taken. Apart from the people engaged in the project, the project manager has to deal effectively with the people at the top because they may jeopardize a project through their lack of awareness of project problems or their decisions are based on personal prejudices, whims and ill advice. Project Management is a "dynamic activity" where major changes are almost the order of the day. Project Manager is also to manage human and non-human resources. There will be conflicts as he manages his project across many vertical organizational lines.

2.14 Management through Committee

Certain problems/activities of the organization/project are carried out by constituting Committees. This management tool is used widely in the Government organizations and also in the private organizations. There are different types of Committees. Some are permanent type, like, Standing Committee of organization (where individuals are changed after certain period). Most of the Committees are of temporary type; these are constituted by higher authorities for specific problem/scope/purpose. The Board of Directors of an organization is a permanent type Committee. There may be Finance Committee, Executive Committee, Bonus Committee, Planning

Committee, Grievance Committee, etc. Committee is a group of persons in an organization for taking decision or for giving recommendation on certain matters. Necessary authority is given to the Committee for going into details of problems and to decide/ recommend on issues. Sometimes, Committee is constituted for deliberating on problems without authority to decide.

There may be Ad-hoc Committee; this is constituted for a specific purpose/to solve a specific problem. Such Committee is dissolved after submitting final recommendation/after solving the problem. Committee may be delegated authority to make decisions. Some Committees have authority to make recommendations to a Manager or other senior authority. Competent authority may accept fully or partially or may not accept the recommendations.

A Committee may be either "line" or "staff" depending upon its authority. If Committee's authority relationship to a superior is "advisory", it is a Staff Committee. Many Committees fall under this category.

Committees may be formal and informal. A formal Committee is constituted by the competent authority as per rules and regulations. A Committee is informal if it is organized without specific delegation of authority and usually by some persons desiring group thinking or group decision on some particular problem.

2.14.1 Advantages of Committee

Under certain circumstances, the Committee has inherent capabilities that enable it to operate more effectively than one person. The Committee has this capability because people in groups react differently from people as individuals. Some advantages are stated below :

(1) ***Pooling knowledge and experience*** – Two heads are better than one. It has been found that results obtained by group judgment are superior to those obtained by individual.

(2) ***Facility of Co-ordination*** – The Committees provide an important "technique" of coordination by bringing together representatives of various Departments/Divisions/Sections.

(3) ***Motivation through participation*** – Due to free participation in a Committee the quality of decision is improved. Participants feel happy to state their submission freely. The decisions can be successfully implemented as if subordinates are made members of the Committee.

(4) Committees are a useful device to give representation to various interest groups. Representation of various groups will feel a sense of loyalty and commitment to the decisions reached.

(5) ***Easy communication*** – All parties and persons interested have simultaneous communication through the proceedings of a Committee meeting. Doubts and ambiguity can be removed on the spot. Joint inspection of construction site can also be done by the Committee for better understanding of site problems.

(6) ***A tool of management development*** – Committee can also serve as a method of management development. In management development, experience is considered more important than formal training.

(7) ***Fear of delegating too much authority to individual*** – To provide checks and safeguards against the abuse and misuse of power, line authority is vested in a Committee.

(8) ***Avoidance of action*** – Sometimes Committees are appointed to avoid any action or problem or to delay action. When the Committee is appointed to avoid individual responsibility, it is called "blurring of responsibility". Such type of Committee is appointed as a strategy for overcoming resistance and pressure.

2.14.2 Disadvantages of Committees

There are some disadvantages of Committee. Extra/high cost is involved for long-term large size Committees. Process is quite long for taking decisions. Sometimes it takes long time. Sometimes Committee is unable to take decision due to different opinions amongst members.

It has been found that people work efficiently and effectively if their responsibility is clearly defined; authority is clearly delegated. Divided responsibility is a problem whenever group decisions are made. Sometimes it is found that there is misuse of Committee.

2.14.3 Measures for making Committees effective

Some drawbacks and limitations are inherent to Committee form; some problems arise due to improper use of Committees. However, there are many advantages. Some measures should be taken for effective use of Committee and same are stated below :

(1) *Appropriate size* – Optimum size differs under different conditions. Suitable number of experts should be there. Size of normal Committee may be 5 or 6; in certain cases it may be 7 or 9. If it is required to include many persons of different interest groups then 2 or 3 Sub Committees may be constituted by the main Committee.

(2) *Selection of Members* – Proper selection of Committee requires a nice judgment as to personality differences, ability in expression as well as status.

(3) *Well-defined authority and scope* – Members should know the exact scope of subjects. Details/Charter may be stated in the order of constitution of Committee; the same should include function and scope of Committee, responsibility, authority and organizational relationship.

(4) *Selecting subject matter* – Committee is not good for research/study or for exercising managerial functions.

(5) *Effective Chairman* – His roles are that of expediting and facilitating the progress of meeting. Members should not be irritated. It is required to take special care for maintaining unity of the group.

(6) Chairman is to decide logical procedure and programme for conducting meetings. Sequence of meeting matters should be decided in advance.

(7) *Follow up action* – Minutes of all meetings should be recorded carefully and to be circulated to all members quickly. If required, necessary modification/correction is to be done in time.

(8) *Review of Committee functioning* – There should be periodic review of effectiveness of Committees.

2.15 Manpower Planning

This is one important part of management in an organization in respect of (i) staff and (ii) line. One part of manpower is for offices and another part is for production/factories/marketing/construction field works. Skilled and un-skilled labourers are also to be deployed and maintained. Specialised authorities are required for manpower planning, recruitment, training, deployment and production/ construction management. Various categories of manpower planning, recruiting, training, deployment are required like, technical personnel/executives/Managers, Specialists/Engineers of different disciplines. Manpower planning is the process including forecasting, developing, implementing and controlling by which an organization ensure that it has the right number of people and right kind of people, at the right place, at the right time, doing things for which they are economically most useful. Inventory of present manpower is to be maintained properly; it should be possible to find out underdeveloped talent present within the organization. It is also required to check the working conditions and relationships in which they work.

There should be short-term and long-term manpower planning. For short term planning details of existing incumbents are checked. Updated manpower inventory is very useful for this purpose. Sometimes problem is faced in the short run for filling up the posts become vacant due to resignation, death or unanticipated transfer. In such time temporary arrangement is made by reorganizing existing manpower. Some incumbents may be strong/ talented and additional duties can be given to them. It has been found that a portion of individual's ability remains unused.

2.15.1 Long-term Manpower Planning

It is more concerned with filling future vacancies. This has following four basic elements :

(1) *Projecting manpower requirements* – It is required to forecast the organizational structure for future needs. Changes occur in the organization. Hence, prediction for future structure and manpower is difficult. It is required to know following factors :

(a) *Rate of loss of manpower* – This is due to resignation, long leave, death, retirement, transfer. Forecast on the basis of historical data can be done.

(b) *Expansion or contraction* – The future capital requirement and sales estimate may provide basis for projecting manpower need for future.

(c) *Job Analysis* – After projecting manpower needs, it is required to predict what capabilities would be required to handle the jobs effectively. In the case of managerial personnel such prediction is more important. If managerial talent is not good then there will be delay in recruitment, selection and training process, standard of selection may not be good. It is required to fix up characteristics which are required in managerial personnel. For job analysis it is required to know several informations like, objective, nature and scope of job, types of plans and projects to be undertaken, types of decisions to be taken by the incumbents on the job, authority (how many subordinates will work under him), etc. On the basis of above position "job description" and "job specifications" would be prepared; the former describes in detail the contents of job while the later describes the requirement of persons performing the job.

2.15.2 Manpower Inventory

Inventory of materials stored in the godown is prepared showing various materials. Inventory of spare parts is shown in respect of machine/vehicle repair workshop. Inventory of equipments, machinery is prepared showing various particulars. For such matters items are tangible.

But in case of inventory of manpower besides number of person (i.e. tangible) intangible factors viz., merit, talents, efficiency, capability, aptitude, etc. are involved. It includes a cataloguing of present and potential abilities and aptitudes. The inventory should include youngmen (for senior managerial posts) who have good potential, even though they are not holding key positions presently. Manpower inventory comprises following four steps :

(a) Who is to be included in the inventory? Executive inventory is more important as compared to operative inventory.

(b) List of personnel is finalized. Detailed information about each individual is gathered.

(c) Appraisal of the personnel is to be included. Rating of each person should be indicated in the inventory.

(d) Inventory will reveal, how many persons are available for various positions.

2.15.3 Training and Development

Employee training is the term used to describe the process by which organizations build the skills and abilities of non-managerial employees. Much of this training is job-centered, rather than career-focused. Management development includes the process by which managers and executives acquire not only skills and competency in their present jobs, but also capacities for future managerial tasks of increasing difficulty and scope. Training emphasizing manual skill at one extreme and development emphasizing philosophy on the other extreme.

2.16 Communication

Communication has been defined as process. Events and relationships are seen as dynamic, continuous and flexible. This involves various elements, viz.,

Sender-Ideas-Encoding-Channel-Receiver-Decoding (Then Feedback).

In an organization, there are a number of channels or paths connecting various positions for purpose of communication. This is known as "communication network". There are formal and informal type of communication channel/path. Formal channel is deliberately created, officially prescribed path for flow of communication between the various positions in the organization. The system is made for filtering of the information to various points to ensure that information does not flow unnecessarily thereby causing the problem of overload. Ordinarily the path is the line of authority linking a position to its superior; this is known as "through proper channel". The informal channel is not the result of any official action. This is related to personal/social type. At least two parties are required for a communication. There are three types of communications in the organisations as under :

(i) Downward communication – This flows from a superior to his subordinates. This has a great force for controlling, influencing and initiating activities.

(ii) Upward communication – There is possibility of upward information being distorted or coloured; colouring takes place primarily in upward communication, because this direction flow carries managerial control information.

(iii) Horizontal communication – It is the flow of information between persons of the same hierarchical level.

2.16.1 Effective Communication

Communication is the "lifeblood" of an organization and without it organization cannot exist. Hence, management should ensure that adequate and smooth communication flow in all directions and it is effective as well. A periodic review of the existing system/pattern of communication effectiveness is made. Communication is one of the most difficult of all managerial activities. There are several elements in communication which can be evaluated to assess directly the effectiveness of communication. These elements are :

(i) *Clarity* – This avoids unnecessary seeking and tendering of clarification.

(ii) *Adequacy* –

(a) One in terms of coverage, types of messages flowing in various directions.

(b) In terms of quantity of various messages.

(iii) *Timing* – The utility of any message to be received is markedly affected by its timeliness. Message is not only to reach to the receiver but also it reaches in time.

2.16.2 Barriers in Communication

Obstructions/barriers impede flow of communication. Any managerial action in this regard can be effective only if it strikes at the very root of the factors. Barriers may be of three types :

(a) External Barriers
(b) Organisational Barriers (i.e. internal)
(c) Personal barriers

2.16.3 External Barriers

These barriers arise from linguistic capacity of the parties. Words are the main for communication but common difficulty with the word is that each word has several meanings. Badly expressed message create barrier/problem. Faulty translation also creates problem/barrier. It may be required to add proper interpretation of message/words. Problem may be due to unclarified/wrong assumptions. Specialist language/words may not be understood correctly.

Some barriers may be due to emotional effect on the persons. Apart from message there is a "meta-message" i.e. what one gets out of a message when decoding. Meta-message is the pungent thing in a sweet word language. The receiver may not be in a wavelength of communicator. These barriers are known as emotional barriers, such as:

The pre-occupied mind of a receiver; people simply fail to react notices, bulletins, minutes, bulky reports, etc. Barrier may be due to loss of transmission and poor attention. Written word is no substitute for sound face to face relationship. Sometimes the Managers fail to transmit the needed messages.

2.16.4 Organisation Barriers

The top management authority laid down many rules and regulations, procedures, policies, norms of behaviour, reward and punishment system etc. in the organization. Some of these inner processes may affect communication system adversely. A few examples are stated below.

(a) *Organisation policy* – If the policy is not supportive to the flow of communication in different directions, communication flow would not be smooth.
(b) *Organisational rules and regulations* – Some rules may restrict the flow of certain messages and may leave many important ones. Communication through proper channel in a specified way prescribed by these rules delay it and works against the willingness of persons to convey the message. Such barrier is strongly operative in Government offices in India.
(c) *Status relation* – The placing of people in superior-

subordinate capacity in the formal organizational structure also blocks the flow of communication particularly in upward direction.

(d) *Organisational facilities* – Organisational facilities like, meetings, conferences, complaint/suggestion box, open door system, social and cultural gatherings, etc. should be utilized effectively.

2.16.5 Personal Barriers

Barriers in superiors – If the attitude of superior is unfavourable then there is a great possibility that message would not flow adequately from/to superiors.

Managers in general try to withhold the information coming down the line or going up as frequent passing of information may disclose their weakness.

Insistence on proper channel – Superiors may not like bypassing in communication. They may block the flow of such communication.

Lack of confidence in subordinates also hampers in smooth communication.

Lack of time – Superiors may feel that they are overburdened with the work and that they have little time to talk to their subordinates.

2.16.6 Methods of Overcoming Barriers

(a) Though communication through proper channel is required for orderly flow of information, it should not always be insisted upon. The system of proper channel serves purpose adequately for routine types of communication. However, when and whatever the situation so warrants, this has to be overlooked and proper channel need not be insisted in such cases.

(b) A successful communication system will only be achieved if top management is determined that it is to be so. It must set good examples itself, clearly expects to follow them.

(c) Communication being an inter-personal process, the development of inter-personal relationships based on mutual respect, trust and confidence is essential for its promotion. Organisational climate should be radically modified to make it more intimate and personal.

(d) The communication should be quite clear about what he wants to communicate. Basic purpose of any communication is to get behavioural response from the receiver. Two-way communication beings two minds together which is the basic core of any communication. The way of effective communication is to be sensitive towards receiver's needs, feelings and perceptions. Appropriate words, language should be used. There should be continuous programme of evaluating and monitoring the flow of communication in different directions.

2.17 Motivation

2.17.1 Management tries to coordinate various factors of productions so that each factor attributes to its maximum efficiency. There are two categories of factor : –

(i) Non-human factors (i.e. materials, machines etc.); there efficiency depends on type of technology & its effective operation

(ii) Human factors who handle and control there non-human factors. Hence it is essential to improve the performance level of human beings in the organisation. This performance is determined by two factors viz. (i) level of ability to do certain work and (ii) level of motivation. These two factors are to be multiplied rather than added. This can be expressed as :

Performance = Ability × Motivation

2.17.2 Motivation is linked with motives which are expression of human needs by a person. Motive or need is important in management of project. The need is due to feeling of lackness for some thing and human being tries to get lackness removed (i.e. satisfaction). Motivation is the process of inducing persons to experience needs for certain desired behaviours so that organisational efficiency is achieved.

2.17.3 Characteristics of Motivation

(1) Motivation is an internal feeling. Such feelings affect the behaviour of the person.

(2) Person in totality (not in part) is motivated. Each individual in the organisation is a self-contained and inseparable unit; his all needs are interrelated.

(3) Motivation causes goal-directed behaviour.

Motion is one of the most important factors determining organisational efficiency. All the organisational facilities will go waste in the lack of motivated people to utilise these facilities effectively.

Every superiors in the organisation must motivate his subordinates in suitable way (which depends on nature of person) for the right types of behaviour.

Motivated employees put higher performance. As per study it has been found that motivated employees worked at close to 80 - 90 percent of their ability.

Due to non-motivation the absenteeism will be high and turnover will be less.

2.17.4 It has been found in many organisations that motivation depends on many factors like, Leadership/management technique from superiors, good behaviour with subordinates, monitory incentives (by paying bonus, proper amount of annual appreciation for good works done by subordinates, approval leave requirement, quick payment of dues including medical facilities etc.), getting due promotion, suitable posting as per employee's qualification and experience etc. Salary level should be good so that it motivates to employees.

2.18 Leadership

2.18.1 As stated earlier manpower resource is most important; other resources also depend or operated by the manpower. Manpower can get the results from the people in two ways: – (i) by exercising authority vested on it or (ii) by winning support of the people. The second method is better as it has a lasting effect. However it is only possible when a manager becomes their "leader". The leadership is the process of influencing the activities of an individual or a group for goal achievement. Certain basic characteristics are stated below : –

(1) Leadership is basically personal quality. This quality motivates the individuals to be with leaders.
(2) Leader tries to influence the individual to behave in a particular way.
(3) There is a relationship between leader and subordinates which arises out of functioning for a common goal.
(4) The situation variables affect the effectiveness of leadership.

2.18.2 Difference between Management and Leadership

Some times it is thought by some section of people of society to treat managership and leadership as synonymous. But in management aforesaid two terminology are different. There is difference in the two concepts in the context of scope of functions performed. A manager has to perform all the five functions of management i.e. planning, organising, staffing, directing and controlling. Leadership functions come under directing through which behaviour is directed to get maximum use of subordinates' ability. Manager can get work of 60-65% of subordinates' capacity. But it may be possible to obtain more output from subordinates by effective application of leadership quality. Thus leadership is a part of management but not all of it. This implies that a strong leader can be a weak manager, because he may be weak in planning or some other managerial duty. The reverse is also possible. A manager can be a weak leader and still be an acceptable manager.

2.18.3 Following are a few aspects of leadership:-

(1) *Motivating Employees*

Motivation is necessary for work performance of the employees. Higher the motivation, better would be the performance. A good leader by exercising his leadership motivates the employees for high performance.

(2) *Creating confidence*

A good leader may create confidence in his subordinates.

(3) A good leader can build morale on his subordinates. High morale leads to high productivity and organisational stability.

2.18.4 **Qualities of Leader**

(1) A leader should have high level of emotional stability. He should be free from bias, is consistent in action and refrains from anger.

(2) He should have adequate knowledge of human relations i.e. how he should deal with human beings.

(3) He should have ability to look at things objectively and understand them from others' point of view.

(4) Leader has requisite quality to motivate his subordinates.

(5) Leader should have good ability to plan, organise, delegate, analyse, seek advice, make decisions, control, win co-operation (which are technical competence).

(6) He should have good skill for communicating effectively Communication has great force in getting the positive response.

(7) Leader should have social skills.

3

Materials Management (Basic Principles)

3.1 Introduction

Major part (more than 50%) of any construction works in Civil Engineering is materials. There are two parts of material management viz.,

(i) Purchase/procurement of materials

(ii) Effective use/management of materials

Cost of construction/economy largely depends on selection, cost of materials and planning/management of materials. Planning, programme, quantities to be procured with time schedule should be based on CPM (Critical Path Method) network of the construction works/projects. Certain materials are purchased/procured by the Government Department/owner for issuing to the contractors and for using day to day maintenance works. In construction works major part or entire materials are purchased by the contractor as per contract provisions. Considerable amount is to be arranged for purchasing these materials. Hence "material management" is very important for management of project construction and other works. Materials related to manufacturing industries and factories will not be dealt in this book.

3.1.1 Certain basic principles and universality of management are being dealt in this chapter. The effective management of supply is continuing to be a major challenge. Economic, political and social changes create both problems and opportunities for the purchase and material management. Many traditional views on materials procurement have been challenged by new views like : multiple sourcing increases supply security has been opposed by a trend of single source policy. Negotiation is receiving increasing emphasis (particularly in private sectors) as opposed to competitive bidding. Longer term contracts are replacing short-term buying techniques.

3.1.2 The specific functions which might be included under the Materials Manager are : material planning and control, production scheduling, material and purchasing research, purchasing, transportation of materials, inventory control, receiving of materials, quality control, storing, scrap and surplus disposal. Communication network is also an important part of materials management.

3.2 Decision making in the Material Management

Some of the issues on this matter are stated as under :

Should we make or buy ? Must we inventory materials and how much ? What price shall we pay ? Where shall we place this order? What the order size be ? When will we require this material ? Which alternative looks best ? Which transportation mode and carrier should we use ? Should we make a long-term or short-term contract? How do we dispose of surplus material ? Who will form the negotiation team and what shall its strategy be ? How do we protect ourselves for the future ? Should we wait or act now ? Do we standardize ? Should we use one supplier or multiple supplier ? If efficiency of supply is not satisfactory then performance of production/ construction progress will be affected adversely. If quality of supplied materials is not as per agreed standard then production/ works will hamper. If the supplier does not meet the agreed delivery schedule, this may require a costly re-scheduling of production/ construction activities; it may result in a shutdown of the production line/construction activities.

3.3 Professionalism in Purchasing

Some factors which have been added and increased efficiency in purchasing materials are mentioned as under:

Personal travel, good transportation system, good strategic planning have to be done by purchasing branch. There should be increased role/responsibility on the purchase branch. More attention should be given on cash flow.

Staff/Managers of purchase wing should possess proper academic qualifications and experience in this field. Training should be given to staff and Manager. Salary levels should be reasonable so that staff/Manager are motivated. They should be associated with relevant Institutes/Associations.

3.4 Objectives

Purchasing of materials is one of the basic and common functions of every organization. Structuring the purchasing/materials management function to obtain effective contribution to objectives is one of the challenges of management.

Purchasing function should obtain the "right materials" (meeting quality requirements), in the "right quantity", for delivery at the "right time" and "right place", from the "right source" (i.e. a supplier who is reliable and will meet its commitments in a time), with "right service" (both before and after the sale), and at the "right place". It may be seen that there are seven "R" (Right) in above stated objectives.

It may not be efficient to buy at the lowest price, if the materials delivered are unsatisfactory from a quality/performance standpoint or the materials arrive too late than schedule date, causing stopping of production/construction activities. On the other hand, the right price may be a much higher than normal price if the item in question is an emergency requirement. The purchasing decision makes attempts to balance out the often conflicting objectives and makes trade-off to obtain the optimum mix of these seven rights.

3.5 Goals

(1) Provide an uninterrupted flow of materials, supplies and services required to keep works in smooth progress as per scheduled time.

(2) Keep inventory investment and loss at a minimum. One way to assure an uninterrupted material flow is to keep larger inventory banks. But this requires considerable amount of money. A balance/optimum amount/quantity is to be decided by the Project/Construction Manager.

(3) Maintain adequate quality standards.

(4) Select reputed competent supplier.
The success of purchasing unit depends on its skill/procedure to locate/develop supplier, analyzing supplier's capabilities and to select a competent reputed supplier.

(5) Standardize item/specifications of materials. Long-term efficiency and economy are achieved by adopting aforesaid principle.

(6) Improve the organisation's competitive position.
It is possible by controlling the cost. Purchase cost of materials is the largest single element in the organisation/works/project.

(7) To achieve harmonious, productive working relationships with various departments within the organization. Production/construction units must provide correct information on material requirements in time. Purchasing unit must work closely with quality control in determining inspection procedures. Payment should be made to the supplier in time.

(8) The objectives of purchasing should be achieved as efficiently and economically as possible, which requires that the Purchasing Manager continually review/monitor the operation to assure that it is cost effective.

It is required to establish relationship between strategies, organization structures and delegation. The work must be assigned and executed in accordance with strategic plans and organizational goals. The organizational planning and delegation procedures are important segments of the integration of strategic goals and organizational design.

3.6 Centralisation/De-centralisation of Purchasing

In large organization, the decentralized system of purchasing may be done so that it can be ensured conveniently for delivering materials at site/store in time and in good quality. In smaller organizations, centralized system is convenient and effective. There are many advantages of centralized purchasing and some of these are stated as under :

(1) It is easier to standardize the items.

(2) It cuts down administrative duplication and save time and cost. Purchasing documents can be finalized on annual contract basis which will be economical.

(3) Requirements of quantities of various items are collected from various projects, works, branches. As the same will be large enough and large supplying companies will take interest and it may be possible to get discount or lesser price.

(4) In periods of material shortage it is easier to tackle the problem.

(5) It is administratively more efficient for suppliers, since they need not approach/pursue several officials.

(6) It provides better control over purchase commitments. Since a large percentage of organisation's cash outflow goes for material purchases, a central control point is effective to monitor various matters of material purchase.

(7) Centralised purchasing system enables the development of specialization and expertise in purchasing decisions.

3.7 Organisational Integration

Purchasing function might be performed by a few key persons who handles scheduling of production/construction activities and watches inventory levels (i.e. items and its quantities with cost) closely so that there will be no problem of coordination or control. In this system the purchasing, inventory control, stores and transportation are grouped together and assigned to one senior person.

3.8 Functions of Material Management

(1) Material Planning and Control – This is planning of material requirements to meet the broad, overall production/ construction plan. Approximate quantities of important materials (period-wise based on works' programme) are obtained/worked out. Role of Production Manager/ Construction/Project Manager/Engineer is important for working out details of material requirement (month/week-wise) alongwith detailed specifications (should be with approved samples of materials).

(2) Specific person is responsible to inspect the quality of materials (alongwith testing), measure the materials and its record on measurement book and to keep in the Store, and to maintain inventory register.

(3) The physical control of Store and maintaining inventory of items are important duties.

(4) Following are some functions of purchasing branch/ department :

(a) Checking the specifications of materials which have been requisitioned by various sections/works' inchargers/ Prtoject Managers. Effort is to be made for standardization of items.

(b) It is required to select the best available sources of supply, inviting quotations/tenders, negotiating the terms of purchase, preparing proper purchase orders, etc.

(c) To take follow up action for ensuring on-time delivery and receipt of the proper quality and quantity.

(d) It is required to carry out search/intelligence process in the market place for new and more effective suppliers and new materials with the objective of reducing costs.

3.8.1 Transportation

This is another important function of material management. Transportation cost has increasing influence on material costs. Type of transportation is to be decided carefully as per local position. Traffic analysis should be done for deciding transportation of materials.

Disposal of scrap and surplus materials

This has become function of material management and carried out by the purchasing unit/department.

3.8.2 Quality Control

Quality control of materials is an important part of materials management. Relevant tests of various materials are to be got tested from approved laboratory as per Codal requirements. Samples are to be collected as per correct method in presence of supplier. Final payment is to be released on the basis of test reports. Quality control is the responsibility of purchase unit/division.

3.8.3 Inventory Control

The inventory control function is responsible for keeping detailed records of all materials supplied by the supplier and surplus materials returned by the user department/ Construction Site Engineer. Handling the physical stocks of maintenance, repairs and operating items to be issued as needed for operations/maintenance is carried out by Stores In-charge. The inventory/Stores In-charge is required to send requisition to Purchase Manager/Engineer when stocks reach the recorder point and for special requirements.

3.9. Steps in the Purchasing System

Essential steps in the purchasing procedure are as under :

(1) *Recognition of need* – Engineer-in-Charge of project construction/works is required to send the detailed requirements of materials (viz., specifications, quantities, estimated cost, approved samples (for certain materials), calendar month, place of delivery, etc. to the Purchase In-charge well in time (i.e. with necessary lead time). There should be effective coordination, discussion, communication between aforesaid two In-charges/units.

(2) Determination and analysis of possible sources of supply. It is very important to select competent and reputed supplier (based on their past performances). It is required to "pre-qualify" suitable suppliers. Quotations/tenders are invited from pre-qualified suppliers in the Government Departments. In private organizations materials are purchased either by inviting quotations or by direct negotiation with pre-qualified suppliers. Price and terms and conditions are finalized during negotiations with the supplier.

(3) Then proper supply order is prepared carefully on prescribed

form and the same is issued to the supplier. The agreement/contract documents are prepared and executed by both the parties carefully. Time is essence of the contract. Penalty/compensation clause is required for delay in supply.

(4) It is required to monitor the progress of supply of materials and related requirements (viz., quality control, testing of materials, recording measurements, payments, etc.). Purchase In-charge should attend the coordination/monitoring meetings conducted by Works Manager/Chief Engineer /Chief Project Manager.

(5) Various records are to be maintained by the purchase unit/department.

(6) Emergency/Urgent Supply Order – Sometimes it may be required to purchase some materials on emergency/urgent basis due to unavoidable circumstances. It may not be possible to follow normal lengthy procedure for emergency requirement. Some short-cut method is to be adopted. However, such orders are costly.

(7) Project Construction Manager deals directly (not through Stores/Purchase Department) in respect of supplying special material like, Ready Mix Concrete. Construction Manager places order to the Ready Mix Concrete supplier who manufactures cement concrete as per specifications of department in his plant and supply through special type of vehicle (Transit Mixer truck) at construction site.

(8) For certain works it is required to purchase manufactured finished items (viz., ceramic tiles, pipes and fittings etc. of certain 2 or 3 manufacturers). It may be required to obtain quotations from these specific manufacturers and then to negotiate with the lowest manufacturing company for placing supply order.

3.9.1 Best Buy

The notion "best buy" puts quality, reliability and suitability into a sound procurement perspective. The best buy assumes, of necessity, a certain minimum measure of suitability, but considers cost and procurability, transportation and disposal as well. Balancing various technical considerations is another aspect. In case of purchase of equipments (viz., pumps, tools for workers, safety equipments, etc.) services to be given by the supplier are also important; services may include installation, training, inspection, repair, advice, etc. This part is also relevant for "best buy".

3.10 Description and Specification

3.10.1 The using/requesting/specifying department must be capable of reasonably describing what is required to be sure of getting exactly what is wanted. The purchasing department has the direct responsibility of checking the description given. It should have authority to insist that description be sufficiently accurate and detailed to be perfectly clear to every potential supplier. The requisitioner may also see the availability of other options in respect of certain special material which might represent better value.

3.10.2 The methods of description normally used may be listed as under :

(1) By brand
(2) By specifications –
  (a) Physical or chemical characteristics
  (b) Material and method of manufacturing
  (c) Performance
(3) By engineering drawings
(4) By market grade
(5) By sample
(6) By combination of two or more methods

3.10.3 There are certain circumstances under which description by brand may be not only desirable but also necessary.

(1) Specifications cannot be laid down in respect of patent item.
(2) When specifications cannot be laid down with sufficient accuracy by the buyer because vendors manufacturing process calls for a high degree of that intangible labour quality some time called "workmanship" or "skill" which cannot be described exactly.
(3) If the quantity is too small then it is cheaper to buy through brand name.
(4) Description of desired material specifications constitutes the best method. It is better to specify the test procedure and results necessary to meet quality standards as part of the specifications. Testing is to be done in approved laboratory.
(5) Specification by physical or chemical characteristic – It represents an effort to state in measurable terms those properties deemed necessary for satisfactory use.
(6) In another method it is prescribed for both the materials and method of manufacture.

(7) Specification by performance – Performance specification is a method employed to a considerable extent, partly because it throws the responsibility for a satisfactory product back to the seller.

(8) Description by engineering drawings – Description blueprint or dimension sheet is common and may be used in connection with some form of descriptive text.

(9) Description by sample – Another method of description is by submission of a sample of the material desired. This method is essential in respect of certain material. As for example, in construction of buildings it is not effectively possible to describe stone slabs/tiles (viz., granite stone slab/ white marble slab, etc.) for flooring/wall cladding finishing items.

Objective of description/specifications of materials is to secure the right quality at the best possible price.

3.10.4 Sources of Specification

There are three major sources from which specifications may be derived :

(1) Individual standards set up by the buyer.

(2) Standards established by certain private agencies (technical institutes/ societies, manufacturers)

(3) Government standards (viz., by CPWD, PWD, Railways authority, National Building Code of India 2005, MES, Rural Engineering Service, Ministry of Road Transport & Highways, Bureau of Indian Standards, etc.). Standard specifications have certain advantages. They are widely known and commonly recognized and readily available to every buyer. Further, the materials of standard specifications have somewhat lower cost. Finally, because they have grown out of the wide experience of producers and users, they should be adoptable to the requirements of many purchasers.

3.10.5 Lists of various tests to be carried out for various materials (raw materials, finished products) are prepared by various authorities (Government departments, organizations authorized by Government, institutes/societies of engineering disciplines, etc.). Particulars of tests and permissible values of test results etc. are stated on the aforesaid codes/standards. Suitable instruments, equipments are manufactured by certain companies. Tests are carried out at site, factories/plants site and laboratories depending on type of tests and materials. In case, certain tests are not available in any records of Indian authorities then the same are taken from international codes, specifications. If necessary, samples of materials are sent to approved laboratory of foreign country for carrying on testing. Certain tests are to be carried out during production of material, like, cement concrete mix at green stage at site.

3.11 Quality Assurance, Testing, Inspection and Quality Control

3.11.1 Before a new vendor/supplier is given an order and even before a vendor/supplier is allowed to quote, the purchaser conducts a capability or quality assurance survey on the vendor's premises. The purpose is to ensure that the vendor is capable of meeting the specifications and quality standards required. The survey is normally conducted by engineering/manufacturing/purchasing/quality control personnel. It is desirable to have continuing involvement with the vendors/suppliers to evolve common quality standards, to agree on inspection methods and to work out ways and means of improving quality so that number of inspection is decreased and overall cost is reduced.

3.11.2 Inspection and testing of material may be done at two different stages i.e. (i) before selection of supplier/vendor and before placing supply order; and (ii) during supply of materials and before acceptance. For certain materials "comparison testing" may be done at different sources before selecting suppliers/vendors. Testing facilities for certain material/manufactured items may not be available with the buyer or in approved laboratories. As for example, certain testing of RCC hume pipes is carried out at factory; testing of neoprene bearings for bridges is carried out at factory. Lists of tests and frequency of testing of various materials are available in relevant Codes, specification books. Manufacturers/vendors are required to maintain records of quality, testing. These records can also be checked by the buyers. After normal checking of quality and testing in the approved laboratory, a test checking of certain percentage should be done by senior Engineers/Managers and same is to be recorded. Senior test checking technical person should collect sample materials from work site and tested in the laboratory after checking calibration of instruments.

3.11.3 During stating, specifications in the tender/quotation documents/supply order, it is essential to include the procedure for inspection and testing as protection for both buyer and seller. The supplier/vendor cannot refuse to accept rejected goods on the ground that the type of inspection was not known or that the inspection was unduly rigid. Vendor and purchaser need to work out both the procedure for "sampling" and the particulars of test to be carried out. Cooperation of both parties is important. It is very important to ensure that inspection, sampling, testing are done correctly in calibrated testing instruments as per correct procedure in approved laboratory. It has been found in many cases that manipulation is done during carrying out testing and preparing test reports. Hence, test checking by senior Engineer/Manager in respect of testing of materials is essential. It is to be ensured that materials collected and tested are genuine and test reports are also genuine.

3.12 Quantities of Materials

3.12.1 Quantities of materials to be purchased is important matter of works/project construction. Considerable fund is required. Inventory control is linked up with quantities of materials. Material cost of a project construction is about 50% to 60% of project cost. There is a dramatic change in management attitudes to both quality and quantity with profound impact on the acquisition process. The need to cut down on unnecessary inventory costs and to increase productivity and competitiveness has led to a new perspective on quantity. Importance evidence is the drive to increase the frequency of deliveries while decreasing the amount delivered at one time.

3.12.2 The decision on "how much" to acquire logically follows clarification of "what" is required. The natural response is to say "buy as much as you need". Such a simple statement is not sufficient. Many factors significantly complicate the decision on quantity. First, Managers take purchase decisions before, often a long time before; actual requirements are not known on this time. Therefore, they rely on forecasts, not only of future demand, but also on lead times, prices and other costs. It may not be possible to get correct forecasts. Secondly, there are costs associated with pricing orders, holding inventory and running out of materials and goods. Thirdly, materials may not be available at a reasonable price. Suppliers may offer reduced price for buying larger quantities. Lastly, shortage of materials may cause serious consequences; like, construction/operation of works may be held up and penalty may be imposed.

3.12.3 In case of project constructions, estimated quantities of materials (period-wise) are worked out by the Planning/Site/Project Engineers and the same (as forecast) is informed to purchase unit in advance. Engineer should work out requirement of materials (month-wise) on the basis of CPM (Critical Path Method) network; this will give better forecast. In case stipulation is made in the contract/agreement about issue of certain materials (like, cement and steel) by the owner/department to the contractor the same are to be purchased by the owner/department in time. Other materials are to be purchased by the contractor as per CPM network programme. Project Managers of owner/Government Department and contractor are responsible to take action on various activities of material management as stated on foregoing pages. Chief Project Manager of the owner/ department is responsible to prepare CPM network of entire scheme/project (i.e. including pre-construction stages and various disciplines). Chief Project Manager of contractor is to take necessary action for various activities related to his contract including preparation of detailed works programme (CPM network), materials requirement (month wise).

3.13 Supplier Selection

3.13.1 The key decision in supply management is supplier selection. Supplier's performance has a greater impact on the productivity, quality. There is trend to buy readymade product instead of making at work site to achieve better quality, lower inventories. Therefore, the supplier selection decision is the "single most important decision". Final decision depends on the nature of purchase (whether it is a repeat or modified repeat or a new requirement), market condition. Present trend is purchasing through a team of concerned persons (viz., purchasing department, Works/Project Managers/Engineers, finance/account department). There is another trend in private organizations for fewer supplier/vendors, longer term contracts and continuing improvement in quality, price, services requires much closer coordination and communication between various people in both the buying and selling organizations. Progressive private organizations are pursuing ways and means of limiting their total number of suppliers and maximizing the results from fewer key vendors. Normally the purchaser's perception of the vendor's ability to meet satisfactory quality, quantity, delivery, price and service objectives will govern the decision of selection; other factors may also be included for large purchase like, past history, facilities, technical strength, financial status, organization and management, reputation, systems, procedural compliance, communications, labour relations. In certain organizations pre-qualification of competent suppliers is made for large purchase. Particulars of large vendors/suppliers are collected/obtained for preparing aforesaid list of suppliers.

3.13.2 Good sources of supply are one assurance of good quality production today, and progressive thinking and planning is a further assurance of improved quality tomorrow. Superior sources of supply are an important asset to any organization.

Sources of information about suppliers :

(1) Catalogues
(2) Microfilm files
(3) Trade Journals
(4) Industrial advertising
(5) Trade Directories
(6) Sales Representation
(7) Vendor Files

3.14 Supplier Management

It is required for extensive communications and communication skills for both parties in the buying-selling relationship. The whole art of supplier management from purchasing perspective is to bring both sides into an effective working relationship. In some private organizations importance is given to team approach to long-term vendor relations. It is required to check a potential supplier's management strength. One key question is : Is the management of this supplier a corporate strength or a weakness ? This will require a detailed examination of the organizational structure, qualifications of Managers, management control, etc. of the prospective vendor. The financial strengths and weaknesses of a supplier obviously affect its capability to respond to the needs of customers.

3.14.1 Purchase from Manufacturer or Distributor

The question sometimes arises whether to purchase from the manufacturer directly or from some trade channel such as a wholesaler, distributor or even a retailer. The real issue involved here is often closely related to buy from local sources. The justification for using trade channels is found in the services which they render. Some large organizations purchase certain type of costly materials (viz., steel, cement, etc.) from original/main manufacturers. Many buyers prefer to buy from local sources due to some advantages.

3.15 Pricing of Item for Sale

3.15.1 As already stated, there are two categories of purchasers :

(1) Government Department/Organisation/Semi-Government organisation

(2) Private persons/organization (contractors in case of engineering works)

In case of Government departments, procedure/rules for purchasing are laid down and the same are to be followed. Like, to prepare notice inviting tenders/quotations (i.e. quotation notice with schedule of materials, specifications, terms and conditions, etc.). For large purchase the pre-qualification of suppliers/vendors/ manufacturers is done before obtaining offer/rates. The criteria of lowest rate may be under following two positions :

(a) Lowest tendered amount considering all items of materials including sales tax, transportation for delivery at specific location etc. Comparative statement of all tenders/quotations for all items of materials are prepared. If any extra conditions, variation in specifications of materials are given by any tenderer then the same are also to be stated in the aforesaid comparative statement; financial and technical implication of such conditions are also to be worked out/analysed for taking decision. It is required to identify – who is the lowest tenderer ? Negotiation may be done with such lowest tenderer. As per result of negotiation the supply order is issued and agreement is executed by both the parties.

(b) There may be another situation. Lowest rates quoted for various items are identified (supplier-wise). Such case is examined as per actual position and necessary decision is taken by the negotiation committee. As per actual position of quantities and items may be possible to split the order for 2 or 3 suppliers.

3.15.2 In case of private sectors (like, private companies/ organizations, private contractors/builders, private societies, private house, factory owners, etc.) procedure for purchasing of materials is different. In case of project constructions steel and cement may be issued by the Government Department to the contractors as per stipulation made in the tender documents. Remaining materials for construction works are to be purchased by the contractor. Necessary set up is made by the contractor for purchasing materials. Various functions, system of material management as stated above are to be followed after detailed planning.

3.15.3 Principles of Price

Determination of the price to be paid is a major purchasing decision. The ability to get a "good price" is held to be the prime test of a good buyer. If by "good price" is meant "greatest value", - this is true. While price is only one aspect of the overall purchasing job, it is extremely important. Major part of project cost relates to materials. The purchasing department/unit should be alert to different pricing methods. Any price quoted should be analysed. The price paid by the purchaser is as much a factor in the final decision on what is the "best buy" as are the technical properties of the item.

3.15.4 Relation of Cost to Price

Rates quoted by the suppliers are the price of the items. Purchasing Manager believes that the vendor/supplier should be paid a "fair price". But what does fair price mean? A fair price is the lowest price that ensures a continuous supply of the proper quality where and when needed. The vendor's total costs including a reasonable profit, should be covered by total sales in the long run. A fair price to one seller for any one item may be higher than a fair price to another. Prevailing price need not necessarily be a fair or excessive one. The Purchasing Manager is called on continuously to exercise judgment as to what the fair price should be under a variety of circumstances. Cost of material item is under two categories viz., - (i) direct cost and (ii) indirect cost. Materials and manpower engaged directly for production of item are direct cost. Indirect cost of a item are those incurred in the operation of a process; some examples of indirect costs are : rent, property tax, machine depreciation, expenses of general supervisions, power, heat, light, overhead cost of offices etc. Analysis of price may be done by:

(i) Cost Approach
(ii) Market Approach (actual market rates vary due to many factors)

Government influence and legislation affect price.

3.16 Negotiation with Supplier

3.16.1 Brief procedure about purchasing materials in Government Departments and private organisations has been stated on foregoing pages of this chapter. For the bid process to work efficiently, certain conditions are necessary :

(1) There must be at least two and preferably several qualified vendors/ suppliers.

(2) The vendors/suppliers must want the business; competitive bidding works best in buyer's market.

(3) Specifications of materials must be clear so that bidder knows precisely/ correctly and the buyer easily can compare the tender/quotations received from various bidders.

(4) There must be honest bidding and the absence of any collusion between the bidders.

Negotiation technique is a sophisticated method particularly for large amount of purchase. By negotiation both parties arrive at a common understanding on the essentials of purchase contract, such as, delivery, specifications, (with or without sample material), warranty, prices of items, terms and conditions, etc. Because of the inter-relation of these factors and many others, it is a difficult "art" and requires the exercise of judgment and tact.

3.16.2 Aspect of Negotiation

While negotiation normally is thought of as a means of establishing price to be paid, and this is the main focus, many other areas or conditions can be negotiated. In fact, *any* aspect of the purchase/ sale agreement is subject to negotiation. A few of these areas are:

1. Quality
   (a) Specification compliance
   (b) Performance compliance
   (c) Test criteria
   (d) Rejection procedures
   (e) Liability
   (f) Reliability
   (g) Design changes
2. Support
   (a) Technical assistance
   (b) Product research and development
   (c) Warranty
   (d) Spare parts
   (e) Training

(f) Tooling
(g) Packaging
(h) Data sharing, including technical data

3. Supply
   (a) Lead times
   (b) Delivery schedule
   (c) Consignment stocks
   (d) Expansion options
   (e) Vendor inventories
   (f) Cancellation options
4. Transportation
   (a) FOB terms
   (b) Carrier
   (c) Commodity classification
   (d) Freight allowance/equalization
   (e) Multiple delivery points
5. Price
   (a) Purchase order price
   (b) Discounts (cash, quantity and trade)
   (c) Escalation provisions
   (d) Exchange terms (monetary rate fluctuation)
   (e) Import duties
   (f) Payment of taxes

3.16.3 Negotiation Strategy

As per CPWD manual (Government of India) there is a guideline that tender/quotation forms (i.e. NIT) should be issued to competent eligible suppliers; in case of large amount of purchase, pre-qualification of suppliers should be done as per laid down procedure. Having complied abovestated procedure the quotation/tender of lowest bidder can be accepted without negotiation (provided quoted rate is reasonable/justified). Unnecessary negotiation should not be carried out.

Success in negotiation largely is a function of the quality and amount of planning done before taking up negotiation. The basic steps in developing a strategy for negotiation are as under :

(1) To develop the specific objectives (outcomes) desired from the negotiation.
(2) To analyse the bargaining position of supplier/vendor. What is the vendor's capacity, backlog and profitability ? How confident is the vendor of getting the contract ? Is there any time urgency?

(3) To gather pertinent data, cost analysis of material item is relevant.

(4) Attempt to recognize the vendor's needs. In successful negotiation, both parties should win.

(5) Determine the fact of relevant points and issues.

(6) Determine the issues of the tender/quotation case. An issue is something over which disagreement is expected.

(7) Set the buyer's position on each issue. What data will be used to support the buyer's position?

(8) Plan the negotiation strategy. Which issues should be discussed first ? Where is the buyer willing to compromise?

(9) Brief all members of negotiation committee.

(10) Discuss various issues amongst negotiation committee members (alongwith brief notes for negotiation) before calling supplier/vendor.

(11) Conduct the actual negotiations with calmness.

3.16.4 Certain Points for Successful Negotiation

These points are stated below in three time phases (i) before, (ii) during and (iii) after the negotiation session :

I. Before the Session :

A. Preparation

1. Establish possible sources
2. Analyze the vendor's position
3. Make a facilities survey
4. Make a financial analysis of the vendor
5. Analyze the vendor's proposal
6. Organise the negotiation team
7. Get a clear understanding of the work statement
8. Determine negotiation objectives
9. Prepare alternative action courses
10. Know your own authority
11. Provide adequate conference facilities
12. Set up the meeting room in advance

B. Basic Ideas

1. Be prepared to compromise
2. You will have to "sell" your position
3. Keep a poker face
4. Do not ever underestimate the seller
5. Never let your guard down
6. Take your time and do the job right
7. Alcohol and negotiation do not mix

8. Be reasonable; do not push too far
9. Regardless of your offer, the seller will want more
10. Nervousness on your part will be interpreted as a sign of weakness

II. During the Session :

A. Size up the Seller

1. Watch for roving eyeballs – and your opponent can read upside down
2. Spot their leader – who really can make concessions?
3. If they are hesitant to discuss an issue, it indicates a weakness
4. If the seller has no data on a key issue, it shows vulnerability
5. Be intense; concentrate; look the other part right in the eye and listed to what he or she has to say

B. Strategy

1. Take command – sit at the head of the table
2. Know names and pronunciation of vendor personnel
3. Establish extent of vendor representatives' authority
4. Assess supplier's minimum position
5. Talk positive
6. Phrase your questions to encourage a positive answer
7. Compromise on minor points when advantageous
8. If you compromise early in the session, you set the stage for the vendor to reciprocate
9. Start with the easy issues first
10. Avoid taking an either-or-position. The vendor may get up and leave.
11. Never give anything away
12. Do not go beyond your mental and physical endurance

C. Tactics

1. Do not reveal maximum position
2. Do not argue without reason
3. Do not lose your cool
4. Do not commit beyond your authority
5. Remember, the supplier has to "win" too
6. Do not interrupt; it is disrespectful and may "turn the other party off"
7. The best reply to a totally unacceptable offer is complete silence

8. Do not go off on tangents – keep the discussion on tract – time is valuable

D. Resolving Negotiation Deadlocks

1. Go on to another point
2. "I see your position – now try to understand mine"
3. "You suggest a solution"
4. "We have come too far to get bogged down now"

III. After the Session :

1. Make sure all items are covered in the final agreement
2. Know how and when to terminate a session
3. Keep complete notes on all points of agreement – both parties should initial or sign
4. Analyze what happened and why

3.17 Disposal of Scrap, Samples or Obsolete Material

3.17.1 Introduction

There are two groups of works :

(1) Manufacturing Plant

(2) Construction and Maintenance works (Engineering works)

This book is mainly for second group of works. However, there are works of plants/factories under Civil Engineering disciplines like :

Factory for manufacturing pre-cast concrete units (Plain CC, RCC, Pre-stressed), Casting yard, Ready Mix Concrete Plant, Hot Mix Plant for bituminous work (including Crusher Plant), Water and Sewage Treatment Plants, Compost Plant for manure from municipal solid waste, RCC Hume Pipe manufacturing plant, Brick manufacturing units, etc.

Dimensions of problems, quantities, type of materials in large manufacturing plants under category of surplus, obsolete scrap, toxic/harmful solid wastes are quite high and complicated. These attracts certain sections of various environmental protection laws. Considerable space is required for keeping such materials within premises of factory. Details in this regard (i.e. for such large manufacturing factories) will not be dealt in this book. However, some brief information will be given. Details in respect of disposal/ management of scrap/surplus/obsolete materials under second group (i.e. construction and maintenance works of civil engineering) will be dealt here.

3.17.2 Following are some examples of materials which are to be disposed off from site of civil engineering works :-

(1) *Construction worksite* – Surplus serviceable materials like, steel bars, steel structural sections (angle iron, channel, I-section, etc.), cement, stone slabs, stone/ceramic tiles, timber of different sizes, plywood, Board, stone aggregates, bricks, sand, paint, various fittings, GI steel and Cast Iron pipes, rejected materials which have not been removed by the supplier, etc.

(2) Various types of scraps like, cut pieces/wastes from steel bars, structural steel sections, timber, boards, plywood, stone slabs, tiles, stone aggregates, sand, bricks/brickbats, GI/Steel/ Cast Iron pipes, paints, etc.

(3) Serviceable materials obtained from dismantling of old structures/ buildings like, old service/utility pipes, old bricks, steel sections, steel bars, doors and windows, GI/Steel/Cast Iron pipes, etc.

(4) Obsolete materials/fittings/spare parts lying in central stores of maintenance works, scraps deposited in central stores by field divisions, surplus new materials lying in stores since long (viz., copper/gunmetal fittings, pipes, spare parts, etc.)

(5) Waste materials like, stone aggregates, sand, cement, steel bars, scraps from factories/plants of pre-cast concrete/RCC/ Pre-stressed units, casting yards, Ready Mix Concrete plant, bituminous Hot Mix Plant, Water and Sewage Treatment Plants, Compost Plants, RCC Hume Pipe manufacturing factory etc. Rejected materials, rejected/defective RCC hume pipes, rejected/defective concrete mix, bituminous mix, waste bitumen/fuel etc. are also to be disposed of.

3.17.3 General

Managers/Engineers are concerned with the effective, efficient and profitable disposal of scrap, surplus, obsolete and waste materials generated within the organization. In recent years, disposal problems have become more complex and important. Recently a new dimension has been added to the overall disposal problem : the need to develop and use new methods to avoid the generation of solid waste and better means of disposing of other wastes. Salvage of all types of materials is also important. As far as possible, re-cycling of certain waste/scrap materials should be done. Effort should be made for keeping minimum quantities of waste/scrap materials. Care should be taken at the stage of procurement of materials. Existence of

above stated materials (which are to be disposed off economically) is the result of a wide variety of causes like : over-optimism in the forecast, changes in the design and specifications during execution/ production, errors in estimating the materials, inevitable losses in processing, careless use of materials, excess stocking of materials at the Stores/Construction sites etc. It is essential to maintain inventory of store properly and to control the Stores efficiently. Systematic, physical stock taking, continuous review of inventory records and occasional clean-up campaigns are necessary. Record is to be maintained about serviceable materials received from demolition works.

3.17.4 Obsolete Material/Equipment

Obsolete material differs from excess stock. Although material or equipment may be obsolete to one user, this need not mean that it is obsolete for others.

Scrap material – Scrap material differs from excess or obsolete stock since it cannot properly be classified as new or unused. Scrap is a term which may be applied to material or equipment which is no longer serviceable and has been discarded. It includes such items as worn out machinery and old tools. Scrap metals normally are separated into ferrous and non-ferrous categories.

3.17.5 Waste

Waste has been defined as material or supplies which have been changed during the production process and which through carelessness, faulty production/ construction methods, poor handling or other causes have been spoiled, broken or otherwise rendered unfit for further use or reclamation.

3.17.6 Material Disposal

Responsibility for the management of abovestated material disposal remains with purchasing in-charge. Reasons for the same are : knowledge of probable price trend, contact with sales people is a good source of information as to possible users of material, familiarity with the organisation's own needs. It is essential that scraps, wastes, obsolete materials be "segregated" by type, alloy, grade, size and weight. Such materials may be sold on auction on "as is where is basis". Brokers can handle the sale of scrap. Their role is to bring buyer and seller together.

3.17.7 Disposal Procedure

(a) Disposing/selling of scraps, wastes, obsolete, surplus

materials is done either through auctioning method or by obtaining rates/offer from scrap buyer. In both methods notices are issued to scrap buyers. In case amount involved is large then short listing of scrap buyers or pre-qualification of scrap buyers is done before auctioning or inviting bids. Bid documents/Schedule of Quantities /Bill of Quantities is prepared before issue of auction/bid notice; and the same is issued to interested scrap buyers. Bidders are required to fill up the rates of various items of materials on the BOQ and put in the bid/tender/ quotation box by specific date and time. The bids/offers will be opened on the specific date and time after checking/receiving earnest money. Comparative statement of all offers alongwith conditions (if any) is prepared. Highest offer after considering financial implication of conditions is identified. If the highest offer is reasonable/ justified then the same is accepted. If auction is carried out, highest offer is accepted.

(b) At frequent intervals, the scrap disposal in-charge should be instructed to clean up the stock and report on the quantities of materials (weight in kg/ton for certain items, in numbers for certain items and by length in metres for certain materials). In case quantities of scrap/wastes is considerable, special arrangement is to be made for weighing. Such scraps are to be loaded on trucks with loader/labourers and the same is to be weighed on weigh bridge. Quantities of various items are to be recorded on the measurement book. Period for lifting these materials is also stipulated in the NIT/ Quotation documents. Proper conditions are stipulated in the NIT/quotation/bid documents about depositing bid amount/auction amount to owner of materials in advance. Stipulation of earnest money of suitable amount is made in the bid document/auction notice. Penalty clause is also stipulated for lapse on the part of scrap buyer.

PART II

★ Details of Engineering Management (Pre-construction/Planning/Design and Construction Stages)
★ Critical Path Method Network
★ Resource Management
★ Total Quality Management
★ Construction Equipment Management
★ Highway Management
★ Safety Measures

4

Approach to Construction Management

4.1 Introduction

The investment on construction projects has been assessed at not less than 50% of the total national plan budget. In India, the construction industry employs a very large labour force probably next to agriculture sector. In India the construction industry has been facing lot of problems due to unsatisfactory system of contract. The contract agreement is one-sided i.e. in favour of owner/ government department. Malafides continue to be attributed and allegations of corruption and nepotism are common at some levels. Efforts are being taken by various authorities to streamline the working system and to reduce the degree of corruption, unfair practices at various levels. When the works are exerted on the basis of the contract system, the right attitude of the owner/client and his engineer/architect is very important for the success of the contract. Contractor should not suffer for wrong specifications and designs of the owner's engineer and architect. It is important to introduce "model contract" documents for construction projects on rational basis this will promote equity in contract relations. This will provide simplicity and improve better relations between two parties of the contract.

4.2 General/Basic Approach

4.2.1 Effective modern management system is essential for proper planning of various resources (manpower, materials, machinery, money, design & drawing etc.). There are various types of projects like, oil exploration, manufacturing industries, building on infrastructure (viz., constructing building, highway, bridges, airport, housing etc.). This book will deal the building industry/infrastructure. Special management techniques are to be applied for emergent/ special types of works.

4.2.2 Main objectives of management are to complete the project within stipulated period and cost by effective/scientific management of manpower, materials machinery and money (cash flow). There is a French dictum : - "When the building industry prospers, everything prospers."

4.2.3 Better co-operation is essential between the workers and management with a review to provide an effective solution to the problems. There is a dire need for rapid and intelligent solution of problems of this industry. The construction industry with all its uncertainties, high risks and pitfalls is most demanding on the construction manager.

4.2.4 Linear scheduling is characteristic of a well planned construction programme. Identification of management process and individuals who carry it out, from an important part of the planning. Contractor's motivations, constraints and decision making patterns are to be kept in view. Manager has to extract better productivity through motivation. Project team should be used for long-range planning efforts. Techniques for tracking the quality of project planning progressively, are to be developed. Initially on paper, projects show many bright promises but later come to grief and some times end up in complete failure. At initial planning stage the environmentalist enters the scene to ensure the delicacy of natures balance during and after commissioning the project. As example hydro-electric projects affect the human and animal habitats. Various problems are faced due to insufficient preparatory analysis, sectoral assessment, feasibility study, technical appraisal. It has been observed that in India over emphasis is given on financial targets and achievements; this may lead to wrong assessment/decision. Proper stress should be given on "physical target and achievement" on an appropriate unit/measurements. Generally proper physical target/ achievement is not done correctly because it takes more effort for measurements after identifying various work break-down structures/ components of project.

4.2.5 Obstructions

There is a basic/essential requirement for taking up a project i.e. to accord administrative and expenditure sanction of the project by the competent authority. In Government department it takes very long time. Clearance is to be taken from many external authorities and internal departments. Bureaucratic delays are normal matter.

4.2.6 The time and cost over-runs in project completion are normal features in India. There are many reasons for the same such as :- communication failures, obstruction in cash flow, contractual ignorance, lack of adequately trained and competent project managers; divided responsibility contributes to project delay; deficient work scheduling, inadequate equipment specification also effect the progress adversely.

4.2.7 External co-ordination at the right level is necessary. There is need for good understanding intra-departmentally as well as inter-departmentally. There is also lot of political interference. "Technology transfer" should be effected properly.

4.2.8 Case studies on various project construction were carried out by some persons and their findings are very useful for management of project constructions. As per these studies it is required to take special care on certain techniques such as :- Managerial strategies and techniques; Networking and Quantitative Management Methods; Plan for Human Resources Management; Philosophy of Project Management itself. It has been revealed during above case studies, such as:- Dual control, confusion of purposes by involving many agencies; lack of co-ordination; change in specifications during construction period; delayed decisions.

4.2.9 In most projects the recruitment is on a temporary basis and competent persons are hesitant to enter in insecure short time assignment. The contractor try to win over the site staff with temptations and compromise in quality is the inevitable result. The site engineers are constantly on the move and due to this reason they are unable to apply correct technology.

4.3 Practical Points

(a) In view of scarcity of some essential materials, non availability of required number of skilled workers, rising trend of labour wages and other similar factors, it is required to adopt new techniques and management in the construction industry. In case of manufacturing industry there is repetition of process for long period. But in case of construction industry each project is unique in itself and manager has to apply suitable management techniques as per new problems which also depend on location and locally available materials and labourers/workers. Project manager moves with his staff, workers, equipments etc. after completion of a project at new place and new environment; he is to set up the unit at new construction site with machinery etc. In this industry prompt and sound independent decisions are called for beyond the usual commercial practices. The manager has to take on-the-spot decisions on various problems as per actual situation and readily available resources about methods of construction. In view of nature of this profession the manager is required to be mobile and constantly moving with new ideas to face a new set of problems with each new project and perhaps with each new client. A good construction manager has to be alive to all the above problems which are normal affair in the construction industry and he still makes the best of what is available to him and his organisation.

(b) There is another aspect of this industry. There are many types/variety/sub disciplines of civil engineering like building, roads/highways, bridges, railway track, irrigation works, dams, airport, public health engineering works etc. Practical problems, design, specifications, construction methods, project planning details etc. are different in each type of works. Engineers and workers are to be trained for each type of project. Effort is to be made to deploy at least some engineers having experience in relevant type of project.

(c) ***Working environment/culture in construction industry*** It is different to work smoothly in this industry. There is no worthwhile code of conduct in Indian construction industry. Lot of misunderstanding is faced between contractors and owner/client on one hand and contractors and the public on another hand. The clients are generally government or quasi – government or large public companies with bureaucratic approach. There is also lack of proper coordination and understanding between builder and the suppliers of materials on one hand and between manager and workers. Redresses of these problems are negligible. There is another gloomy/unrealistic position of contract conditions between the owner and contractor. The agreement is one sided. There is always reluctance on the part of government officers to change the old methods even though there are convinced about the drawbacks. However contractor has to sign one-sided contract agreement which gives the owner unlimited powers; and contractors face lot of problems. The owner/government department takes very long period (1 to 2 years) for finalising final bills of smaller works. The owner/department take lot of time for taking over the completed project from the contractor. Even after taking over the project it takes further long period to occupy/use the project/building due to faulty planning, design etc. Contractor and users face lot of problems due to above stated situation. Project manager is required to take action in advance so that above stated problems are not faced.

4.4 Corporate/Head Office Planning for Construction/ Development Works

4.4.1 Building construction projects owner/departments, are Government Department/ Organisation, Quasi-Government/ Organisation, Private Organisations. The construction works are carried out through contractor (private company or quasi-government organisation). These organisations should have proper long-term (say 5 years) and short term (say 2 to 4 years) planning. Short-term planning is also known as "rolling plan". Short term planning is generally prepared by certain type of smaller organisation (like, Govt. Undertaking, contractor, private contracting organisation or other private organisation).

4.4.2 In government departments (State Govt. or Central Govt. like PWD, CPWD, Irrigation Department, Railway Engineering Department, MES etc.) there is a regular system for preparation of Annual Plan and Five Year Plan. There are four different types – viz., Plan Schemes, Non-Plan Scheme, Loan Scheme and Deposit Work. These plans are prepared and controlled in the head office of above stated government departments. Various head of the departments are associated for preparing aforesaid plans in due time. Finance Department is the nodal department who co-ordinate with various heads of the departments. Certain policy guidelines are issued from time to time for preparing annual plan and Five Year Plan. Necessary scrutiny of these plans are carried out by the competent authority. Annual Plans and Five Year Plan are checked and approved for specific amounts (which become budgetary fund) by the Planning Commission in respect of Govt. of India fund. Similar planning is done by, various State Govt.

4.4.3 As per above stated approved plans preliminary schemes are prepared as per priority fixed by the authority. Top priority is accorded for schemes of annual plans. Actually preparation of preliminary schemes (particularly medium and large projects) are started a few years back on the basis of Five Year Plan.

4.4.4 In case of Quasi-Govt./Private organisations, who take up contract works, the short-term planning (i.e. annual plan and advance planning of 2 or 3 years) is done. This is also known as "rolling plan". During annual conference of the organisation review is made about position of old dues, actual turn over etc. and then target amounts are fixed for next years; reasonable projections for next 2 or 3 years are also made.

4.4.5 From above stated corporate/Head Office planning documents it is required to prepare preliminary schemes and resource planning. Projected amounts are made on the basis of past achievements. Contract organisation is to take action for acquisition of work as per target/annual plan. Necessary monitoring is to be done at Head Office with all concerned departments about progress achieved periodically (at least quarterly) so that budgetary allotted fund is not surrender at the end of financial year. It can ensure a systemic working and planned growth in the environment in which it is placed and within the limitation to which it is subjected. Participative planning should be encouraged. Middle level managers should also be involved in above stated corporate planning system.

4.4.6 A typical corporate/Head Office planning is depicted in the Flow Chart given at figure 4.1 below :

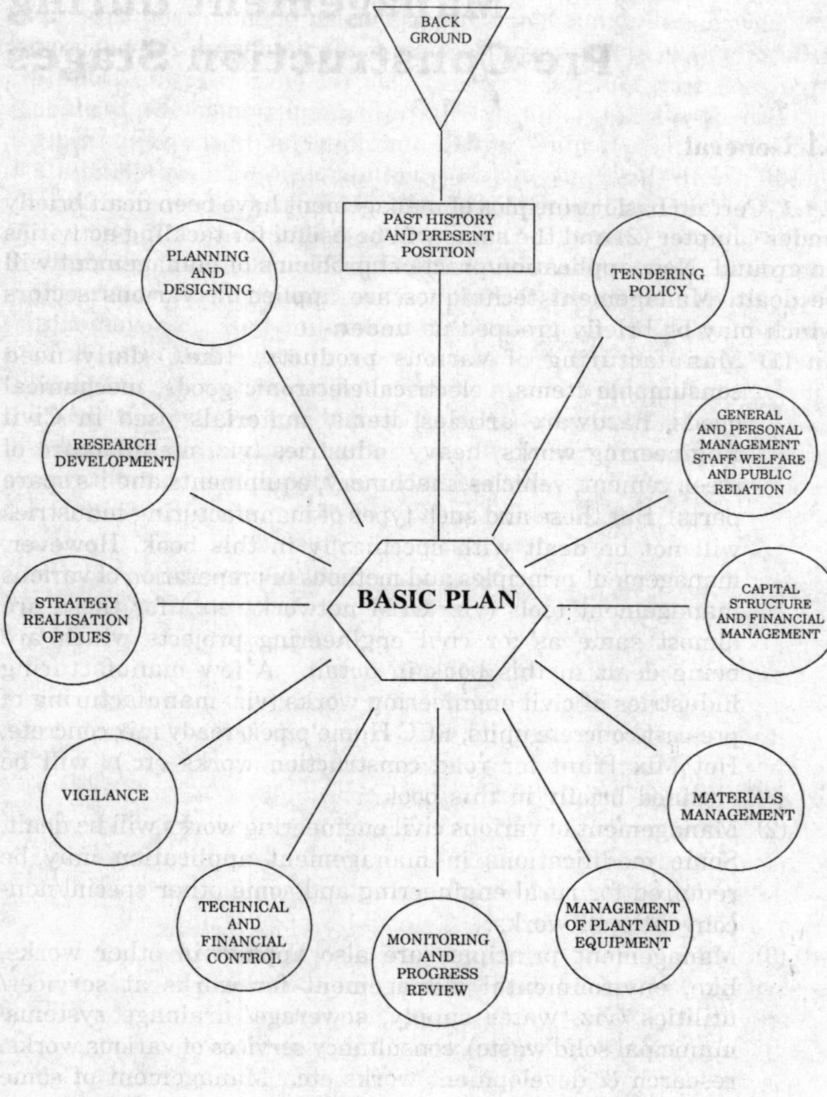

Fig. 4.1 : Flow Chart for Corporate/Head Office Planning

5

Management during Pre-Construction Stages

5.1 General

5.1.1 Certain basic principles of management have been dealt briefly under Chapter (2) and the same will be useful for tackling activities on ground. Now application/practical problems of management will be dealt. Management techniques are applied in various sectors which may be briefly grouped as under:-

(1) Manufacturing of various products like, daily need consumable items, electrical/electronic goods, mechanical goods, hardware articles, items/ materials used in Civil Engineering works, heavy industries (viz. manufacture of steel, cement, vehicles, machinery, equipments and its spare parts). But these and such types of manufacturing industries will not be dealt with specifically in this book. However, management principles and methods of preparation of various management tools (viz. CPM networks etc.) for these are almost same as for civil engineering projects which are being dealt in this book in detail. A few manufacturing industries of civil engineering works (viz. manufacturing of pre-cast concrete units, RCC Hume pipes, ready mix concrete, Hot Mix Plant for road construction works etc.) will be outlined briefly in this book.

(2) Management of various civil engineering works will be dealt. Some modifications in management application may be required for rural engineering and some other special/non-conventional works.

(3) Management principles are also applied in other works, like, environmental management for works at services/ utilities (viz. water supply, sewerage/ drainage systems/ municipal solid waste), consultancy services of various works, research & development works etc. Management of some of these works will be dealt. But works of defence sector will not be dealt, except for the management in construction and maintenance of border roads, which will be discussed briefly.

5.1.2 The first group (i.e. manufacturing, factories) is vast and effective management application is important. The same basic principles of management are applicable for manufacturing goods and materials as for other disciplines. However, there will be some difference from management of civil engineering construction and maintenance works.

Material management has been dealt under Chapter (3). This is applicable for various sectors/group of works.

5.1.3 One example is given here. There is a proposal to construct a multistorey building in a city. Broadly, there are three main stages of this project, viz. i) Pre-construction stage, (ii) Construction stage and (iii) Post-construction stage (i.e. operation and maintenance). The government department/owner of the building will engage contractors for construction purpose and, thereafter, operation and maintenance will be carried out by deploying necessary skilled and unskilled manpower departmentally. Necessary planning of the entire scheme/project is to be done by the organization of owner. The first stage (i.e. pre-construction) is also to be carried out by the office/organization of owner. Details of project management inputs will be dealt with.

5.1.4 The contract agreements for various components/sub-heads/ disciplines of the project are finalized and works are awarded to the main contractor and other contractors. During first stage, owner's organization is also required to prepare master programme charts (in the form of Critical Path Method (CPM)/Programme Evaluation Review Technique (PERT)/Bar Chart of the entire project (i.e. pre-construction and construction stages). Thereafter, the contractors take up the construction works and they are required to prepare detailed programme of various activities of project (week/month wise), including procurement of materials, equipments, machinery, arranging fund etc. Thus during the construction period, there will be two groups for control/management i.e. one group from the side of owner's organization/department and another group from the side of contractors. There will be Project Manager/Engineer and his subordinates of various disciplines from the side of owner/ department, as also a similar organisational structure from the side of contractors. In case, owner/department engages Architect and other consultants, then they will be controlled by the owner's Project Manager. In case the project is large, then owner may deploy Chief Project Manager/Chief Engineer. All the above stated managers/

engineers (Civil, electrical, mechanical, horticulture) of owner and Project Manager of contractors, architects and consultants engaged by the owner/department are to work under control of Chief Project Manager/Chief Engineer. There should be one Chief Project Manager/ Chief Engineer (Civil) of the owner department in respect of all disciplines including electrical, mechanical etc.

5.1.5 Necessary management inputs/techniques are also to be applied by contractors to achieve the goals. In fact, task of contractors is more difficult and important. Success of project mainly depends on efficiency, capability, resourcefulness of contractors. The main contractor engage some sub-contractors and specialized agencies. However, contractors for other discipline works (viz. internal electrical installations, A.C. system, Lift works, Fire Safety installations etc.) are to be engaged by the owner/department.

5.1.6 Suitable organisational structures are framed for various other civil engineering works/projects such as construction/maintenance works of roads (city road, highway, hilly roads), railway embankment and track construction, bridge construction, water supply & sewerage systems with treatment plants construction/operation/maintenance, drainage works, irrigation project, factory building, municipal solid waste management, environment management, civil engineering factory productions etc. Though basic principles of management for aforesaid works are same, yet certain suitable technique/strategy/ management are required for various types of projects/ works.

5.1.7 Contract management is part of project management. This is a specialized subject. Hence contract management including arbitration matter and laws related to construction works will be dealt in a separate book.

5.1.8 Stressing particularly, it is required to focus on the control of two highly influential factors: i.e."time and cost". CPM is used for time control. Planning, scheduling, project shortening and resource management will be dealt in detail. Contractors are more concerned for effective management techniques because they have to arrange adequate manpower, materials, equipments/machinery economically and to obtain maximum output, maintaining proper quality. Most difficult and important resource is money for executing the works smoothly. Economy and cost controls are more important for the contractors. Hence they have to prepare various management tools in detail and to carry out in-depth monitoring daily. Unfortunately, management of most of the contractors in India is not satisfactory. They do not apply various management tools, techniques. They face lot of problems to arrange adequate fund early enough and are unable to put together the necessary resources (manpower, materials, machinery/ equipments) well in time, thus hindering implementation of the planned programme.

5.1.9 Pre-Construction Stage is divided in following three sub-stages:-

A Preliminary/conceptual planning stage
B Detailed planning/design stage, and
C Tendering stage.

These stages are very important like foundation of a structure. If these stages are weak/defective, then construction stage will be affected adversely. Unfortunately, in India adequate care is not taken for management during the aforesaid three stages of pre-construction period. Suitable management techniques/tools are not applied during planning/pre-construction stage. Due to this reason, it has been observed that it takes considerable time for completing various activities of the pre-construction stage; for example project (i.e. construction of a multi-storey building in a city), it may take about 18 to 30 months. Administrative authorities also feel concern for the same. In many organizations/government departments, the engineers do not get proper freedom to implement technical management requirements. Top management support is essential, particularly during pre-construction stage. One important factor is that adequate time should be given for various activities which are stated under subsequent paragraphs/sections.

5.2 Preliminary/Conceptual Stage of Planning

5.2.1 Various activities of this stage are carried out by the Engineer-in-Charge/Project Manager of the owner's organization/department. As first stage, structure of project team for carrying out various activities during pre-construction period, which is also known as planning period, is to be decided by the competent authority. Then specific engineers having necessary experience are to be posted. Initially, fewer persons are required at senior level. Number of engineers is required to be increased at the stage of detailed planning, designing and tendering. Suitable ranks of engineers are also to be decided. Another important policy decision is to be taken about engaging private Architect/Consultant for preparation of building plans including obtaining approval from various statutory authorities, structural design & drawings and other specialized services (like fire suppression and safety system, air conditioning systems etc.). If sufficient competent engineers and architects are available in the owner's organization/government department (like CPWD, PWD, MES, Railways), then it may be possible to carry out all or some of the activities by owner's/department's engineers/ architects. If necessary, private architect consultant may be engaged for preparing plans (including obtaining approval from various statutory authorities) and to prepare detailed drawings, detailed structural design & drawings, detailed estimates, detailed specifications, NIT (tender documents) etc. If architect consultant and other consultants are engaged, then detailed scope of services, terms & conditions are to be prepared very carefully. Detailed procedure about engaging architect/ consultant will be dealt in a separate chapter. Engaging architect/consultant also depends on magnitude and intricacy of project.

5.2.2 In case architect-consultant including structural designer is engaged, then number of owner's/department's engineers will be less. Policy decision on this basic issue is to be taken by the top management of owner's organization/department after considering detailed proposal of the head of engineering department. Thereafter, detailed requirement of owner's/departments technical personnel are to be worked out and approved by the competent authority. These personnel are to be posted in phased manner. The most important factor is selection of competent planning engineers and architects; they must possess suitable and sufficient experience to carry out such planning duties in respect of building constructions. They should have aptitude in planning and design works. In reality, it has been found that personnel having experience on irrelevant works and having no aptitude, expertise or experience in planning works are posted. Due to this reason, not only the quality and progress of planning/designing activities suffer badly, but a lot of serious problems also crop up later during construction stage. Due to aforesaid wrong selection of technical personnel, project cost is increased considerably; quality of works (particularly finishing items of building) are also affected adversely due to stipulation of wrong specifications.

5.2.3 After finalizing proposed structure/set up of technical personnel (technical persons of department/owner, private architect-consultant and other consultants), it is required to post suitable persons without delay. Project Engineer (P.E.)/Project Manager (P.M.) is posted first and then other persons report to him. Office along with various facilities (viz. computer, telephone – (mobile and landline), furniture, vehicles etc.) is to be set up. A circular about setting up of this office should be issued to all concerned offices for better communication and coordination.

5.2.4 Basic broad proposal for constructing the multi-storey building, indicating the available land, total covered area for offices, source of fund etc. is to be prepared by the P.E. and the same is to be examined by all concerned branches/departments including finance department and, thereafter, the same is to be placed before the competent authority (top management) for principle approval. Most important element is "availability of suitable land with valid title and permissible land use for constructing office complex or for other use". After examining this aspect and clearance of the available plot, further step is to be taken. Total permissible covered area of the building is worked out on the basis of area of the plot, considering permissible FAR (Floor Area Ratio) as per records of Development Authority of the city. For knowing budgetary fund, rough cost of the project is worked out on plinth area rate basis. Owner/department will check the sources for arranging fund during relevant period through the finance department (in case owner is a government organization). Sufficient period will be available for this purpose because considerable time is required for awarding the construction work to the contractors.

5.2.5 As next step, P.E. is required to discuss with user department and concerned persons who can give detailed requirements, facilities etc. for the project. This part of project planning is very important because building plans, specifications, provision of various facilities are made on the basis of above particulars. These particulars are to be given to the architect through Project Engineer for preparing building plans, detailed drawings, specifications etc. If the building is to be used by multinational companies, then standard of buildings, facilities, specifications, safety provisions etc. are to be envisaged accordingly.

5.2.6 Thereafter, architect is to prepare conceptual/preliminary buildings plans on the basis of above stated preliminary data. Architect is to discuss the same with P.E./Chief Project Engineer and other authorities of the owner/department and take cognizance of their observations/suggestions suitably. Accordingly, the modified conceptual plans and model are to be prepared and the architect will discuss with P.E./Chief Project Engineer and then with top management authorities. At this stage, broad clearance/comments are to be obtained from the structural designer as the final technical feasibility of a proposed building depends on the clearance of structural design. Thereafter, final modified conceptual drawings and its model are prepared. The drawings and model are to be sent to the laboratory for "Wind Tunnel" test, which is important, particularly for multistorey/high rise building. After that, the scheme, including project conceptual building plans, report showing various data, FAR (Floor Area Ratio) calculations, floor-wise areas, plot area etc. are to be submitted to all concerned statutory authorities. First, it is required to obtain clearance /approval from main building plan approval authority (i.e. local municipal authority/ city development authority/city urban art control authority). It may be required to modify the building plans as per observations of above stated statutory authorities and to be resubmitted to them.

5.2.7 Engaging Architect Consultant

5.2.7.1 Immediate after principle approval of the proposed project by the owner, it is required to fix up architect and structural designer for preparing conceptual/ preliminary building plans. In the present example, project (i.e. construction of multistorey building), it is assumed that competent authority of the organization has decided to engage architect consultant who will engage a structural design consultant. It may take two to four months or more to engage architect consultant.

5.2.7.2 For intricate large building, it may be required to invite architectural competition from reputed architects. Guidelines are issued to the architects about submission of drawings, models, project particulars etc. Stipulation is also made in the guidelines that certain amounts will be paid to first, second and third prize winners by the owner/department. Notices for the same are issued in the newspapers. Board of assessors is constituted by the top management after considering proposal of Chief Engineer. Senior/reputed private & government department's architects and owner's Chief Engineer/ P.E. are made as member and Secretary of the board for conducting this architectural competition. Marking system should be adopted on various elements of the submission made by the architects. Presentation is given by all the architect competitors before the board of assessors. If any observations are raised by board members, necessary clarification is provided by the competitors. Thereafter, necessary markings are conferred by the board of assessors on the submissions made by the architect competitors. On the basis of above, the board decides the names of competitors/architects who won the first, second and third prize. Necessary prize money is paid to the three winners. The first prize winner architect is selected as Architect Consultant. Their fees are to be decided by negotiation on the basis of prevailing market trend. For conventional building work, some guidelines on fees have been given by the Council of Architects (India). The detailed scope of services to be rendered by the Architect Consultant and Structural Design Consultant are to be laid down under one agreement. Various terms & conditions are to be outlined in the agreement after checking some agreements of similar works. Unfortunately, proper standard agreements for such consultancy services are not available in most of the organizations/ departments. If required, it may be possible to enlarge the scope of services of architect-consultant; like - consultancy services in respect of certain utilities (viz. water supply, sanitary installations, internal electrical installations, lifts installation). Day to day supervision is also to be done under such agreement. Joint measurements and checking of contractors' bills should also be part of consultancy agreement. Entire coordination with all contractors, his own associate consultants, various departments/branches of the owner's organization is to be done by the architect consultant. He is also to attend all weekly site meetings, inspections and meetings with other organizations for sorting problems and obtaining clearance from various authorities (viz. fire service authority, local body, city development authority etc.).

5.2.7.3 From management point of view, it is important to state here that the decision for preparing building plans and structural design is to be taken immediate after principle approval of the proposed project. The project can be processed only after finalization of conceptual/preliminary buildings plans. It may take 3 to 5 months' period to finalise conceptual plans for approval of various statutory authorities. If it is decided that private architect-consultant and structural design consultant (who will be associate of architect consultant) are to be engaged, then immediate action is to be taken to engage them as per procedure stated above. Considerable period (3 to 6 months) is required to engage a competent architect-consultant. Success of the project also depends on competent/ reputed architect consultant. As per practical experience, progress of work was hampered in many large/medium size projects due to poor/

unsatisfactory services rendered by the architect-consultant. Hence special precaution must be taken by senior engineers for selecting a reputed, competent architect-consultant. Past performance of architect and other consultants should be checked from other organizations before selection/pre-qualification. Thus it may be seen that considerable period is required for engaging architect-consultant (including his associate consultants) and to finalise the conceptual buildings plans.

5.2.7.4 Next step is to submit conceptual building plans to various statutory authorities i.e. local body (municipality), city development authority/city urban control authority. Their observations are to be complied on top priority (if required by preparing modified building plans). All these authorities are to be pursued by the architect for obtaining approval. Approval is accorded by these authorities after their observations are complied with. But this approval is subject to obtaining clearance from other various authorities viz. fire service authority, Civil Aviation Department (for height clearance), Archeological Department (if any archeological structure/ruins exist at nearby to the proposed building) etc. Obviously, it would take considerable period (about 4 to 7 months) for approval of conceptual building plans from the local body as well as city urban control/ development authority. Further period will also be required to obtain clearance from other authorities. For multi-storey building, clearance of fire service authority is most important. Besides fire safety installations, it is also required to make provisions for other fire service requirements like, refuse areas at all floors, suitable staircases, external emergency staircases, circulation pavement at ground level etc., architect is required to submit the building plan for these requirements as first stage approval of Fire Service Dept. Second stage approval is to be taken for detailed design and drawings of fire safety installations later on.

5.2.7.5 On the basis of above stated preliminary building plans, preliminary/project estimate is to be prepared for obtaining administrative approval and expenditure sanction from the competent authority. In case owner is a government department, 'competent authority' depends on the amount of project estimate. The Project Engineer must send preliminary building plans to all concerned branches/departments of the organization like, electrical department for obtaining preliminary estimate in respect of electrical work (internal & external), lift works, A.C. installation, electronic/ telephone installation etc. Project Engineer is to incorporate these estimates in the project estimate. The project estimate is to be submitted to Finance Department for clearance and then to the competent authority for according Administrative Approval and Expenditure Sanction which are important basic requirement to process further.

5.3 Land Position

5.3.1 Matter related to purchase of land for government/private project is a very long process, particularly in a city. Lot of difficulties are faced. Land acquisition matter has been dealt under another chapter.

5.3.2 Sufficient time is available after first step of principle approval of project till approval of building plans from various statutory authorities. During this period, it is essential to check thoroughly all documents related to legal title, ownership, land use as per records of master plan/city development authority. If there are encroachments and other hindrances on the plot, then details are to be surveyed and superimposed on the building plan. Necessary action is to be taken to remove above stated hindrances. It has been observed that considerable period is required to remove encroachments/ unauthorized structures. Encroacher may obtain temporary stay order from the court. Such problems are to be pursued personally and sorted out before actual commencement of the project.

5.4 Soil Investigation of Plot

This is an important part of project management. Soil investigation is to be carried out at this preliminary stage of planning. Detailed technical requirements of soil investigations by making bores are to be obtained from the structural designer. Soil investigating agency must be reliable, competent and reputed. Numbers of bores are to be marked on the layout plan of the plot by the structural designer. Correct locations of bores are important from technical point of view. Date of soil boring is important because underground water level varies considerably during a year; effort is to be made to obtain historical data of underground water level. Maximum thrust on foundation will be at the time of highest water table. Proper selection of agency is important to ensure correct test reports and correct analysis of tests.

5.5 Surveying and Levels of Plot

After taking decision to construct a building (example project), it is required to carry out field survey of plot including surrounding/ adjacent areas. Existing physical features are also to be plotted on the survey plan. Levels are also to be taken in respect of plots and adjacent roads and other features (including ground and road levels). Benchmark (as reference point) is to be shown on the survey plan/ contour plan/ sectional drawing. Above survey plan and levels are to be sent to architect. Plinth level/floor level of ground floor, levels of internal roads etc. are to be decided carefully.

5.6 Monitoring

5.6.1 CPM network for above stated activities of preliminary planning stage should be prepared by the P.E. Generally, it is not prepared and effective monitoring is also not done. Due to this reason, it has been found that prolonged period (12 to 24 months) is taken to obtain approval of building plans (conceptual stage) from various statutory authorities after clearance given by the owner. It may be mentioned here that it is difficult to predict likely period to be taken by statutory authorities for approving the building plans because it depends on certain factors viz. type and number of observations raised by above stated statutory authorities and compliance period taken by the architect, whether building plans committee meeting is called fortnightly or monthly and when the plans are placed before the meeting etc.

5.6.2 However, it may be mentioned here that after approval of preliminary building plans by statutory authorities, it takes much less time to obtain approval for the detailed building plans from statutory authorities (i.e. second stage approval). For small buildings, one time approval is required. Minutes of the weekly/fortnightly monitoring meetings to be held under chairmanship of Chief Engineer/P.E. are to be sent to all concerned persons including top management. During monitoring, reasons for delay/bottlenecks are to be ascertained. It may be required to meet concerned authorities alongwith architect. As illustration, a CPM network programme for pre-construction stage planning is given at figure No. 5.1 (under para 5.11). Time required for various activities is assessed by senior experienced engineer on the basis of completed similar projects. During monitoring, if it is found that there has been considerable delay on account of any activity/activities, then remedial measures are to be taken and revised CPM programme is to be prepared. This CPM network is put on the overall CPM master network of the total project activities (i.e. planning, construction activities of main contractors and other utilities viz. electrical works, water supply, sanitary installations, fire service installations etc.). Various activities in respect of CPM network planning for pre-construction stage is given in Table (under para 5.11 as illustration of example project).

5.7 Detailed Planning, Structural Design Stage

5.7.1 After approval of the building plan or preliminary building plan action on next stage for detailed planning, preparation of detailed architectural drawings, detailed structural design & drawings, detailed estimates etc. is started. If required by the statutory authorities, then final modified building plans (in case of large projects) are to be submitted for approval. Generally, at this stage, it does not take much time to get approval. Detailed planning, preparation of detailed architectural drawings, detailed structural design & drawings, detailed specifications etc. are started without waiting for further approval of building plans. As stated earlier, further approval of building plans is not required for small or certain type of building works from statutory authorities.

5.7.2 Detailed Estimate

Detailed estimate for proposed main contract/tender is prepared on priority. Many major management decisions regarding the project must be made during this stage. Management attention is required at this stage for following:-

(i) Project supervision structure (i.e. staff & line)
(ii) Major construction methods to be used
(iii) Time schedule of operations
(iv) How the project will be equipped.

Future success of the project depends on above.

5.7.3 The detailed estimate is prepared on the basis of architectural drawings which include drawings of toilets, kitchen, doors, windows, floors, wall claddings, railings, grills, glazing, finishing on external walls etc. Quantities of structural works are worked out on the basis of detailed structural design and drawings.

5.7.4 Nomenclature/specifications of various items are prepared on the basis of above stated architectural & structural drawings and also as per special specifications. Nomenclature/specifications of many items are available from Schedule of Rates and Specification books published by various authorities (viz. CPWD, PWD, MES, Railways, National Building Codes 2005 etc.). For items not available in above stated publications, nomenclature and specifications are to be framed as per approved proposals. Rates for such non-schedule items are to be analyzed on the basis of market rates.

5.7.5 Analysis of Rates for Items of Works

Rates of most of the items are generally available in the schedule of rates. If required, approved cost index due to increase of market rates of materials & labour is to be added on the schedule of rates. If items are non-schedule, then analysis of rates are to be prepared on the basis of prevailing market rates or partly taking data from relevant schedule item. Co-efficient of labour (i.e. labour requirement) is taken from standard analysis of rates published by competent authorities. If the same is not available in standard analysis of rates, then it is to be found out by taking site observations of sample works.

5.7.6 Above stage (i.e. D.E.) is very important because NIT (tender documents) are prepared on the basis of approved detailed estimate (i.e. technically sanctioned D.E.). If there is mistake in any items or specifications of D.E., then problem is faced during execution and for making payment to the contractor. Changing/modification of agreement items during construction may create contractual and payment problems. Contractual disputes may hamper the progress of works; such disputes should be resolved on priority basis. Prevention measures (i.e. good planning and design process) are more important.

5.7.7 Before preparing the D.E. and detailed architectural drawings, it is essential to discuss with all concerned persons/users/branches & departments of organization and to obtain their detailed requirements. It may be required to visit some similar buildings alongwith architect, representatives of user departments for finalizing finishing items, type of floor and wall cladding from functional points of view. Since the present example project is a multistorey office building, it is essential to decide specifications of doors, windows, glazing, glass panes of windows/glazing areas, floor, wall cladding in toilets, lift wall cladding/archives etc. as per functional and maintenance requirements, also considering the market innovation. If building is used by multinational companies, then specifications/ finishing should be at international standard. It has been found that many architects do not consider the functional & maintenance aspects effectively and problems are faced by the occupants of the building and maintenance engineers. As an example, due to wrong specifications/ materials of external wall finishing; many problems are encountered after completion of the building. This has been dealt in another book of the author (Practical Problems and Solutions in Civil Engineering Works). Many serious problems like, chronic seepage/leakage in lift pits, ceiling walls of toilets, basement floor & walls etc. crop up due to faulty architectural design/drawings/ specifications. Sometimes safety aspect is overlooked by designing faulty railing, grill or due to non-provision of grill in windows, glazed areas etc. Many such problems encountered during construction period and during use of building have been dealt in the aforesaid book of the author. Root cause of many such problems vests in faulty planning, design and drawings, wrong selection of specifications/materials etc. at detailed planning stage. It is stressed again to study this important matter in depth from user point of views.

5.8 Certain Management Matters

(1) ***Wind Tunnel Test***

Serious problems were faced in one high rise building at Delhi as wind tunnel tests were not carried out at the conceptual stage. Conceptual building plans were approved by all concerned authorities (viz. Development Authority, Urban Art Commission and Local Body) without wind tunnel test report. When wind tunnel tests were carried out after finalization of detailed architectural and structural drawings it was found that the proposed building failed structurally. The shapes of two towers were like reverse curves. The velocity of wind was increased by six times at the shortest gap between of two blocks. Besides delay in preparing the revised building plans, extra cost on account of manpower/ extra fees of the consultants were also involved. Hence it is essential to make suitable models of the building and to carry out wind tunnel tests at conceptual stage itself to know adverse effect.

(2) ***Proof Consultant for Checking Structural Designs***
If the height of the building is too tall or the building is of intricate type (due to special shape or other reasons), then planning should be done for engaging a competent, experienced structural designer as "proof consultant". Practically, it has been found that due to long process of planning, pressure is put on engineers by higher administrative authorities for inviting tenders without proof consultant. If it is required to check the structural design and drawings then proposal for engaging a competent structural design should be initiated at preliminary planning stage.

(3) To avoid subsequent problems, it is essential for civil engineers to scrutinize the detailed architectural & structural drawings and specifications, particularly from the functional/maintenance angles and economy. As example, size of shafts for various utility lines should be adequate so that maintenance person can enter in the shaft. If necessary, drawings and specifications are to be modified for preparing D.E. It has been observed that generally these important requirements are not looked into properly.

(4) ***Clearance of project from various statutory authorities***
Considerable period is required to obtain clearance from various statutory authorities at planning stage, such as:-

(i) Principle approval of the proposed scheme considering budgetary aspects, availability of land etc. by the top management authority of the organization.
(ii) Approval of conceptual drawings/scheme from statutory authorities viz. Development Authority, Urban Authority, Local body etc.
(iii) Height clearance from Civil Aviation Dept. and clearance from Archeological Department.
(iv) Approval of Fire Safety scheme from Fire Service Authority.
(v) Approval of final building plans from above stated statutory authorities.

Timely action must be taken for seeking approval of these and other specific items. In certain projects, Archeological and Civil Aviation departments' approval were not obtained and as such, the work was held up due to this lapse. It is essential to pursue above stated authorities at higher level for obtaining approval at the earliest time.

(5) ***Detailed requirements of User departments***
It is essential to obtain detailed requirements (like, sizes of rooms, broad specification, electricity points, A.C. etc.) from various user departments. Architectural drawings and details of various service facilities are to be prepared and the same is to be discussed with the users. Clearance is also to be obtained from user departments in writing at "preliminary planning stage". Problems are faced during and after construction due to proposed changes from the side of users. If necessary, it may be required to visit some existing buildings alongwith user department representatives at preliminary planning stage.

(6) ***Making approved samples as part of NIT***
Provision of huge quantities of granite stone, marble stone, ceramic/vitreous tiles is made in the NIT. Dispute arises with the contractor on the quality of white marble stones; the cost of white Makrana marble (first quality) is about double of the white Rajnagar marble. Hence it is required to make approved samples of stone slabs as part of NIT. Regarding ceramic/vitreous tiles, it is better to stipulate names of two or three reputed manufacturers in the NIT. Similarly, for ensuring good quality of flush door shutters (bearing ISI mark), names of a few reputed manufacturing companies can be stipulated in the D.E. /NIT.

(7) ***Seepage/Leakage Problems***
Unfortunately, problems related to seepage, leakage, dampness in walls, floors, roofs , lift pits, basements etc. are found during use of the buildings; life of buildings are also shortened. It is, therefore, essential to decide correct detailed specifications of water proof treatments of all relevant locations of the building by experienced engineers. Details in this regard have been dealt in other two books of author. Appropriate grade of pipe (like, "C" grade G.I. pipe which is heavy grade) should be stipulated in the D.E./ NIT. "Brick Coba" water proof treatment may be done instead of "Mud-phaska" treatment on the roof.

(8) ***Problem related to protection measure for deep basement works***

(a) Generally, in mega city, adjacent buildings (very old or new tall buildings) are very near to the proposed tall building. In large high rise buildings, it may be required to construct 3 basements to fulfill the approved norms of parking and other service equipments (electric substation, central A.C., pump rooms, water tanks etc.). It may be required to excavate upto 14 metres below ground level. It is not possible to provide side slope or stepping/benching on excavated basement area due to shortage of space. This safety problem is generally solved by constructing RCC retaining round piles at suitable interval as the same is economical. But there is some risk of collapse of RCC retaining piles. Slides of excavated area may occur during heavy rain; such incident had occurred at a certain project site during construction of RCC raft, which created havoc. Another safe method is by constructing diaphragm wall along entire length of periphery of the deep basement area. But this is very costly. Cost of diaphragm walls may be 4 to 5 times of the RCC round retaining bore piles.

(b) Aforesaid safety arrangement is essentially to be made for construction activities of basement floors. Final decision on method of safety arrangement should be taken by a group of senior technical experts/engineers as conscious decision after considering various factors, site conditions, soil conditions etc. For large deep basement areas (three floors with extended areas), cost of diaphragm walls is very high (may be to the tune of Rs.6 to 9 crores). This important part of structural design is to be finalized at the stage of detailed designing and provision made in the detailed estimate accordingly, otherwise adjacent property can be damaged due to construction activities and it will be required to pay damage/compensation (may be huge amount). Accident can also occur if above design is not made correctly.

(9) ***Completion Period of Construction***

Completion period of construction is stipulated in the NIT (tender documents). As per CPWD manual guidelines, the period for constructing high rise buildings becomes very long. On the other hand, administrative authority desires a shorter period. Both positions create problem. If it is too short period, then actually the same is not practically feasible. It is required to grant extension number of times. The CPM construction programme is prepared on the basis of stipulated period. It is required to revise the network several times. Several problems are faced during construction period for reporting to higher authorities, payment on account of escalation, too many changes about deployment of machinery, equipments, adverse effects on other conected contracts (viz. electrical works of lift, A.C., internal installation, substations etc.). Contractors of above stated services claim extra amounts due to delay of main civil work. Thus due to unrealistic stipulated period, many problems crop up during construction period. Hence realistic completion period should be stipulated in the NIT in consultation with experienced senior engineer.

(10) ***Earth work and Dewatering of Underground Water***

(a) It is required to excavate earth for a depth of 13 to 14 metre (depending on two or three floor basements) on entire area of basements. Huge quantity of earth is of good quality for which lot of demand exists in the market. As per normal method, large payments are made for earth work in excavation and cartage of earth to dumping ground, as per relevant items of CPWD Schedule of Rates.

(b) Trucks are not allowed to ply in the city during day time. Problem is faced for locating dumping ground for such huge quantity of earth. It may be required to make payment by the department/owner for using dumping ground. It is also difficult to measure the carting of earth including lead during night (after 10 p.m.). Another problem is about measuring and making payment for dewatering of underground water (to keep water table below bottom of raft i.e. 14 to 16 metre below G.L.). It may be required to carry out pumping throughout the day and night for very long period (may be 6 to 18 months). As per requirement of structural designer, above pumping is required till 3 or 4 floors are constructed to create sufficient downward load/force so that upthrust on the bottom of foundation/raft is counter balanced. As per normal method, measurements are carried out on the basis of relevant item (i.e. in the unit of kilolitre per hour) of CPWD schedule of rates. It is not practically possible to measure correctly throughout day and night for long period. Huge amount is involved on this item. Following method has been found to be beneficial to the department/owner.

(c) Additional/special condition be stipulated for above stated three items in the NIT (tender documents) i.e. earth work in excavation, disposal/carting of excavated earth and dewatering of underground water. These items will be executed by the contractor. Entire quantity of earth (except certain quantity required for back filling) will be property of contractor. No separate payment will be made to the contractor for executing above stated three items. Contractor would be required to quote rates of other various items of NIT considering the above position.

(11) ***Upgradation of Technology***

(a) For achieving good quality and expediting progress, it is required to upgrade technology in respect of certain items of works. One most important item is RCC work, particularly in basement floors (raft, outer vertical/retaining walls), exposed concrete finishing work etc. Serious problems are faced on account of seepage, leakage in the basements, lift pits. It is not only on account of faulty specifications of waterproof treatment, but also on account of average/fair quality of concrete and its laying operation without using good equipments. Structural design method should also be correct for water tight structure. Hence "Ready Mix Concrete" should be used after finalising proper mix design of appropriate grade of concrete. Appropriate water-cement ratio is to be achieved for correct workability. It is required to use suitable plasticizer in the mix. In case quality of concrete mix and its laying operation are very good, and construction joints are made properly then RCC raft, wall should be water tight. Leakage/seepage is also due to defective construction joints in the RCC walls and raft slab. These joints should be done as per details given in the code.

(b) There is another example of upgradation of technology. All floor slabs of high rise building are constructed by providing "precast slabs". RCC columns and beams are constructed as cast-in-situ. Floor slabs of large sizes are cast on ground at project site after making special type shuttering (using plastic fiber glass lining on strong wooden frame). Good exposed concrete finish on the bottom of slab can be achieved. Then the precast slabs are placed on the beams with cranes. A considerable time is saved and good exposed concrete finishing is also achieved. Ready mix concrete should be used.

(c) *Slip Form Shuttering*

As sequence of construction, it is required to construct the lift cores independently at superstructure. For expediting progress of lift-core construction, "Slip Form Shuttering" technique should be used. The form is lifted mechanically upward after casting of one metre height of lift core. This is a specialized work. Special agency with experienced, competent operator should be engaged. High operation skill is essential to maintain accurate verticality of lift core. Verticality has to be checked carefully at every one metre height and also accuracy of verticality at every 5/10 metres interval. Accuracy of verticality checking also depends on skillfulness, correct type/size of plumbob and string.

(12) ***Finishing on external walls***

(a) *Exposed Concrete Finishing with pattern*

Problem is faced if exposed RCC wall surface (external walls) is done with special type of pattern to be made with vertical strips/beadings. Practically, it is not feasible to make good finishing with normal shuttering and concrete. It is also difficult to maintain straight vertical lines of such type of beading/strips on external walls of 100 metres or more height. Skilled workers for such work may not be available locally. Ready mix concrete should be laid with pump and compaction should be done with special type vibrator equipment. There is another problem, viz. to maintain uniform colour of exposed concrete finishing on entire external wall surface because it is not possible to procure same colour of coarse sand and cement during long period (3 to 4 years).

It may be required to apply "silicon" coat on exposed concrete surface to protect from rain/weather effects. It is, therefore, recommended for not adopting above type of special pattern. Civil engineers should examine such proposal of architect before stipulating in the NIT. Suitable specifications, details of pattern, ready mix concrete, special type of shuttering with lining etc. should be stipulated in the NIT.

(b) Practical problems have also been seen for finishing on external walls with stone slabs (fixed with metal hooks/clamps). Some slabs fall down and fatal accident/injury may occur to person who walks near the building. Further, it is not easy to repair/maintain such type of finishing on a tall building. Hence special care should be taken for deciding specifications of external walls.

(13) ***Fire Prevention/Fire Suppression Works***

Main problem is faced for obtaining NOC from Fire Service Authority after completion of building. Hence it is essential to design and prepare details, specifications, building plans etc. as per building bye-laws/codes. Compliance is to be done in respect of "12 point" of fire suppression system as laid down in the Municipal Bye-Laws, including proper approach roads for heavy fire fighting vehicles, equipments (viz. snorkel, long ladder, tender etc.). Sufficient pavement space is required around the building for aforesaid heavy equipments, vehicles. The authority also requires certificate from competent engineer about structural strength of certain load for movement of snorkel vehicle. In many buildings, the plaza/pavement around the building are constructed on roof of the extended basement, but structural strength of aforesaid roof is less than the requirement of Fire Service Authority. Due to this reason, problem is faced; issue of NOC is held up for long period.

(14) ***Cutting of Tree***

Architect should prepare layout plan of the entire project in such a way so that cutting of trees are avoided as far as possible. It is very difficult to get approval for cutting trees. If cutting of some trees is absolutely essential, then the same should be processed in advance alongwith relevant preliminary drawings.

5.9 Monitoring and CPM Programme

CPM network for various activities of pre-construction stage/ planning should be prepared. Monitoring of all activities should be carried out by conducting meeting fortnightly under the chairmanship of Chief Engineer. All concerned persons, including architects, structural designers, civil and electrical engineers, consultants are to be present. Reasons for delay, problems/bottlenecks are to be looked into and remedial measures be taken. The CPM network diagram of pre-construction stage is at Fig. 5.1. Activities are stated in Table No. 1.

5.10 Bidding/Tender Stage (Pre-Construction Period)

5.10.1 Next is the most important stage of planning i.e. tender stage. Tender documents (also known as NIT – Notice Inviting Tender) become contract/agreement documents after acceptance of tendered/negotiated offer. Success of project depends on competency and reliability of the contractor. NIT consists of: - Standard/General conditions, tender notice, Schedule of Quantities/Bill of Quantities, specifications of each item of BOQ. Specification books and Schedule of rates published by the competent authority are also part of the NIT. Approved samples of certain items (like stone slabs for floor, wall cladding item etc.) should be part of NIT. Certain drawings should also be part of NIT. Additional/Special conditions and Specifications are also prepared and placed as part of NIT. If any materials (like cement, steel) are to be issued to the contractor by the owner/department, then necessary stipulation is made in the NIT. The NIT is to be prepared as per department's rules, regulations and procedure. Completion period is stipulated; this is very important. Realistic completion period is to be decided by senior engineer/Chief Project Engineer after considering many factors. If it is too short, then the same is not practicable and can create various problems during construction period.

5.10.2 Schedule of Quantities/BOQ are prepared on the basis of approved D.E. Many problems are faced during construction period due to faulty NIT. This matter has been dealt in two books of the author (viz. (i) Practical Problems and Solutions in Civil Engineering and (ii) Modern Technology in Civil Engineering). If these guidelines are followed, then many problems encountered during construction period will be solved. At this stage, first NIT of the proposed main Civil Engineering contract is prepared on top priority basis. This is framed on the basis of D.E. It is important to add work of certain utilities/service lines (e.g.. Water supply system, sanitary installation, drainage) in the NIT of the main civil contract. It may be mentioned here that normally D.E.s of various utilities are prepared separately; but certain utility works are added in the aforesaid main tender document. It ensures responsibility of one main contractor. However, tenders for electrical works, lifts and A.C. system are invited separately. If the multi-storey building (present example project) is above 10 storey and soil bearing capacity is poor, then it may be required to construct "deep bore piles". Such deep bore piles are specialized work and separate tender should be called from pre-qualified specialized firms. Accordingly, separate NIT should be prepared for pile work. Since this work is to be started first, top priority should be given for preparing D.E. and NIT for pile work.

5.10.3 Prequalification of Contractors

One main factor for successful completion of project construction depends on reliable, competent and reputed contractors having experience on such type of works. We know that once tenders are opened, it is required to award or negotiate with the lowest tenderer. Hence it is essential to pre-qualify (PQ) the tenderers for various works of the project. To save time, processing of PQ of contractors should be started alongwith preparation of D.E. For this purpose, PQ documents (like, NIT) are prepared by laying down the eligibility criteria. Applications are invited by issuing notices in the newspapers. The tenderers are required to obtain PQ documents from the department and to submit the same alongwith various supporting records by specified date. Technical Committee finalises the pre-qualified list of tenderers by marking method. Subsequently, NIT (tender documents) are issued to these pre-qualified firms/ contractors for submitting their tenders. Thereafter, pre-bid meetings are held with all the pre-qualified tenderers (who purchased the tender documents). Various issues/observations from both sides are discussed and final decisions taken by the owner's organisation/ department and the same are communicated to these firms; this letter is made part of NIT/tender documents. It is also essential to verify registration documents of the firms (as registered in various departments of constructions).

5.10.4 Tender Process

Details on tender matter have been dealt in another book of the author. A few important requirements/procedures are stated briefly as under:-

5.10.4.1 There are various types of tender (16 types of tender), each having certain types of advantages/disadvantages. Selection of type of tender depends on the nature and magnitude of works. For large building work (like, present example project), "item rate" tender is best. Rates of items are filled up on the BOQ/Schedule of Quantities by the tenderers.

5.10.4.2 If the works are not large and it is required to invite tenders urgently, then tenders are invited from contractors who are registered in appropriate class in certain approved government departments (like, CPWD, PWD, MES, Railway etc.). In such cases, it is essential to verify the eligibility of tenderers (i.e. class of registration, validity, list of similar works executed showing cost of work, period taken, performance certificate etc.). In case tender documents are issued to non-eligible contractor and he becomes the lowest tenderer, then it is a problem to reject the lowest tender; in such position, it would be required to reject all tenders and to re-invite. It is essential to verify minimum numbers of similar works executed during the last five years or the period as stipulated in the NIT, as also the monetary value of these works. In case a tenderer executed the works in different capacity (viz. under joint venture), then relevant records (joint venture/shareholder on partnership basis etc.) are to be verified and decision is to be taken about issue of tender documents by the competent authority. Hence it is essential to stipulate in the NIT about details required to be submitted by the tenderers "within specified date". Difficulties are faced due to non-submission of proper records or late submission of supporting records by the tenderers. Time is also required for verifying their records and obtaining approval of the competent authority for issuing or not issuing the tender documents. For important/large works, it may be required to visit a few completed works or checking from the organizations under whom the contractor executed the works. If it is essential, the date of submission of tender may be extended by issuing notice in the newspapers and to various offices and contractors who purchased the tender documents.

5.10.4.3 As per detailed instructions/procedure of tendering stated in the NIT, the tenderers are to submit their tenders in three sealed envelopes (as per one procedure for large works or intricate type of works). First envelope contains earnest money and is to be opened first, on specified date & time. In case earnest money has been submitted in correct mode and amount, then only the second envelope containing technical bids is opened on the same date/time. The third sealed envelope contains financial bids and is not to be opened on the same day; tender opening officer will sign on the cover of sealed envelopes and record on the register as record. The technical bids are examined and eligible contractors are identified. Then intimation is sent to all eligible tenderers about date and time for opening the third envelopes (i.e. financial bids). The third envelopes are opened and rates are read and recorded in words (if not written by the tenderers) by the tender opening officer. Cutting/over-writings/additions etc. are to be attested suitably on each page of tenders by the tender opening officer. Above stated procedure must be followed strictly, otherwise serious problems may crop up and it may be required to cancel the entire tender and to re-invite. In two envelope system, the first envelope contains earnest money. The second envelop contains bids on original tender documents issued by the department; rates are filled up against each item (or percentage rate in case of percentage rate tender) and the same are opened after opening and checking valid earnest money. Under this system tender documents are issued only to eligible (registered contractors or pre-qualification done separately before issue of tender documents).

5.10.4.4 Tender Evaluation

(a) In case of three-envelope system, there should be no condition (technical/financial) in the financial bid. Thus it becomes easy to evaluate the tenders and to prepare comparative statement of financial bids. It is also easy to identify the clear lowest tenderer (L-I). In this system it is also possible to select reliable competent through P.Q. method.

(b) In case of two-envelope system, the tenderers may put some conditions and suggestions and there may be some financial/technical implications of the same. These implications are to be worked out carefully. Comparative statement of financial and technical implications is also to be prepared, besides comparative statement of rates quoted by the valid tenderers. Financial implications of conditions (which may be plus or minus) are to be effected on the quoted rates on the comparative statement. Thereafter, it is required to identify the real lowest tenderer. The apparent lowest tenderer (i.e. without effecting financial implications) may not remain as actual/absolute lowest tenderer. It is very important to identify the correct/absolute lowest tenderer. If negotiation is to be conducted, then the same is to be carried out with the actual lowest tenderer (not with apparent lowest tenderer). In case negotiation with actual lowest tenderer fails, then the entire tender is to be rejected and re-invited. As per government procedure, it is not permissible to negotiate with second lowest (L-2) tenderer. However, in case of private works/organization, the owner can take decision suitably as per his benefit.

5.10.5 *Detailed Justification of Rates*

5.10.5.1 This is another important part of tender evaluation. Following methods, factors are relevant to work out justification of rates. Lowest tendered rates and justification of rates are compared to ascertain the reasonability of tendered rates.

(1) Working out of detailed justified rates considering market rates of labour and material:-
Rates of each item are prepared by analysis, considering market rates of labour and materials. Then total amount of all items is worked out. In case some elements are not included in normal analysis of rates of schedule items, then suitable extra amounts are added. Considerable time is required to prepare such voluminous records after verifying prevailing market rates of material and labour. Percentage (above/below) of tendered amount over the above justified amount is worked out.

(2) Trend of market rate based on other awarded works of such type is to be ascertained. A list of such awarded works with awarded date is prepared. If some works were awarded 3 or 4 years back, then suitable cost index is added.

(3) Additional amounts on account of certain special/additional conditions of NIT (which have not been considered in the above stated detailed justified statement) are to be worked out and added in the justified amount as worked out under para (1) above.

(4) It may not be possible to work out extra specific amount on account of certain factors like, movement of heavy vehicles are not allowed from 6 AM to 10 PM in city, night shift working due to short stipulated period/crash programme work, double handling of materials, earth due to very less available site, space etc. Some time extra amount on account

of electricity is known only after completion of works. These factors are also to be considered for justification of rates. Certain reasonable percentage may be added on above factors.

(5) Percentage on account of profit/overhead is more than normal value of 10% in respect of pre-qualified high standard big firms. Extra cost is considered for deploying special/large machinery, plants, ready mix concrete, interior decoration works etc. as per NIT conditions. These factors should also be considered for justification of rates.

(6) *Healthy competition of tender*

It is required to consider healthy competition aspect i.e. there should be at least 3 or 4 tenders from competent firms. It is also to be seen that the tender is not "pull tender" (i.e. all tenderers have a common understanding in malafide way). If it is "pull tender" case, then all tenders should be rejected and re-invited.

5.10.5.2 Other Points

Before taking final decision on tender (i.e. to accept or reject the tenders), it is required to check various factors. Reasons should be recorded for acceptance or rejection of tender. There may be problem if tenders are rejected and recalled and, thereafter, much higher rates are received. Assessed position should be recorded. If any letter is received by the department from any tenderer after opening and before acceptance/rejection of tenders, then the same is deemed to be as "post tender development" and is not to be considered. The department/organization is required to finalise the tender case within validity period as stipulated in the NIT. In case it is not possible to decide within the validity period, then it is required to obtain letter containing extended validity period from all the tenderers, particularly from the lowest tenderer.

5.10.5.3 Tender case is scrutinized at various levels/branches (i.e. field division, accounts branch, planning branch) and final recommendation is to be given by the Chief Engineer/Chief Project Manager and the case is sent to Finance Department. Thereafter, the entire case is considered by the Tender Committee for decision/ to carry out negotiation with the lowest tenderer. If the tender/ negotiated offer is to be accepted, then "Letter of Acceptance" is to be issued within the validity date. Mode of dispatch/delivery of letter of acceptance is important. Legally, the "offer" and "acceptance" have to be very clear i.e. without any ambiguity/condition. Legally, contract is not formed if there is any ambiguity/condition in the letter of acceptance. Thereafter, contract agreement is to be signed by both the parties. Various documents i.e. complete original tender document of first, second and third envelopes, letters issued by the owner's organization/department, negotiated offer, letter of acceptance/work order etc. form part of the agreement. Certain points and check list related to negotiations have been stated under chapter 3 (paras 3.16.2, 3.16.3 and 3.16.4).

5.11 Monitoring/CPM Programme

Tender process should be monitored closely at senior level (viz. S.E./G.M.). It must be ensured that Letter of Acceptance is issued within validity date or decision is taken for rejection of tender and to re-invite. For larger work, validity period of tender stipulated in the NIT is generally 90 days from the date of opening of financial bids. CPM network programme of above stated activities for pre-construction stage should be prepared. Regular monitoring meeting should be taken on the basis of CPM network. This CPM network should be reflected suitably on the master CPM network of total project. The CPM network diagram for pre-construction stage is at figure 5.1. Activities are stated in table No. I.

Table No. I

Name of Project - Construction of High Rise Building
Activities during Pre-Construction Period

Sl. No.	Name of Activities	Activity arrow (Fig. 5.1)	Duration in week	Remarks
1.	Initiating proposed project including land matter and its principle approval by the owner	1 – 2		
2.	Constituting project team and their posting	2 – 3		
3.	Decision on engaging Architect consultant with Associate Structural Design Consultant and other services consultants and their appointments	3 – 4		
4.	Preparation of conceptual/preliminary building plans by the Architect to the owner (including Wind Tunnel Test).	4 – 5		

Sl. No.	Name of Activities	Activity arrow (Fig. 5.1)	Duration in week	Remarks
5.	Submission of conceptual building plans to various statutory authorities and their approval.	5 – 6		
6.	Preparation of project estimate and its administrative approval and expenditure sanction.	5 – 7		
7.	To carry out soil Investigation and its approval.	5 – 8		
8.	Engaging proof consultant for review/ checking of structural design.	7 – 11		
9.	Preparation of detailed architectural and structural design/drawings including specifications and its approval by the project Chief Engineer including report of proof consultant.	6 – 9		
10.	Preparation of detailed design, drawings, specifications of various services/utilities (water supply, sanitary, drainage) and its approval by the project Chief Engineer.	9 – 10		
11.	Preparation of detailed Estimate in respect of main Civil Contract and its approval (technical sanction).	10 – 15		
12.	Engaging consultants for A.C. system, Lifts, Internal Electrical Installations, Electric Sub-Station.	8 – 13		
13.	Preparation of detailed design, drawings, specifications, estimates of electrical works.	13 – 14		
14.	Pre-qualification of main Civil Constrictors.	11 – 12		
15.	Preparation of NIIT (tender documents) for main Civil Contract works and its approval.	15 – 16		
16.	Inviting tenders of main its receipts, evaluation, process, negotiations, decision and issue of letter of acceptance of offer.	16 – 1		

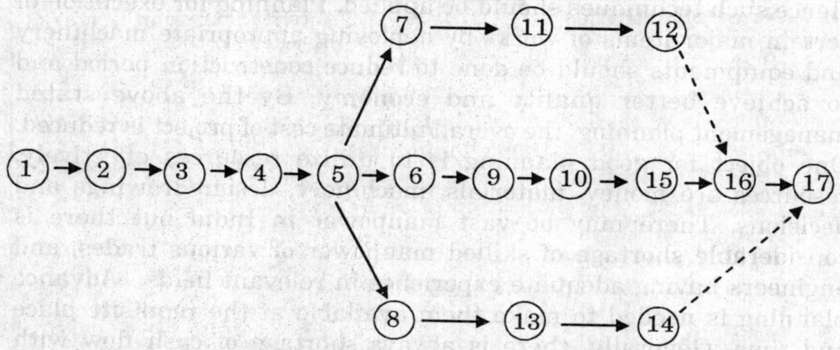

Fig. 5.1 : CPM Network for Preconstruction Period

5.12 Other Common Matters Related to Pre-construction (Planning) Period

5.12.1 On foregoing pages under this chapter, the matters related to management of pre-construction planning period have been dealt considering one example project (viz. construction of multistorey building in a city). Though management principles/ methods/ techniques in respect of other civil engineering works are same/ similar, yet there is difference in respect of activities, working details etc. Management of following works will be dealt under separate chapters:-

(1) Public Health Engineering/ works related to environments viz. water supply system, sewerage and drainage system.

(2) Highway Project/Road works.

(3) Municipal Solid Waste Management

(4) Special structures/Building

Management on Finance, Materials, Total Quality, Consultancy services etc. will be dealt under separate chapters.

Certain common management matters related to planning are stated as under:-

5.12.2 As per conventional manual method, it takes considerable time for finalizing structural design & drawings, particularly for large multi-storey/high rise building, special buildings/structures (viz. chimneys, towers etc.) It is easy to design the same by using appropriate computer programmes (software); preparation time is reduced considerably in designing and preparation of drawings. It is also easy to modify the design/drawings in future (if required).

Hence such techniques should be applied. Planning for execution of certain major items of works by deploying appropriate machinery and equipments should be done to reduce construction period and to achieve better quality and economy. By the above stated management planning, the overall/ultimate cost of project is reduced. One object for good planning is to utilize resources effectively; resources are money, materials, machinery, design/drawings and decisions. There may be vast manpower in India but there is considerable shortage of skilled manpower of various trades, and engineers having adequate experience in relevant fields. Advance planning is needed to make them available at the requisite place and time. Generally, there is always shortage of cash flow with contractors due to which execution of works suffers. To avoid such obstructions, advance planning is required. It is also essential to plan various activities considering season (certain items of work cannot be executed during rainy season or under very low temperature). Hence there should be flexibility of planning. It may be required to prepare contingency plans. It is essential carry out planning in advance.

5.13 Various Types of Tenders/Contracts

(1) Item Rate Tender
(2) Percentage Rate Tender
(3) Lump Sum Tender (There are variance under this category)
(4) Turnkey contract (including design, drawing, execution & commissioning)
(5) Build, Operate and Transfer (BOT) contract (Entire fund is arranged by the contractor)
(6) Partial BOT (like Joint Venture). Under this category, part fund is provided by the government department and rest is arranged by the contractor who is known as Concessionaire.
(7) Service Contracts
  (a) Running, operation and maintenance contract
  (b) Operation of parking space in city
  (c) Cleaning/sweeping of buildings
  (d) Security arrangement for buildings, industrial complex
(8) Consultancy contract
(9) Labour contract
(10) Foreign contract
(11) On actual cost plus percentage basis
(12) Rate contract/Running Contract
(13) (a) Contract for purchase and demolition of old buildings.
  (b) Contract for demolition of building

(14) Supply contract
(15) Consultancy contract for Highway project

Item Rate tender has been dealt under this Chapter. A few other types of contracts will be dealt under other chapters alongwith certain types of project management.

5.14 Work Break Down

(a) Considering practical aspects of planning, the "Work break down" of the project is important. There are large complex/campus projects like, University campus, industrial/factory complex, township of industries etc. consisting of various buildings/blocks, lot of external development/utility lines, landscaping etc. From practical consideration, it is difficult to control one large contract of various disciplines. Hence break down of components/buildings of total project is made, like.

(b) In a large complex of Asian Games Village, New Delhi, following components/work break down were made:- different four blocks of houses, central dining hall building, administrative block, cultural centre, external development works etc. Tenders were invited for above stated components of the project separately. CPM networks were prepared separately for each component for effective control. However, all the components of total project were shown as milestones on the master CPM network.

(c) Works break down of major items/group of items of each building/component are also prepared for detailed planning and detailed monitoring.

5.15 Planning of Contractor for Submitting Tender

(a) Planning is also to be done by the contractor. He purchased the tender documents from the office of department/owner's organization and the same is studied. The work site should be visited. Climate, availability of labour (skilled/unskilled), materials etc. are enquired and noted. Distance for transporting materials is also recorded. Market rates of materials, labourers are ascertained from local markets/sources. Duration of rainy season is noted particularly. Availability of electricity and its reliability (i.e. load shedding) are important. If electricity is not available or not reliable, then diesel generator set is to be arranged. After studying tender documents, all the drawings are to be studied in the department's office. Various clarifications are obtained by discussing personally with engineer/architect. After checking the position of other works which are being executed by the contractor, final decision is taken to submit tender.

(b) Various planning aspects are ascertained from the owner's engineer/architect. Construction methods, procedure, as planned by the department, are also ascertained because costing depends on the same. Sequence of construction is broadly decided by the contractor. Requirement of construction equipments, machinery is to be worked out/ planned. An improperly or under-equipped job is apt to be loser from the beginning, one that cannot be salvaged by any amount of project management.

(c) Estimated cost for the purpose of tendering (i.e. to quote rates) is to be prepared carefully. Specifications of items and its quantities are available in the BOQ issued by the owner/department. Labour and material cost of each item is worked out on the basis of past experience and prevailing market rates of similar works. Generally, such information is maintained by the contractor, including awarded rates of other works/contractors. Necessary plus/minus is made for the present tender case as per site location and other factors. Above cost is made as "direct cost". Equipment cost is also worked out for relevant items. A statement (i.e. abstract of cost) showing direct cost of labour, materials and equipments for each item is prepared. Thus total direct cost is worked out.

(d) Many other costs are to be added over the aforesaid total direct cost. There are some elements of indirect cost such as:- employer's contribution to social security, workmen's compensation, insurance, property damage insurance, fringe benefits, health & welfare funds, paid vacations etc. These items of indirect labour cost are added by certain percentage on total labour cost.

(e) Equipment cost is worked out on the basis of hire charge. Expenses associated with move-in, set up, operation, dismantling and move-out are also to be added in the cost estimate. If equipments are owned by the contractor, then "depreciation cost" is worked out and the same is considered for preparing estimated amount.

(f) Amount on account of "project overhead" (related at site set up) is assessed on account of various points related at project site, such as:- Site/project manager, supervisor, security guards, electricity, water, telephone, sanitary facilities, storage, first aid, dewatering , temporary shed/hut, photographs, barricades, job travel expenses, clean up etc. Overhead on account of head office is also to be added.

(g) Tax as applicable is to be included. Contractor's profit/mark-up (in percentage) is to be added last (i.e. on the grand total of direct cost plus indirect cost as stated above). Mark-up is also known as marginal amount (which may vary from 5 to 20% of project cost as worked out above). Deciding the markup/margin figure depends on many factors like, size of project and its complexity, location, evaluation of risk, difficulties inherent in the work, competition and other intangible factors. Contractor may think to mark the job as high as possible and still be the lowest tenderer. The usual selection of markup is largely based on historical perspective of past works. This is a "bidding strategy"; there is no rational procedure for deciding the markup/margin figure.

6

CPM Network Method of Works Programme (Construction Stage)

6.1 Introduction

(a) Certain basic management principles which are universal and applicable in various types of management have been dealt very briefly in Chapter – 2. Practical aspects of management in civil engineering works are dealt in this book. Basic management principles are applied in such construction works also as done in case of pre-construction period (also known as planning period) of project. Now management of construction period is dealt under this chapter. Here also basic management principles are applied on ground during construction period from where actual result/achivement on ground will be known.

(b) It may be seen from Chapter – 5 that the construction work has been awarded to contractor at the end of pre-construction period. Now further management action is to be taken by the contractor. His project engineer is to work out working details for execution of construction work. For this purpose "job planning" for management details is to be done. Proper management tools, various schedules etc. are to be prepared for day to day works/activities. Such work execution planning is to be done by the engineers of contractor initially and the same is to be submitted to the owner of the work. This job planning and working details are to be done on the basis of various particulars, items of works, specifications, detail design & drawings, terms and conditions, period of construction etc. stipulated in the contract agreement. The owner's organization/department is required to hand over hindrance-free land and various working drawings (good for construction) to the contractor immediately after awarding the work. Time is essence of contract and if there is any hindrance or delay from the side of owner's organization/ department then contractor will not be responsible for such period of delay. It has been found in many works actually that owner/department is unable to comply with their responsibilities in time, like, to hand over hindrance-free land, detailed working drawings (good for construction), soil investigation report, materials to be issued to contractor (if stipulated) etc. Department is also required to assist for obtaining approval of electricity and drinking water connections, permission for plying trucks from traffic police etc. Mobilisation action is to be taken by the contractor (viz., fencing the work site with security gate, post, temporary site electricity, installing drinking water pipeline, drainage arrangement, temporary urinals, latrines, bathing place, disposal of sewage/water, constructing site office, approach road to worksite etc. Mobilisation period which may vary 15 to 30 days or more depending on magnitude/type of works/ project is stated on the letter of award and the same is counted from the date of letter of award/acceptance of tender.

6.2 First Part of CPM Network – Job Planning/Construction/ Management Planning

6.2.1 Workable scheme of operation for day to day field activities is to be prepared. Construction planning is concerned primarily with establishing the sequential order in which various operations will be carried out on the job site. Such planning is the most important for the job management system. Inter-relationships of large number of job activities are intricate and complex. Hence it is required to apply effective management tools to manage these activities. "Criticial Path Method" is one such tool and is the basis for planning and scheduling methods. This is also known as "arrow diagram planning". CPM planning involves a depth and thoroughness of study that gives the contractor.

6.2.2 A project plan at this stage is usually the result of a joint study made by a number of contractor's engineers and other persons. The network diagram and the management data obtained from it will be realistic and useful if the job plan is produced and updated by those who understand the job to be done at site. Participation by sub-contractors is also important for preparing job plan.

6.2.3 Construction Planning

This consists of :

(1) determination of the job steps/activities;
(2) ascertainment of the sequential relationships among these activities;
(3) presentation of this planning information in the form of a network diagram (viz., CPM network).

6.2.4 "Work breakdown" of the construction work (as stipulated in the contract documents, drawings) is to be done as beginning. If the project consists of more than one block/segment/reach then CPM networks, schedules etc. are prepared separately but simultaneously. These are interlinked suitably on the master CPM network (through "interface" events which will be dealt later on. Detailed breakdown of each block/ segments are done by identifying various activities of works. Above stated three steps are carried out almost simultaneously (i.e., including preparation of initial preliminary/ rough sketch of network). It is important that the plan which is prepared be the one that the contractor actually expects to follow; because the entire objective is to plan the work and work i.e. implement the plan.

6.2.5 Job Activities

The extent to which a work is divided into activities depends upon a number of practical considerations such as :

(1) By area of responsibility
(2) By category of work (workers)
(3) By category of work (by type of equipments)
(4) By category of work (materials – concrete, steel, timber)
(5) By structural elements (foundation, walls, beams, slabs)
(6) Considering owner's breakdown (for payment purpose)

6.2.6 Job Logic

Start of many activities depends upon completion of others. Some activities are independent of others. Some activities are independent of one another and can proceed concurrently. It depends on approach to the work. It may be required to try to draw a separate CPM network for alternative approach. Here lies one of the major strength of the CPM planning method. It is a versatile and powerful planning/ management tool in the time evaluation of alternative construction method. The network diagram itself is nothing more than a graphical display of job logic, presenting in pictorial form all of the dependencies among activities. Project/ work Manager (Engineer) is to implement the aforesaid planning on ground. He has to identify various problems, hindrances in advance through proper management and effective system/tools.

6.2.7 Activity Restraints

The restrictions of job logic itself might be called physical restraints. This arises from the necessary order of construction operations. As example, placing of steel reinforcements for RCC column cannot be done unless shop drawing is received and approved fabrications are done and available at site. Practical limitations of this type are called "resource restraints" or "material restraints". Other instances of job restraints are non-availability of special skilled labourers, non-delivery of owner-provided material (viz., cement, steel bars), non-availability of approved structural working drawings etc. There may be "safety constraints" (on the sequencing of structural operations) on a multi-storey building. The recognition and consideration of job restraints is an important part of job planning. Failure to consider such constraints can be disastrous. It is usual to treat restraints much the same as activities and to represent them as such on the network diagram.

6.2.8 Draw of CPM Network Diagram

(a) Detailed lessons, method on drawing CPM network diagram are not dealt in this book as it is assumed that readers have elementary knowledge on the same. This is being dealt briefly.

(b) Activity is shown with a arrow notation; tail of arrow is the starting point and head of arrow represents its completion. In initial diagram length and orientation of arrows have no significance. Subsequently it can be converted to draw to a "time scale". The sequential relationship of one activity to another is shown by their relative position. Number is put in a small circle at tail and head of arrow for identifying the activities. The letter "i" is used as a general symbol to indicate the tail of arrow (i.e. start) and the letter "j" to designate the head of arrow (i.e. finish). The circles of junctions of the arrow (which are numbered) are called "events". The activities are named (like, i-j i.e. 1-2 (placing steel bars in column), 2-3 (pouring concrete) etc. If there are some events that are particularly important, they can be named and these events referred to as "milestones" (to indicate important intermediate goals). This matter with "interface" will be dealt under subsequent chapter.

(c) Another example project, "construction of a small bridge (one span) across a stream" has been considered for CPM network planning. The network is given at Fig.6.1. Various activities (as arrows), events, dummy activities, etc. have been shown on this figure. Following are main items of the work :

(1) Excavation and back fill
(2) Piling below two RCC abutments
(3) RCC Abutment No. 1
(4) RCC Abutment No. 2
(5) Steel girder
(6) RCC Deck
(7) Finishing items.

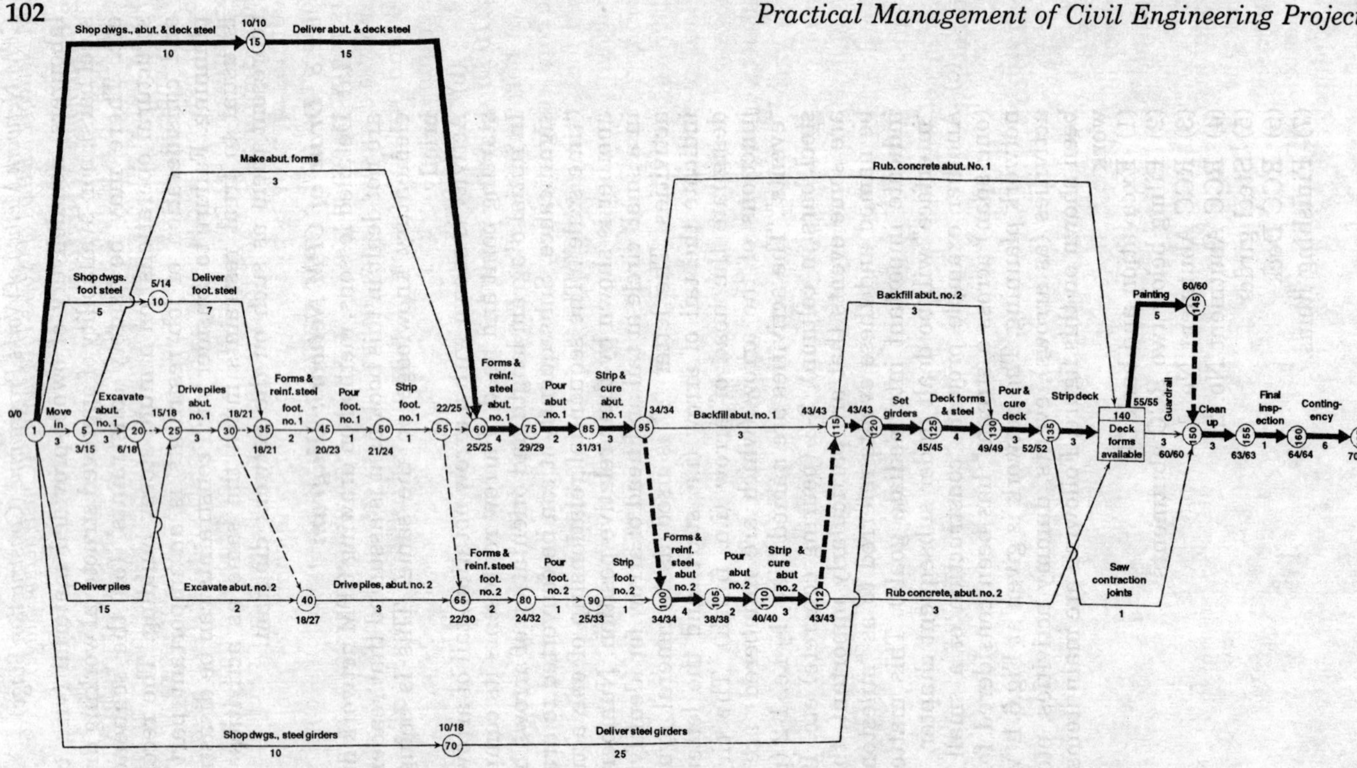

Fig. 6.1 : Arrow Diagram — Event Time Computations (Example Project)

6.2.9 Dummy Activities

In Fig. 6.1 the dashed arrow 20-25 is not a time-consuming activity but is an example of a logical connection called a "dummy" or "dependency arrow". The logic is : Drive piles abutment No. 1 (25-30) can be started only after completion of activity (5-20) i.e. excavate abutment no. 1. There is another type of dummy activity when computer is used for preparation of network. The computer identifies an activity by its (i-j) numbers and computer programme requires that each activity has a unique (i-j) designation. To achieve this requirement of computer it may be required to create dummy activity (i.e. dashed arrow) between two event numbers. The accurate use and interpretation of dummy is difficult aspect of arrow diagramming.

6.2.10 Diagram Format

A horizontal diagram format has become relatively standard in the construction industry; sometimes vertically oriented diagram is used for multi-storey buildings. It is useful to have a central chain of activities serve as a network backbone with branching activities looping above and below. It may be seen in Fig. 6.1 that construction activities of abutment No. 1 and 2 will be executed simultaneously (i.e. as parallel); activities of abutment no. 1 have been shown on above the central line and abutment No. 2 has been shown below. The network is intended to be an everyday tool used and consulted by a variety of people. Emphasis here is not on drafting elegance but on contriving the simplest and most intelligible form of network possible. Crossovers occur when one arrow must cross over another in order to satisfy job logic. These are not arrow intersections, because activities can come together only at the nodes/events. If desired colour can be used on network diagram like, red line for critical path, another colour for sub-contractor works etc.

6.2.11 Value of CPM Network Diagram

CPM network diagram (Fig.6.1) is not only a lucid job model, it is also an effective management tool for day-to-day direction and control of the work. The network diagram is the job plan for the example project. If the contractor does nothing else beyond this point, the time has been well spent because he has gained valuable insight into the project. Preparing the network has forced him to think the job through completely from start to finish. Decisions have been taken about equipment, construction methods and sequence of operations. The network diagram is an expedient medium for communication between field and office persons (of contractor and owner's organization and Architect-Consultants). The diagram makes effective job condition with material suppliers, sub-contractors, owners, architect-consultant and engineers much easier.

6.2.12 Interfaces/Interface Events

In certain large projects it is required to make "work breakdown" in 2 or 3 or more separate entities for effective control/management of the entire project. As example, the project is construction of highway. It is required to construct a bridge along the highway at Kilometre 5 to cross a river. The work breakdown is made in two parts viz., (i) construction of the bridge at Kilometre 5 and (ii) construction of highway from Kilometre 0 to 5. The detailed CPM network diagram of the bridge has been prepared as stated under this chapter. The completion event number of this network is (165). A separate CPM network has been prepared for construction of highway (Kilometre 0 to 5); the completion event number of this part of highway work is 286 as per network. Now it is required to interlink above stated two networks for representing the entire project (i.e., bridge and highway of 5 Kilometre length). Above mentioned two events (i.e., 165 and 286) are called "interface" events of the entire project. Arrow diagram of a large project is commonly drawn as a number of separate networks, each depicting the job plan of major segment of the project; the dependencies among the diagrams are shown through interface events. Another application of interface arises when "sub-networks" are used. As example, an important component of different discipline like, providing and installing a large boiler (which is mechanical engineering) in the building complex. A specialized contractor is engaged for this component either by the owner or by the main contractor. A sub-network for the boiler job is prepared. This is interlinked with CPM network of main project through two relevant interfaces of above mentioned two networks.

6.2.13 Master CPM Network/Summary Programme

Initially detailed CPM network of many activities for day-to-day field works is prepared. Various schedules are also prepared with long list of activities of field operation on the basis of above detailed CPM network. But it is difficult to study such detailed network programme at the level of top management in head office. Hence, a smaller CPM network programme consisting of main/important/critical activities in condensed form is prepared. This is also known as "summary programme" of the project. However, this is prepared after retaining the essence of the original logic.

6.3 Second Part of CPM Network – Activity Duration, Critical Path/Time Schedule

6.3.1 General

This is second part of CPM network which is for computation of activity duration, earliest start and latest start and earliest finish and latest finish of each activity, computation of floats at each event, drawing "Critical Path", calendar/time scale network diagram etc. In the first part of network (i.e. foregoing paragraphs) the work breakdown, listing of activities of works, placing various activity arrows suitably on the diagram keeping in mind of sequence/ dependency/operation etc. This is the first version of the network.

6.3.2 Following are the steps in sequence for this second part of CPM network :

(1) Estimated time/duration of each activity (without adding contingency allowance)

(2) Network computation i.e.

(a) To determine "earliest start" (ES)

(b) To determine the earliest finish (EF)

(c) To determine latest finish (LF)

(d) To determine latest start (LS)

(3) To compute value of Float/Total Float/Free Float

(4) Early Start Schedule

(5) Schedule of Total Float

(6) To draw Critical Path on the network diagram. Crucial activities are identified from the Critical Path for timely completion of the project.

6.3.3 Estimating of Activity Duration

(a) List of all activities has been prepared as per first part of CPM network preparation. Now it is required to assess/ estimate the time/duration required to complete each activity considering normal availability of resources (manpower, machine, materials, etc.). This should be done by senior engineers having sufficient practical field experience. Estimating of period for completing each activity is done on practical experience for such item works. Historical data on this matter should be maintained by the contractor and the same is useful for estimating activity duration. The duration also depends on many factors, like : Method of construction, type of technology adopted, efficiency/performance standard of skilled workers, operators, weather/ climate, altitude (plain/mountainous area) type, size (capacity), and number of suitable construction equipments, number of shifts, of working etc. As example duration of activity "earthwork in excavation" depends on size (capacity) and number of Bulldozers; if soil is hard/soft rock/soil mixed with stone boulder then it is required to deploy larger capacity (viz., D-80 or D-120) Bulldozer. Number of Bulldozer depends on quantity of earthwork in excavation. If the technology of "Slip Form Shuttering" is used for casting RCC chimney/ Lift Core then activity duration will be less than conventional type shuttering. Progress of casting of RCC units (structure members) as pre-cast unit (i.e. by adopting pre-cast technology) will be more and duration period will be less. It may not be possible to execute bituminous items during rainy and very cold winter season. It may not be possible to carry out earthwork in excavation in deep pit for basement work or in deep trench for drain/sewer construction items during rainy season. Duration may be more if risk is involved to execute the item. In view of above stated position, estimating for working duration/time for executing various activities is done on normal deployment of labour, equipments etc.

(b) During estimating the time period required for various activities, the completion date of the project should be kept out of mind. Otherwise there is apt to be, consciously or unconsciously, an effort made to fit the activities with the total period available. This is one of the serious drawbacks of the bar chart as a planning and scheduling device.

6.3.4 Forward Pass Computation

(a) The earliest finish (EF) of an activity is the earliest possible that it can be completed and is determined by adding the activity's duration to it early start (ES) of activity (5-20) is 0+3 = 3 days. EF of activity (5-20) is 3 + 3 = 6 days (of Fig.6.1); this (i.e. 6 days) is written at event (node) No. 20. EF at event No. 35 is to be worked out as : first path is along (10-35) and it works out to 5 + 7 = 12 days. As per second path (25-30-35) it works out 15 + 3 + 0 = 18 days. Hence, earliest finish time at event 35 is 18 days (which is the highest). Hence 18 is written at event No. 35. This part of computation is known as "forward pass" (i.e. left to right of network diagram). The concept of "event time" is stated here. An event time is defined to be that instant of time when "all" of the activities merging into that event have been completed. In above example, the completion of activity (25-30) controls the event time at event No. 35 i.e. 15 + 3

= 18 days. The "early event time" is the earliest possible time that the event can occur considering the time requirements of those activities that precede it. As first step computation of earliest event times at each node (event) is done starting from extreme left event/node No. 1 (i.e. zero time) to right side direction and to reach at last event/node No. 165 (of Fig.6.1). The earliest event time of node/event No. 165 is 70 days. This process is known as "forward pass".

Backward Pass Computation

(a) Activity (5-20) :
LF = 21 days – 3 days = 18 days
Activity 20-25 = 0
Hence, LF = 18 – 0 = 18 days (at event No. 20)
LS = LF-3 days = 18-3 = 15 days

(b) Another example, Activity (5-60) :
LF = 25 days (at Event No. 60)
LS = 25 – 3 = 22 days (at Event No. 5)

Thus for computing LF and LS of activities, it starts from extreme right side event (i.e. completion of project event) towards extreme left side event (i.e. starting event of project) EF and LF of each activity are written at each event like – EF/LF

At event No. 5 = 3/15
At event No. 20=6/18
At event No.60=25/25

For Activity (5-60):
EF = 6 days
LF = 25 days

For Activity (10-35):
EF = 12 days
LF = 21 days

It may be seen that earliest time at event 35 works out to 12 (5+7) along path (10-35); but it works out to 18 (15+3+0) along another path (25-30-35). In such a situation, "E" (earliest) is to be taken as 18 (which is highest) at event No. 35.

(b) From above stated position, it is seen that there are two computed figures (in days) at each event/node. This is called earliest time (indicated as "E") and latest time (indicated as "L"). Time/duration (in days) is computed at each event as "forward pass" which is known as earliest ("E") time. Thereafter, time/duration is also computed at each event as "backward pass" and the same is known as latest time ("L").

Hence, it is better to indicate value as (E — days) and (L — days) at each event of the network.

6.3.5 Float of Activities

(a) *Total Float*

The "total float" of an activity is the difference between the earliest start (ES) and its latest start (LS). Similarly, difference between earliest finish (EF) and latest finish (LF) is also total float. The float of an activity indicates that the same (in days or other time unit) is the available spare time for completing the activity. An activity with zero total float has no spare time. If more than this amount (i.e. spare time) is given to the activity in order to expand itself, it may result in the change of critical path and may increase the overall project time already calculated.

(b) *Free Float*

In calculating the total float, only a particular activity has been considered with respect to its head and tail event times. It is also necessary to find out how much an activity can move or float without affecting the flexibility of movement of the "succeeding" activity. This amount of constrained float is called "Free Float" (FF). The term "free" indicates that the use of float does not affect the succeeding activities. Let it be checked as example of free float in case of activity (10-35) of network (Fig.6.1) :

Total float of this activity is 14 – 5 = 9 days

But if this total float is utilized then succeeding activity will be affected. Maximum float of 6 days can be used without affecting succeeding activity. Hence "free float" in respect of this activity (10-35) is 6 days. This free float of activity (10-35) is worked out as under :

(i) Total float of activity (10-35) = 9 days

(ii) Total float of immediate succeeding activity (which is 35-45) = 3 days

Hence, free float = 9 – 3 = 6 days

Formula is :

Free Float = Total float of concerned activity minus total float of immediate succeeding activity

(c) *Independent Float*

In some cases, the absorption of float affects neither the predecessor nor the successor activities. The float is then said to be independent.

(d) *Significance of Float/Critical Path*

The critical path in a network is that path which has "zero float". Often it is desired to increase the duration time in order to release effort elsewhere. Under these conditions, independent float or free float can be used without necessitating any re-planning. But total float, if used completely, may affect both previous and subsequent activities in certain cases.

Hence, it can be concluded that :

(i) Float represents under-utilised resources,

(ii) Float represents flexibility of an activity and disappearance of float is loss of flexibility for the activities.

6.3.6 Now, it is seen from the network (Fig.6.1) that estimated activity time (in days) of various activities have been indicated on the arrows (quantum in days), earliest start, earliest finish, latest start and latest finish times (in days) have been indicated on the network diagram. All these amounts (in days) have been shown at each event/node points. Serial number of all events/nodes have been shown inside the event circles. Latest and earliest amounts (in days) have been shown as ES/LS or EF/LF at each event. Difference between these two amounts is the available float. Critical path has been identified along the path/route where float is zero (shown in bold line). This critical path line may be shown in red colour. In case more activity time is taken in any activity (one or more activities) then overall project completion period will be increased (i.e. completion period of the example project will be more than 70 days at event No. 165.

6.3.7 At this point the contractor has completed the most important document i.e. detailed working job plan//works programme (activity-wise) through CPM network diagram. Contractor's job planning is completed. This CPM network diagram is sent by him to all concerned persons (viz. his all field engineers, accounts branch, sub-contractors, material suppliers, vendors and also to concerned engineers of owner's organization, architects, consultants, etc.).

6.3.8 Since it is not possible to prepare above stated detailed working CPM network diagram within 2 weeks after receiving award letter from the owner, the contractor starts preparation when he comes to know that he is the lowest tenderer as advance action. Then it will be possible for him to mobilize manpower, materials, machinery, etc. within mobilization period (viz., 3 weeks).

7

Resource/Manpower Management (Construction Stage)

7.1 Introduction

As stated on last para of preceding Chapter (i.e. immediate after mobilization period), it is essential to start field works activity-wise as per CPM network programme. For this purpose, contractor is to deploy necessary manpower (skilled and unskilled), arrange materials at works/project site and to deploy equipments with efficient operators and also engineering manager, technical supervisors, etc. Requirement of manpower, engineers, materials and equipments are worked out on date/week wise requirements as per scaled/calendar CPM network diagram.

7.2 Basic Points on Resource Management

(a) Project management is the judicious allocation of resources to accomplish project completion at maximum efficiency of time and cost. It is a management responsibility to avoid or minimize the adverse effects on account of shortage of materials, skilled/ unskilled labourers (including labour disputes), breakdown or shortage of equipments, etc. The real key to the efficient usage of labour and equipment is the consistent achievement of favourable production rates in the field. In this regard, project management and field supervisors play complimentary roles. The supervisory technical staff are responsible for getting maximum out of their men and equipment on a day-to-day basis. The project manager works mostly in the future tense as he is mainly concerned with advance job planning and scheduling.

(b) Duration of each activity/item of work depends not only on the quantum of effort required, but also on the availability of required quantum of resources viz. man, materials, money, equipments/machinery, drawings etc. Main aspect of planning/management is the proper utilization of resources. Progress/ completion period of particular item of works/activity depends on resources. Hence resource planning which consists of (i) planning for finalization of drawings & decision, (ii) materials planning, (iii) manpower planning, (iv) equipment planning, (v) finance planning/cash flow chart is most important; this should be done on the basis of CPM networks. Control/management of construction activities is carried out by continuous monitoring on the basis of above planning records/CPM network. Actual achievement on ground is compared with above stated planning records during construction period.

(c) There are several facets to the management of project labour and equipment related to contractor. If it is assured or planned for the present project only it may not be practically correct. The contractor may have some other works at other places/with other clients. Sometimes it may be required for the contractor to shift some labour and equipment for short period to other works due to some urgency. Management action in this regard cannot be stereotyped but must be based on judgment and economic factors.

(d) If it is found that total labour requirement during a certain short period exceeds the original labour requirement then it may be possible to "stretch out" of non-critical activities or the use of overtime or sub-contracting on critical activities. Regarding equipments, if there are conflicts among activities for the same equipment, then re-scheduling of non-critical activities will often solve the problem. If there are conflicts for the same equipment then it may be possible to solve the problem by working overtime, holidays (Sundays), multiple shifts. Job materials in the proper quantity and specified quality must be available at the right place at the right time. Certain resource constraints are kept in mind during activity planning through network diagram. But it is not possible to identify various resource constraints during job planning. In case network diagram is modified during construction period, then various resource constraints and other problems are also to be considered. It is assumed that necessary planning (through schedules and network diagram) has been made for required quantities of manpower, materials and equipments. But during day-to-day activities at field it is essential to prepare list of various resources for next day/week's and to check the actual availability at work site. If there is some shortage then remedial measures are to be taken urgently. Various resource planning/management are now being dealt as under. Money is another important resource. It is very difficult for contractor to arrange adequate money at right time. Effective use of money is also important.

7.3 Manpower Planning (Construction Stage)

7.3.1 Manpower planning is to be done by the owner's organization (Government/Private) and contractor. In case the owner's organization is Government Department then deployment of engineers of different ranks are planned with necessary delegation of powers (technical, financial and administrative) and suitable experienced persons (in-house persons) are deployed by distributing various segments/functions of the project. In case the project is large/intricate then department may engage Architect-Consultant, specialized consultant for fire suppression works, structural consultant, etc. In case of private owner/organization it may be required to engage architect-consultant, supervisory engineers, engineer-in-charge (Project Manager/Chief Project Manager) and other consultants.

7.3.2 In case of contractor, manpower is decided in two categories viz., (i) Managerial persons (in engineering) and supervisory engineers and (ii) skilled and unskilled workers/ labourers. The manpower planning and deployment are different than the owner's organization. Contractor has to work out detailed skilled and unskilled manpower for each activity on day-to-day basis. This is worked out on the basis of CPM network calendar diagram.

7.3.3 It is required to prepare schedules of different skilled workers and unskilled labourers. Activity-wise manpower is prepared on table form as per proforma given below.

Table No. II
Tabulation of Manpower Need

Activity	Activity Arrow	Number of Workers/Labourers			
		Mason	Carpenter	Beldar	Blacksmith
(1)	(2)	(3)	(4)	(5)	(6)

7.3.4 Then another table is prepared for each working day i.e. 1st working day to 63rd working day which is completion period of the example project as per network. This table has been prepared on the basis of CPM network diagram for example project i.e. construction of a bridge on the basis of network at Fig.6.1. Total manpower (date wise) of the project (i.e 1st working day to 63rd working day) has been prepared on the basis of CPM diagram and the same is as per table No. III given below.

Table No. III
Example Project, Daily Manpower Compilation

Craft \ Working Day	1	2	3	4	5	6	7	8	9	10	11	12	13	14	15	16	17	18	19	20	21
Pile crew															1	1	1	1	1	1	
Carpenters		1	1	1															2	2	
Iron workers																			2	2	
Cement masons																					1
Operating engineers		1	1	1	1	1									1	1	1	1	1	1	
Truck drivers		2	2	2	2	2															
Labours		4	4	4	2	2												2	2	3	2

Craft \ Working Day	22	23	24	25	26	27	28	29	30	31	32	33	34	35	36	37	38	39	40	41	42
Pile crew																					
Carpenters		1	2	2		2	5	5	5	1	1	2	2	2	5	5	5	5	1	1	2
Iron workers		2	2	4	7	7	7	7							4	7	7	7	7		
Cement masons				1						2	2	1	1	1	3	3	3			2	2
Operating engineers					1	1	1	1	1	1	1				2	2	2	1	1	1	
Truck drivers																					
Labours	2	2	3	6	4	4	4	6	6	4	4	4	7	7	7	4	6	6	4	4	4

Craft \ Working Day	43	44	45	46	47	48	49	50	51	52	53	54	55	56	57	58	59	60	61	62	63
Pile crew																					
Carpenters	2	2				5	5	5	5	1	1	1	5	5	5						
Iron workers			6	6	6	6	6	6											6	6	6
Cement masons	1	1	1	3	3	3				4	4	4	4	1							
Operating engineers				2	2	2	1	1	1	1	1	1	1	1	1						
Truck drivers																					
Labours	4	4	7	4	4	4	6	6	6	6	6	6	1	1	1				4	4	4

7.3.5 It may be mentioned here that during estimating activity duration approximate number of workers/labourers are considered. But now during executing at actual field work final position is to be thoroughly checked. Generally manpower is worked out on the basis of early-start schedule of operations (as shown on schedule of activity duration and CPM network diagram). Table No.II is prepared on the basis of early-start schedule of various activities. It is seen from the table No. III of manpower that there are several days of gaps in labour requirements. It is required to carry out such "manpower levelling" or balancing. This may be done by different methods, like, to stretch certain activities which possess some floats. If there is shortage of workers/labourers during certain short period or peak period it may be required to work on overtime or on Sunday or by engaging labour contractor. Practically it is difficult for piecemeal hiring and laying off of labourer on a short term basis. Manpower levelling is the process of evening out labour demand over the course of the job. Perfection in this regard can never be attained but the worst of the inequities can often be removed by rescheduling selected activities. Peaks may be levelled off by utilizing the floats of non-critical activities. Manpower levelling is essentially a trial and error process. Small daily fluctuations may not really occur because the same is deployed on some other miscellaneous works. Special care is to be taken for activities which fall on critical path and other path having low float activities. If special care is not taken then activities having low floats will become critical and new problem will crop up.

7.3.6 There are acute shortage of labourers during certain season/ period when crops are cut or certain time of harvesting in the field. Majority of labourers go to their villages for aforesaid purpose (may be two times in a year for 3 to 5 weeks). This is happening throughout India. Due to this reason, contractor is unable to arrange labourers through labour contractors or sub-contractors. The basic implication of a labour shortage is that the duration of certain activities inevitably will be extended.

7.3.7 During shortage of labour the first action is to "stretch out" the duration of activities having sufficient amount of floats; this will reduce the requirement of labour per day. When the extension of non-critical activities becomes equal or exceed the available floats, the contractor must seriously consider the use of overtime or engaging sub-contractor. Otherwise, new critical path and "sub-paths" will be formed, possibly superseding the original critical path and duration of completion period will also be increased. It is a trial and error process that involves allocating the available labour to relevant activities. While maintaining the necessary job logic so as to achieve project completion in the shortest possible time.

7.4 Certain Practical Aspects of Manpower Management (Construction Stage)

7.4.1 With the advancement in the construction technology and with ever-increasing labour legislation, personnel management in the construction industry has become increasingly difficult. The two of the worst evils of this industry are the exploitation of labour and the drudgery to which the workers are subjected to in the highly labour intensive construction works. While various Labour Legislations (viz. Contract Labour Act, Minimum Wages Act, Payment of Wages Act, Industrial Disputes Act, etc.) do provide some element of relief by way of improved working conditions and minimum wages, yet the construction workers who are basically illiterate or very little literate and untrained, continue to live in uncertainty of the job and in inhuman surroundings. Project/ Site Engineer of department/owner and contractor do not take interest for implementing various provisions of Acts/Laws and Contract conditions related to labour.

7.4.2 In private and Government construction organizations (as contractors) realistic yearly turnover target should be fixed by top management. This turnover target should also be considered for manpower planning on a rational basis. It is a cardinal function of the personnel management to place the right men on the right job at the right time to achieve the desired end results. A good deal of appreciation is needed in the placement of personnel and if any wrong posting is done particularly at a critical place, the damage to the organization can be irreparable. At the unit level, it must be ensured that the project manager gets adequate support on the financial and material management including posting of competent accounting and storekeeping personnel. Recruitment programme has to be properly planned and personnel branch has to work in very close liaison with other branches. The position should be reviewed from time to time.

7.4.3 There should be continuous facilities for training employees (on or off the job) to achieve higher level of efficiency and productivity. Organisations should have their own courses and expose their employees to various short-term/long-term courses including practical training. In India, it is unfortunate that there are no worthwhile training facility for construction workers who form the bulk of construction manpower. There should be "induction" training at the time of joining in the organization. Motivation is a management philosophy to keep an employee always trim in all respects to perform his tasks with optimal efficiency. In the construction industry, group incentive schemes sometimes become unavoidable for timely/faster completion of projects.

7.4.4 Performance appraisal has come to be regarded as a necessary tool in personnel management. In Government departments the annual performance report is prepared in secret way; but in present system private management carry out this annual appraisal after discussing with each individual. If there is any adverse remark in the report then the same is communicated to the individual. Annual salary hikes depend on performance report. In the construction industry, job requirements are comparatively more exacting, particularly as the working conditions are more rigorous and results have to be achieved under severe constraints, over which even the contractor may have a little control. Functioning in such an environment, therefore, demands initiative, courage, perseverance and a higher precision and judgment in decision making.

7.4.5 Wages are, by and large, governed by various Minimum Wages Acts which vary from State to State. Private contractors pay less wages with insufficient facilities. But Government undertakings (as contractors) pay fair wages to the workers and provide better facilities. This creates a wage differential in some categories of workmen employed in the private sector and the public sector. Hence, it is required to standardize wages throughout India.

7.4.6 Industrial relations in the construction industry have also assumed greater importance because of continued exploitation of the labour in the past, expanding activities of building industry and in the wake of ever-increasing labour legislations. Strict enforcement of various labour law provisions, rules, conditions stipulated in the contract agreement are to be done by all concerned authorities particularly by Engineer-in-charge of owner's organization/ Government Department who is one party of contract. There is a tendency that "More you ask, the more you get". There is industrial unrest on account of demand, disputes from the side of labourers. To solve such problems, representatives of workers should be allowed to participate in Workers' Committees, etc. A pragmatic approach is required from both the management and the workers. The construction industry which is essentially a field industry, is generally spread over far and wide, needs a result oriented approach in dealing with its personnel and administrative problems. Implementation of administrative policies/decisions in relation to its personnel should be prompt.

7.5 Welfare Measures

7.5.1 The welfare of staff and workers in the constructions industry attains importance particularly because of the disorganized nature of industry. The life of construction worker remains most unsettled and mobile because of the very nature of these works. The workers have to work under hazardous and difficult working conditions, putting in long hours of work, often in remote and inaccessible areas. Like other industries, importance should be given in building industry in respect of welfare measures. Tendency for exploitation to workers should not be there. Government welfare measures improve the productivity.

7.5.2 Nature of manpower resource is different than material management and equipment management. Clear specifications, quality particulars are laid down in case of raw materials and materials which are manufactured. But manpower is different. There are more possibilities of variation in manpower. They may leave their existing job and join in other organization/project for better place, salary/wages etc. They may go to other countries for earning considerable amount. Persons retire from their jobs due to age or illness. Hence, planning of manpower is more difficult than planning of other resources. Productivity of labour (i.e. output per labour per hour) in respect of various items of work (viz., earthwork, brickwork, concreting, etc.) depends on conventional or specialized works. Such information are published by certain organizations like, CPWD, National Building Organisation, Bureau of Indian Standards, etc. Output of labour depends on various factors like, age, health, weather, etc.

7.5.3 Another important aspect of manpower management is relation between manpower (managerial, supervisory engineers, skilled and unskilled labourers) and the contractor. Contractor is required to take various welfare measures, to pay good salary/wages, bonus, medical facilities, leave facility, etc. Labour disputes should be sorted out on priority basis. It is required to see how workers, labourers will reach at work sites i.e. on foot or through vehicles. If they are transported with vehicle then time is saved and output will be more. Hence, output of workers/labourers will depend on above stated factors. Contractor is required to follow various laws, Acts related to labourers. Realistic output may vary from project to project and on contractor's labour management etc. CPWD and certain organizations have worked out labour coefficient in respect of various workers and items of works on the basis of site observations. These coefficients are quantity of workers (in numbers) for 8 hours' working. Total number of workers are worked out for relevant quantities of items. On the basis of these theoretical standards, total quantities of workers (in number) for total quantities of various schedule items of works are worked out. Table No.III has been prepared for total labour requirement of example project of this chapter.

7.5.4 Manpower/Labour management is a specialized subject and lots of difficulties/ problems are faced by senior in charge/supervisor. There is large variation in nature, behaviour, aptitude, capability of workers, skilled and unskilled labourers, supervisory staff. It is required to study each person on various factors and suitable technique is to be applied on each person separately. Same technique may not be applicable on all persons. It is important to create good working atmosphere and personnel relations amongst various ranks. It is important to appreciate the good workmanship/performance of persons. In case of inefficient person, it is required to train or motivate. Human resource is the most important resource for timely and economical completion of work. After observing their performance, the deficiencies of each person are to be identified and suitable training for removal of these deficiencies should be given. In case labour union or staff association exist, then the same is to be handled suitably.

7.5.5 After working out total requirement of supervisory staff/ engineer, accounts staff etc. for entire work (month-wise/week-wise) contractor is required to arrange/recruit/ deploy as per job planning/ labour requirement schedule/CPM network diagram. Contractors may have other construction works. Availability of various persons/

workers/labourer from his other works at different time are checked by the contractor and the same are considered for shifting from other works to the present new work (stating date/period). Then programme/planning is made to recruit some workers stating period and remaining labourers may be engaged through sub-contractors to execute certain activities and labour contractor (i.e. hiring some labourers on hire basis for certain period). Details of above stated manpower arrangement from different sources are planned by the contractor through different manpower schedules. These are done in such a way so that necessary supervisors, workers, skilled and unskilled labourers are deployed in required numbers for specific period as per calendar CPM network diagram/manpower schedules. Thus it may be seen that some labourers are deployed on hire basis (for specified period) and they are relieved after the specified period; this method is also known as "hire and fire". However, effort should be made to retain good/hardworking/efficient supervisors, workers, skilled workers on long term basis because they will be useful for completing the work in shorter periods and in an economic way. Recruiting agencies are available for various categories of human resources and these agencies may render such services. Lot of time is required for recruitment of personnel of different categories and proper advance planning in this regard is essential so that correct number of persons are in position in correct time. It will also be seen that certain minimum numbers of labourers will be required throughout the project construction period and this minimum number of labourers are engaged for entire period. Remaining labourers are arranged through other methods as stated above.

7.5.6 It has been observed at construction site that some accidents occur due to negligency or faulty action on the part of labourers/ site supervisors. This will be dealt under chapter of "safety".

7.5.7 Deployment of various managerial and supervisory engineers and staff is to be done correctly. If higher qualified person is deployed under lower qualified person then there will be friction and work will suffer. If manpower/labour management is poor then staff, labourers may leave the organization suddenly and it may create serious problems for smooth progress of work. Besides, considerable delay in completion of project, there will also be cost over-run due to above stated situation/shortage of labour.

8

Materials Management (Construction Stage)

8.1 Introduction

Material Management (Basic Principles) has been dealt under Chapter – 3. Material management during construction stage of works/project is being dealt under this chapter. This is application of basic principles on ground for construction works. During construction stage the contractor's organization is required to manage materials for effective control, planning, procurement, storing, inventory control, issue of materials, maintaining accounts, recording scraps, cost control etc. Stipulation is made in many contracts that cement and steel will be issued by the owner (Government/Private organization) to the contractor at certain issue rate. In such works, the owner's organization (purchase branch/unit) is required to prepare plan month-wise for purchasing required quantity of stipulated materials as per laid down specifications and quality. As per procedure it is required to invite tenders from approved main manufacturers of steel and cement. Cement is generally purchased in bags. If silo is installed at large construction project site then cement in bulk is procured and programme is prepared in consultation with the contractor for supplying in the silo. Manufacturers are required to submit test reports of materials alongwith consignments of cement/steel. A few samples of materials should be collected on random basis from site and get it tested from approved laboratory. Purchase process is to be monitored by SE/GM of the owner's organization for ensuring supplying necessary quantity of cement/steel in time to the contractor. Co-ordination between purchase unit/store division and other concerned project engineer, finance/accounts departments should also be carried out by SE/GM. The contractor should also pursue the owner's organization in this regard so that progress of work is not hampered due to shortage of aforesaid basic important materials. If progress of work is hampered due to shortage of aforesaid materials then this hindrance is recorded in the hindrance register and considered for granting extension of project completion period. Hence, owner's organization is to take special care for procuring above stated stipulated materials in time. Lead time required for planning, inviting tenders, finalizing tender, obtaining supply in the store from the supplier as per schedule time (on calendar month basis) etc. is to be considered for finalising the programme. All other materials for constructions are to be arranged at work site by the contractor. In case there is no stipulation in the contract agreement for issuing cement/steel from the side of owner, then the same is to be purchased by the contractor; there will be extra heavy financial burden on the contractor for purchasing these main basic materials. Cost of money on this account will be added. Contractor may quote increased rate due to this factor. Considering various practical aspects it is recommended that stipulation should be made in the NIT (tender documents) for issuing cement and steel by the owner's organization to the contractor at specified rates.

8.2 Scope

(a) Materials of construction constitute a bulk of construction cost. Depending on the type of work, it may vary from 40 to 60% of the contract cost. Proper material management can contribute substantially to the success of project. Mismanagement in this field may lead to overall delay in project completion and to a financial disaster. Proper planning for material procurement and other factors is required before start of project. Material management in construction industry involves planning, timely purchase, inventory control, storage of materials. Good material department can contribute substantially to the efficiency and the profit making potentials of the contracting organization. Precaution is to be taken to avoid wastages. Use of sophisticated materials can contribute to saving in time and improvement in quality of construction. Purchase unit is required to keep close track of the market trends in terms of rates and availability. Purchasing department can be centralized or de-centralised or can be a combination of the two. There are some advantages and disadvantages in respect of early and late procurement of materials. Standardisation is possible in centralized system. Urgent procurement is generally done by decentralized system (i.e. by field purchase unit). Quantities to be procured are to be worked out month-wise based on CPM network programme of construction.

(b) There are huge number of materials in building industry.

These are put in different categories/groups considering cost, type of materials, seasonal/off-seasonal, perishable, inflammable, etc. One type of classification is known as "A.B.C."(i.e. High-Medium-Low consumption in value). There are other types/tools of management like, VED (Vital-Essential-Desirable) items in terms of criticality. There are other type like, SDE (Scarce-Difficult-Easy to procure items), SOS (Seasonal-Off-seasonal nature of supplies). Inventory control by ABC analysis should be done. In construction project, it has been found that 70% of the cost of materials falls within 10% of the total number of items classified as "A" material (viz., cement, steel, stone aggregates, bricks, door, windows (wood/aluminium), sand, pipes, etc.). 20% of the cost of materials are represented by 20% of the items which are classified as "B" materials (viz., tiles, electrical fittings, water proofing, etc.). 10% of the cost of materials is represented by 70% of the items (viz., nails, screws, various fittings, valves, soft wood, etc.). As "A" category materials are important and required to execute various items, it may be required to procure little excess of limits. Efforts should be made to keep administrative lead time and procurement lead time to barest minimum.

8.3 Inventory Control

8.3.1 Efficient inventory management of the project implies the availability of the right quantity of stock at the right place and at the right time (neither more nor less). Inventory control is very important in the overall material management of a project. Such control eliminates the rush and emergency purchases which adds unnecessary additional cost. A typical format for inventory control is as per Table No. IV given below. As materials constitute 40 to 60% of the project cost, it is imperative to apprise the status of procurement to the management periodically. It is necessary to exercise adequate control on inventory of materials based on actual consumption requirements and lead time on procurement to avoid excess blockage of funds of the organization but at the same time ensure that the project is not starved of materials when needed. Inventory control may vary from project to project or place to place. In construction industry the holding of two month inventory is considered satisfactory. Optimum inventory helps releasing scarce capital to other important activities; cost of money is also reduced. Cost of storage, handling, deterioration and obsolence will be reduced

by optimum inventory. Excessive inventory/excess stock will reduce profitability of contractor. Proper planning and control on material purchase, proper storage facility are essential. An effective inventory management can contribute to improvement in efficiency, profitability and overall performance of the project construction.

Table No. - IV
Construction Organisation
A Typical Format for Inventory Control

NAME OF WORK :....................

INVENTORY CLASSIFICATION
A/B/C

Sl. No.	Description of material	Weekly consumption in works etc.	Lead time in procurement in weeks	Recommended Stock of material Maximum	Recommended Stock of material Minimum	Actual stock of material on date of reporting	Remarks and proposed remedial action

Purpose : It is necessary to exercise adequate control on inventory of materials based on actual consumption requirements and lead time on procurement to avoid excess blockage of funds of the organisation but at the same time ensure that the project is non starved of materials when needed.

Frequency : Monthly – For A materials
Quarterly– For B materials
Half-yearly– For C materials

8.3.2 During planning for purchase of materials priority is to be accorded for various materials. Generally it takes about 50% of total project construction for constructing structures of the building. Hence, priority is to be accorded for procurement of certain materials viz., cement, steel, stone aggregate, coarse sand, bricks, etc. Materials required for finishing items stone slabs for flooring/wall cladding, fine sand, fittings for water supply works, ceramic tiles timber, paints etc. are to be purchased at a later stage i.e. after completion of structural works.

8.3.3 There is another important aspect of material management. Proper accounts of materials at site are to be maintained by Site Engineer/Material Manager. It is essential to work out theoretical consumption of materials for various items on the basis of standard norms and permissible variation/wastage. Then actual purchase and consumption of materials are to be compared. If variation/actual consumption (on higher or lower side) is much more than theoretical requirement then there are some faults which are to be found out. Over-consumption of materials may be due to: (1) Improper collection and storage or (2) Bad workmanship. It is essential to prepare proper site layout for storage (shed, room, open yard) of materials including cut pieces and waste materials. It is also to be ensured for proper circulation of vehicle for unloading materials and transportation to exact location of items of work. Proper storage arrangement can go a long way in reducing wastage and ensuring accounting. Effective watch and ward is to be arranged for preventing theft/pilferage of materials. Periodical stock verification of materials including waste materials should be done and compared with store/stock register. Rejection of bad works/concrete mix is also to be considered. If there is shortage of materials after above stated verification then it is required to fix responsibility against concerned persons. Maintenance of proper account is important for correct valuation of materials issued and consumed. This should be done periodically during progress of work. Method of storing depends on type of materials. Issue of cement should be regulated on the principle of "first come-first out". Steel bars are to be stacked diameter-wise. Fire fighting precaution is to be taken for inflammable materials (viz., fuels, paints, etc.). Measurements of materials (like, sand, stone aggregate, etc.) should be made on stacks instead of truckloads. It is always better for the inspection of materials before the materials are received at site. Approved samples of materials should be part of supply orders to avoid disputes and to check quality of materials.

8.4 Scrap/Surplus Materials

Material Engineer is also responsible for maintaining accounts in respect of scraps/cut pieces of steel, wood, tiles, stone slabs, waste/un-serviceable materials, surplus serviceable materials, obsolete equipments, etc. All these materials are to be collected/stacked properly (category-wise) and the same are to be measured. Weighment is to be done in respect of scraps, cut pieces of steel bars etc. Survey report on prescribed form is to be prepared for above stated materials. Then these materials are to be disposed of suitably like, scraps, junks, obsolete items may be disposed of by auctioning. Surplus serviceable materials may be shifted to some other work sites or may be disposed of by auctioning.

9

Project Time Control, Monitoring, Progress Reporting and Changes (Construction Stage)

9.1 Monitoring and Rescheduling

(a) Until its present stage of development, the management system has been concentrated on work planning and scheduling (pre-construction and construction stages). Contractor has also prepared detailed job planning through CPM network method before starting actual construction activities physically at site. Contractor has also completed manpower and material planning/scheduling which are based on CPM network. Equipment management will be dealt under a separate chapter. The contractor can now proceed with the comforting assurance that the entire project has been subjected to advance detailed study and analysis.

(b) It is known that as per ground reality no plan can be perfect nor can the planner anticipate every future job circumstances and eventuality. Contingencies arise everyday during construction period that could not have been foreseen. Adverse weather, delivery delays, labour disputes, changes made from the side of owner/ department of the project, delay in supplying approved working drawings (good for construction) from the side of owner's organization/ department, delay in supplying stipulated materials (viz., cement, steel) from the side of owner's organization, delay in payment and conveying various decisions from the side of owner's organization etc. Above adverse factors can disrupt the original plan and schedule. Due to these reasons the construction activities deviate from their developed programme and schedule. Hence, considerable time and effort are required to check the time progress and to take suitable action that is appropriate and necessary to bring the work back on schedule. These actions constitute the monitor and re-scheduling of the job management system.

9.2 Steps for Time Control

(a) Following are the steps for time controlling of the project constructions :

(1) Recording job progress (on schedule and CPM diagram) monthly/ fortnightly and reporting

(2) Compare work actually done (as on particular date) with that planned earlier

(3) Take appropriate action to correct schedule slippages

(4) Update CPM network diagram and schedules for works yet to be done

(b) There should be some system of feedback of various matters (viz., progress, problems/bottlenecks/hindrances etc.) on weekly/fortnightly basis. CPM activities are good basis for progress reporting and recording. Periodic network recomputations that incorporate accumulated plan (CPM network) and schedule changes are called "updatings".

9.3 Early Start Schedule (Calendar Dates)

CPM network programme and schedule for construction stage have been prepared as stated under chapter 6; the same are based on "early start schedule" or a "normal schedule". There project times have been expressed only in terms of "expired working days". For purposes of project monitoring and control, it is necessary to convert these times to "calendar dates" on which each activity is expected to start and finish. This is easily done with the aid of a calendar on which the working days are numbered consecutively, starting with number on the anticipated start date and skipping Sundays and other holidays. This is shown as per Conversion Calendar (Fig. No.9.1). Starting working day has been numbered as (1) on calendar date of 14th June.

(Calendar sketch)

JUNE

Su	M	Tu	W	Th	F	Sa
		1	2	3	4	5
6	7	8	9	10	11	12
13	14 (1)	15 (2)	16 (3)	17 (4)	18 (5)	19
20	21 (6)	22 (7)	23 (8)	24 (9)	25 (10)	26
27	28 (11)	29 (12)	30 (13)			

JULY

Su	M	Tu	W	Th	F	Sa
				1 (14)	2 (15)	3
4	5	6 (16)	7 (17)	8 (18)	9 (19)	10
11	12 (20)	13 (21)	14 (22)	15 (23)	16 (24)	17
18	19 (25)	20 (26)	21 (27)	22 (28)	23 (29)	24
25	26 (30)	27 (31)	28 (32)	29 (33)	30 (34)	31

AUGUST

Su	M	Tu	W	Th	F	Sa
1 (35)	2 (36)	3 (37)	4 (38)	5 (39)	6	7
8 (40)	9 (41)	10 (42)	11 (43)	12 (44)	13	14
15 (45)	16 (46)	17 (47)	18 (48)	19 (49)	20	21
22 (50)	23 (51)	24 (52)	25 (53)	26 (54)	27	28
29 (55)	30 (56)	31				

SEPTEMBER

Su	M	Tu	W	Th	F	Sa
			1 (56)	2 (58)	3	4
5	6 (60)	7 (61)	8 (62)	9 (63)	10	11
12 (64)	13 (65)	14 (66)	15 (67)	16 (68)	17	18
19 (69)	20 (70)	21	22	23	24	25
26	27	28	29	30		

Fig. 9.1 : Conversion Calendar (Example Project)

9.4 (a) During making a "job calendar", the true meaning of elapsed working days must be kept in mind. Considering practical aspect contractor may prepare above stated job schedule/network/job calendar two months in advance; it may not be required to prepare voluminous schedules/CPM network for entire large project. In reality it will be required to prepare revised schedules/CPM network number of times. After converting elapsed working days to calendar dates it is possible to associate activities with calendar times. A statement/Project calendar schedule is prepared as per proforma given below (Table No. V):

Table No. V

Sl. No.	Name of Activity	Arrow/Activity Number (i-j)	Activity Duration (Working days)	Scheduled Starting Date	Scheduled Completion Date
(1)	(2)	(3)	(4)	(5)	(6)

N.B. Critical activities are to be marked in red colour/ Asterisks

(b) It is possible to know from the above stated calendar/date statement about season/climate when activities will be performed. If it is found that certain activities cannot be performed during particular month/season then it is required to modify the schedule. As for example, it is not possible to execute bituminous items of works on roads during rainy season and very cold weather.

9.5 Events

(a) Activities (i.e. i-j) have been considered for preparing CPM network diagram under Chapter 6. One activity starts from one point/node and end (i.e. completed) at another point/node after consuming some time (unit may be in days). Their early start and early finish (as forward pass) have been indicated on the network. Latest start and latest finish of activities have been worked out through "backward pass". These have been shown on the CPM network diagram and floats for various activities have been worked out (vide Fig. 6.1).

(b) Now it is required to know about "event". The starting point and finishing point of activity as shown on the CPM network diagram are known as "event". The event point is shown as a circle. Event number is written inside the circle. Event times are also to be worked out. Early event times are worked out first. An event occurs (i.e. event is completed) when all incoming activities have been finished and the early event time is the earliest that an event can occur. At one event there may be one incoming activity or more incoming activities. The early event time is the earliest that an event occurs. Values of "early event times"(the symbol for which is "T_E") are computed by completing a "forward pass" through the network. It proceeds from event to event. Event times are obtained manually on the arrow diagram itself. As on CPM network diagram (Fig.6.1) there are two figures with slash mark; the left digit (in the unit of day) is the early time and right digit is the event late time. The T_E (i.e. early event time) of a merge event is the latest (i.e. largest value) of the incoming activity value (i.e. Early Finish value). The T_E value for a given event is the length of the longest path from project start to that event.

9.6 Late Event Times (T_L)

In case of late event time, every activity starts as late as possible without disturbing the original proposed completion period of work/project. This is worked out through "backward pass"(as done in case of late start/late finish of various activities). In this case when working backwards to a "burst event", the smallest of the late activity start prevails.

9.7 Critical Events

(a) It will be seen from Fig.6.1 (CPM Network) that at some events, the values of T_E and T_L are the same. These nodes are called "critical events", because zero time elapses between their earliest possible and latest possible occurrences. These critical events form a continuous path through the network from start to finish. This is the same critical path already located by means of critical activities with zero total float (under Chapter 6).

(b) It is stated that during actual management of construction certain important events are identified as :

(i) Milestone

(ii) Critical events

(iii) Interface events (completing two segments of one project)

Floats can also be calculated from T_E and T_L.

9.8 Milestone Events

Milestone events are those events or points in time, which have been identified as being important reference points in accomplishing the work/project. Milestone events include dates imposed by the Department/owner for the accomplishment of certain tasks as well as target dates set by the contractor for completion of certain segment of the work. On large projects owner/contractors establish number of milestones throughout the project and use these reference points for project monitoring.

9.9 Interface Event

This has been dealt under Chapter 6. Examples of laying pipeline and crossing across a stream/river has been given there. Two separate CPM networks diagrams are prepared for above stated two parts/segments of project. Both the parts of project are to be completed by interlinking completion dates of two segments of project. These two dates (i.e. events) are known as interface events. This is required to be monitored carefully. If one segment is completed in time but other segment is delayed then entire project will be delayed.

9.10 Time-scaled Networks

A time-scaled network is an arrow diagram drawn to a "horizontal time scale". There are two time scales, one in terms of working days and the other as calendar dates. The calendar dates are shown on a horizontal line drawn at bottom of the network diagram. The time-scaled network diagram of the same example project (i.e. construction of a bridge) has been drawn considering early start schedule. The arrow diagram (CPM network without time scale) of the same example project has been dealt under Chapter 6 (vide Fig. No.6.1). The time-scaled diagram has been drawn considering early start and early finish times vide Fig. No. 9.2. The length of the connecting arrow is the duration of that activity. It may not be required to show arrow head as the time flow is drawn horizontally from left to right side. Events are also plotted and numbered in the circles considering T_E (i.e. earliest time). Free floats are shown as horizontal dashed lines in Fig. 9.2. A time-scaled network has a distinct advantage over the other normal arrow type network (without scale) diagram. Time-scaled diagrams are very convenient devices for checking daily project needs of labour and equipment and for the advance detection of conflicting demands among activities for the same resource. Such diagram is also much more convenient for making cash flow studies. This is also very useful for recording field progress. "Summary diagrams" for management, owners should be drawn in time-scaled way.

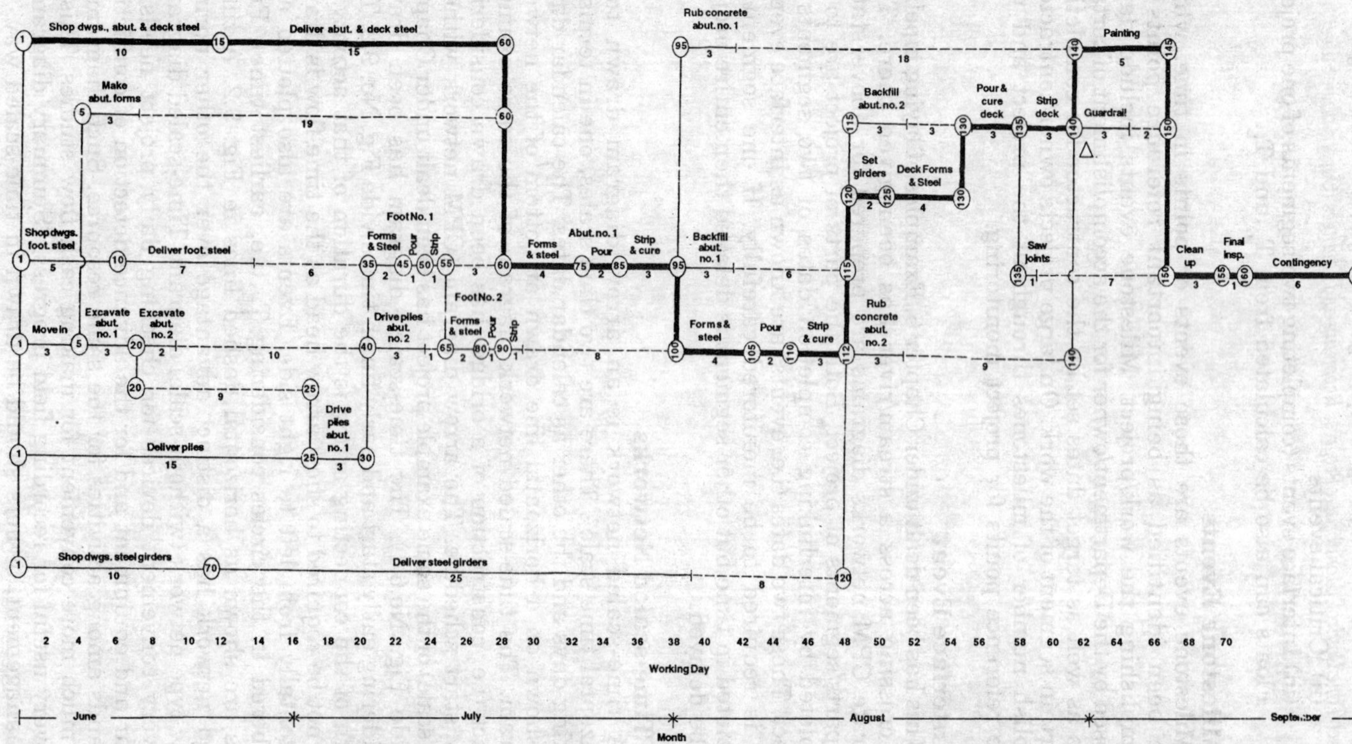

Figure 9.2 : Time - Scaled Network (Example Project)

9.11 Monitoring

9.11.1 Master CPM network/summary diagrams are prepared showing certain important events (milestones, interface, critical events) and critical path for senior executives indicating progress of works, shortfall, main bottlenecks/hindrances etc. On the basis of such reports and summary diagrams necessary action is taken by senior executives/management authorities to sort out various problems/bottlenecks. Corrective steps are taken by contractors and architect-engineers of owner's organizations. Updated/modified CPM network diagrams in respect of main contract and other contracts related to various disciplines, utility works are prepared. If the project completion period is to be extended then it is required to know how much period is to be extended ? It is also required to take necessary steps (by changing logic of CPM network, by deploying extra manpower, equipments/machinery, cash flow etc.) for compensating the backlog by certain extent. If necessary, the main contractor can engage sub-contractors to increase resources. Monitoring and co-ordination meetings are taken at project site weekly/fortnightly by the Chief Project Engineer/ Manager when all concerned persons of contractors and owner's organisation, architect, structural designer etc. are present. Site inspections are also carried out during these site monitoring and co-ordination meetings. Special care is taken for coordinated action plans in respect of various services/utility works, various vendors/ suppliers etc. It is most important to know problems faced by various contractors and to discuss the same during aforesaid meetings and decision be taken to solve the same. Minutes of these meetings are to be issued to all concerned persons "within 3 days". Review/compliance reports on the aforesaid minutes of meetings are taken during next monitoring/coordination meeting. It is also required to take review meeting at suitable intervals (after 2 or 3 months) by the top executives of owner's organization with concerned Senior Managers, Architect, Main Contractor, etc. for sorting out unsolved problems and to take decisions on policy matters.

9.11.2 Progress of works is to be marked in different colours on the CPM network diagram every month. After comparing the shortfall is worked out. Reasons for shortfall in respect of various activities are found out. Then remedial measures for the same are to be taken. Effort is to be taken to make up shortfall upto maximum extent. Hindrance/bottlenecks are to be removed on priority basis.

As per position it may be required to revise the CPM network diagram so that updated CPM programme diagram is available and monitoring is done on the basis of same.

9.11.3 It is also required to prepare monthly progress report of project for perusal of top management/Managing Director/Board or Council Members who are generally non-technical persons. CPM network diagrams may not be suitable for such persons. This is to be prepared in a different way/format for easy understanding at that level. Bar chart of certain main items of works may be useful for such progress report. Original programme and progress (physical and financial) achieved as on date of progress report are to be shown on the bar chart. Quantum of physical and financial progress in percentages are to be stated. Quantum of shortfall in percentage is also to be stated. Main bottlenecks/hindrances, pending decisions on important issues etc. are also to be stated in the progress report. Summary of progress report including deployment of manpower, construction equipments, machinery, target for next month/quarter, etc. is part of progress report for executives, top management and Board/Council members. This summary should be brief with relevant sub-heads. Executive Summary is very important and the same (in 3 or 4 pages) is stated at the beginning of monthly progress report.

9.11.4 Bar Charts

For smaller work the bar chart may be prepared. The bar chart is prepared for site staff (viz., foreman, junior supervisor) on the basis of CPM network for weekly works. This is useful for non-technical persons in respect of certain main items of work. The bar chart does not show dependencies among operations/activities. It does not show critical activities and floats. Consequently each activity receives the same consideration with no indication of where management attention should be focused. However, it is a valuable medium for displaying job schedule information. It affords an easy and convenient way in which to monitor job progress. This is a time-scaled diagram. "Each bar shows the early start and early finish calendar dates". The bar chart of same example project (viz., construction of a bridge) is given *vide* Fig. 9.3. In India, the bar chart is very popular for recording job progress because it is very easy to prepare and easy to understand. Non-technical persons and lower level site supervisors can understand the same. It provides historical record as well as current status. As per field procedure a vertical line is drawn (or a colour movable string) on the bar chart for a particular date of review of progress.

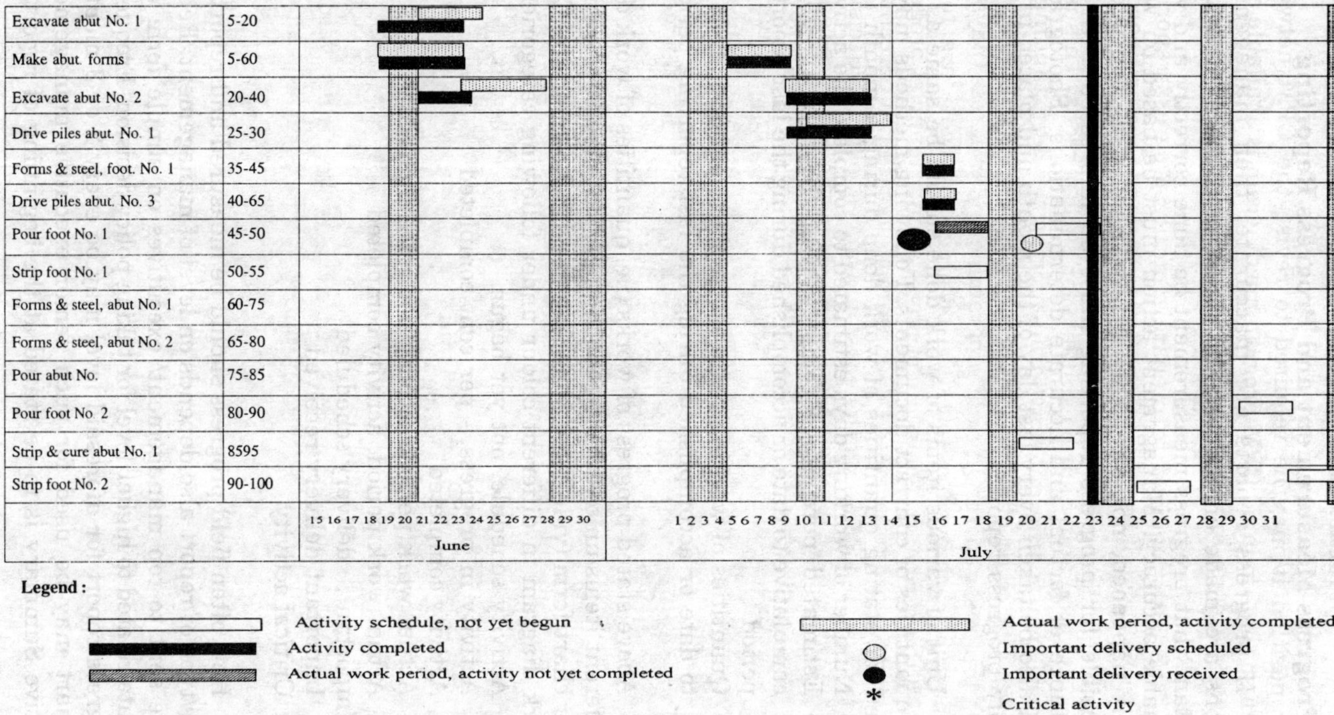

Fig. 9.3 : Progress Recording on Bar Chart (Example Project)

9.12 Progress Measurement and Progress Reporting

9.12.1 If progress reports are inaccurate, then management decisions are made on the basis of incorrect position. It is very important that progress measurement be done correctly and with reasonable accuracy. Management action must be based on what actually happened, not what should have happened. The person responsible for progress reporting must be impressed with the importance of factual and accurate determinations. Photographs should also be taken every last day of the month and placed in the monthly progress report.

9.12.2 Unit of various items of work done should be same as per bill of quantities of contract documents. Following methods may be followed for stating quantities of work done during a month :

(1) Number of working days estimated to complete the activity.
(2) Estimated percentage completion of the activity, either cumulative to date or accomplished during the last reporting period.
(3) Quantities of working units set in place, either cumulative to date or accomplished during the last reporting period.

9.12.3 Above stated progress of works (i.e. quantities of work done for different items/units during reporting period) may be shown on the bar chart form (vide Fig.9.3) or may also be shown on the CPM network diagram in different colour under following categories :

(a) Activity schedule, not yet begun
(b) Activity in progress, - per cent completed
(c) Activity completed
(d) Actual work period, activity not yet completed
(e) Actual work period, activity completed
(f)Important delivery scheduled
(g) Important delivery received
(h) Critical activity

9.12.4 How often field progress should be measured and reported? Details/type of report also depends on level of management. Report may be sent to top management/ executives on simple form and with data related to higher level for taking policy decision. Frequency of progress report for aforesaid level may be every 2 or 3 months. Bar chart may be used for such senior executive management. Executive Summary is to be stated at the beginning of report.

9.12.5 However for technical authorities progress report should be prepared with various technical data alongwith CPM network. Various technical problems are to be brought out in the monthly progress report. Reasons for shortfall are also to be recorded. Proposed action for accelerating progress and making up backlog is to be stated. Depending upon the size, complexity and characteristics of the work, the frequency of progress report may vary from week to monthly or even longer. The project manager is responsible for correct and timely progress measurement and submitting report to all concerned authorities in time. The important thing is not how often the network is recalculated but how well the plan still fits the actual conduct of the work. The requirement for revision depends more on the severity of the schedule deviations than on their number. He provides an independent review of work accomplishment and how the job is coming along. Position should be stated about critical activities. Independent reporting is most important. Supervisor in charge may not state adverse points in the report. Hence an independent engineer who is not directly involved in construction activities may prepare progress report on random basis quarterly or half-yearly and submit the report to higher authority (viz. Chief Engineer). If progress of work is hampered due to shortage of resources (manpower, materials, equipments, cash flow) then details in this regard should be brought out. Action taken for arranging such short resources should be stated in the report.

9.13 Progress Monitoring on CPM Network Diagram

This method of monitoring/review of progress (weekly/monthly or on any particular day) is very effective and technically sound for engineers. A vertical pencil line is marked on relevant day (i.e. working day) or date (if time-scaled calendar network is available). The colouring is done along activity arrows which have been completed or partly completed (indicating percentage of work completed in respect of particular activity which is in progress). Actual time (in days) taken to complete the activities are written by crossing out the originally planned duration time (in days). Similarly actual event times are written by crossing out the originally planned event times.

Above system can be of great value in the settlement of claims, assessment of liquidated damages and requests for extensions of time.

9.14 Progress Analysis

(a) Progress review/monitoring is done weekly/monthly. Frequency depends on magnitude of project/intricacy of project/type of project/urgency of project (i.e. to complete the project on crash programme basis). Bar charts are used due to simplicity and as it takes very less time for preparation. But there is serious fault because bar chart does not show operational dependencies. If a bar chart is to be prepared, it must be used in conjunction with the CPM network diagram.

(b) The status of the work on the date of review of progress can be readily determined by noting the event times. Position of critical activities is also known easily from the network (i.e. whether behind or ahead). Status of activity floats is also known from the network. Actual available total floats on the day of review of progress can be worked out easily from the network and the same can be compared with originally planned total floats. However, it may be stated here that the most comprehensive understanding of current job status (on the day of review of progress) can be gained from the time-scaled CPM network diagram. It is easy to assess the impact of schedule or logic modifications on other activities and on the project as a whole.

9.15 Corrective Action

When it has been observed during monitoring that certain work areas are lagging behind schedule, it must be decided what to do about the situation. It is required to investigate the cause of delays of certain activities. Whether the same has affected or likely to affect any critical activity? If critical activity is affected/delayed then original overall projection is likely to be affected/extended. Evaluation of delay must include the determination of how the delay affects activity floats. Corrective action can take two forms. One possibility is by shortening certain selected activities which are still to be done. A study of network will identify certain activities whose shortening will gain some time. After shortening, a revised project schedule/modified CPM network is to be prepared and the same is to be followed further. The second possibility involves an alternate job logic/operation plan. This line of action is taken when it is not desirable to follow original job logic/operation plan. If original network is done skillfully then CPM job logic is quite durable. Minor modification may help for controlling time overrun.

9.16 Site Management

From practical points of view, contractor should organize certain system of site operation. Project Manager of contractor will find it good practice to provide concise, short-range summaries along with bar chart (which is prepared for a week's construction activities on the basis of CPM network). This will serve both to inform and remind site engineers/supervisors about current job operations. These might be called a "weekly reminder list"/check list. This is merely a list of activities and requirement of resources which will continue for next one week (viz., schedule to start or finish during that week). Site supervisors can concentrate on priority works. Another list contains activities whose progress is not in accordance with minimum scheduled requirements. Project Manager of contractor advised supervisors about potential trouble before it gets out of hand. If there is continuous delay and situation is alarming then Project Manager may issue specific notice directing for taking appropriate action and to take corrective steps to the site supervisors/ Asstt. Manager/sub-contractor/material dealer or other concerned party. Project Manager may call special meeting with all concerned persons.

9.17 Project Changes

(a) This is an important matter in project construction. It has been found in many project/work constructions in India that many changes are made in respect of drawings, specifications of items, site conditions, etc. during construction period. Due to these reasons many variations/extra items of works with reference to agreement provisions crop up. Site conditions are also changed after carrying excavation work and due to special weather/natural calamities. There are financial effects on this account and extra time is involved due to these unforeseen factors/developments. Certain changes (viz. change of specifications of certain items, additional requirement, etc.) are due to subsequent requirements of users; users did not submit detailed requirement to the Chief Project Manager/Chief Engineer of the owner's organization before detailed planning/ designing. As stated earlier, detailed planning stage is very important. Frequent changes occurred during construction period is due to faults or unsatisfactory detailed planning and faulty NIT (tender documents). Due to such changes the contractor may suggest certain construction changes in accordance with the contractor's value engineering. Besides financial effect (i.e. extra payment to be made to contractor), extension of time without penalty/compensation is involved. Project Manager of contractor has an important responsibility in evaluating the effects of such project changes and taking all steps necessary to protect the contractor's interests as per terms and conditions of contract agreement. Negotiations between contractor and owner's organization are involved to sort out various effects of above stated changes. Contractor's Manager is required to keep detailed job records, documentation of extra cost, and timely notification to the owner. The time and cost consequences of project changes are difficult to analyse. The CPM network diagram is one of the most powerful and effective management tool available for such analysis. As changes occur, network analysis is very effective in determining the time effects produced. Such time effects are also often of great value in determining the cost consequences of project changes. It may be stated that if major changes like, change of use of building, change of specifications of major items, increasing number of seats in auditorium/ conference hall, structural change, etc. are made at a late stage then serious problems may crop up and contractor may stop the work. In case major contractual disputes related to contractors, architect consultant and other consultants are not solved then it may be required to close or rescind the contracts and serious consequences are faced. "Rescission" of contract means that the balance work will be executed at the risk and cost of contractor. "Closing" of contract means that balance work and rectification of defects will be executed without risk and cost of contractor. Hence, it is essential to accord importance to various aspects of project at detailed planning/design stage so that major changes are not made during construction period.

(b) Following information is required for re-computation and to prepare revised CPM network during construction period :
   (1) List of activities which have been completed.
   (2) New activities which are to be added.
   (3) Original activities which are to be deleted.
   (4) Any changes in job logic to be made.
   (5) Any correction to original material delivery or resource arrival date.
   (6) Estimated times to complete remaining activities.
   (7) Any changes in estimated activity durations.

10

Reduction of Project Time Duration

10.1 Introduction

CPM network planning and scheduling have been dealt under previous chapters and "normal" project time duration has been established considering estimated normal time of each activity. One basis/reference is completion period of work/project as stipulated in the contract documents. The contractor may wish to achieve completion by a certain date to avoid undesirable weather (rainy season/very cold weather) or to free men and equipments for other works. In case delay occurs initially due to certain reasons then it will be required to expedite by shortening duration of certain critical activities so that project is completed on original period as stipulated in the agreement. If there is delay on the part of contractor then liquidated damage/ compensation may be required to be paid by the contractor. Sometimes it is required to advance certain activities for achieving milestone/interface events. This is the time to find out about such potential problems rather than being suddenly confronted at a time too late to take corrective action.

10.2 Reduction Procedure

There are different methods for reducing time duration. For certain situation extra cost is involved for reducing duration of certain activities. It is also required to see that amount of extra cost is minimum. It may also be possible for reducing time duration without extra cost. Longest path is relevant for completing any milestone or interface events or to complete the project (from starting event of the work). This can be done by "shortening the longest path" leading to that event. Critical path of the project is also the longest path. In absence of CPM network, when a work is known to be falling behind schedule, the usual reaction is to expedite, willy-nilly, all the on-going activities in an attempt to make up the lost time. Hence, it is essential to establish the longest path on the CPM network diagram and then duration of some or all activities of this longest path is to be reduced. In the case of advancing a milestone or interface event, the network critical path of project may not be involved.

10.3 Cost of Work/Project

At this stage it is relevant to state about cost of work. The cost consists of :

(a) Direct cost (viz., manpower, materials, equipments, sub-contractor's amount)

(b) Indirect cost (viz., overhead cost of work)

When activity duration is shortened then extra cost is involved on account of overtime, extra shift of working hours, deploying extra labour and equipments (efficiency of which will decrease) etc. If the project period is reduced, the job indirect overhead expense is reduced. The contrary behaviour of these two costs has resulted in some interesting position about striking a balance between the two with possibility of optimizing project duration to achieve least total cost.

10.4 Shortening of Construction Period by Re-study of Network

Shortening of period can be done by incurring extra cost or without incurring extra cost. Initially the contractor should re-examine the critical path/longest path between two events of the project. At initial stage it is better to make a careful study of the longest path/critical path with the hope of detecting possible ways of shortening it. The intent here is to take advantage of any factors previously overlooked that will result in greater time efficiency with no increase in direct cost.

There are some methods for shortening of critical path with no increase in direct cost. The second look of CPM network may help to locate some lacuna for shortening time. Any network re-examination should be conducted with practical attitude that modifications have to be realistic. It is also required to see about effect of change on non-critical activities. Shortening of critical activity duration reduces the float values of other activities. There is a possibility of conversion of non-critical activity to critical activity. Most convenient way for aforesaid process is by the use of a time-scaled CPM network diagram (as illustrated in Fig. No.9.2). Network re-computations can always be performed after each shortening to determine its effect on activity floats.

10.5 By Reviewing Activity Duration Estimate

Before preparing original CPM network, the estimated duration time of various activities is done. This estimated time depends on individual and vary from person to person. Some persons are optimistic and some persons are pessimistic; due this reason the estimated duration time may be underestimated or overestimated. After examining above situation, rough allowances to compensate for it can be applied. Care should be taken for not reducing times arbitrarily.

10.6 Parallel Critical Activity Path Method

Generally critical path is in series. After re-examining it may be seen that there is a possibility to make two critical paths for some activities in parallel. Main contractor may deploy sub-contractor to carry out certain activities of one parallel critical activity path. It is required to ensure that time duration of both parallel critical paths is reduced. It may be possible that a critical activity can be sub-divided and a portion of it can be done in parallel with another.

10.7 Shortening by Change in Network Logic

(a) Before awarding the work to main contractor, overall CPM network programme is prepared by the owner's organization. Completion period as stipulated in the contract document is considered in preparing aforesaid CPM network. After awarding the work to the main contractor the "detailed CPM network" diagram is prepared by the contractor considering certain logic. Based on this logic the breakdown of work structure, activities, sequence of construction, dependencies of various activities, consideration of manpower, types/sizes of various construction equipments etc. are prepared initially by the contractor. The completion period of project as shown on the detailed CPM network may be shorter than the completion period as stipulated in the agreement.

(b) It may be required to review the logic and re-working/ refinement of the original logic in respect of selected portion of the project. The object is to reduce the length of critical path with no appreciable increase in project direct cost. Innovative thinking and fresh approach can sometimes result in improvements to work method which may help to reduce duration time of certain activities. "Brain-storming" session may produce some excellent ideas concerning new approaches. In certain contract the design is carried out by the contractor (like – turnkey project); contractor may change design for shortening the project construction period.

10.8 Project Shortening by Expediting on Extra Cost

(a) Certain methods for shortening project duration have been discussed on foregoing paragraphs and the same are without extra cost. If it is not possible to shorten the project duration of required number of days without extra cost then there is no other alternative for contractor to "buy time" (i.e. extra cost involved); this method is known as "expediting by extra cost". Project expediting can involve a change in job logic, shortening of individual critical activities or both. Project cost is now a variable with time.

(b) Expediting of an activity can be achieved in a variety of ways depending upon its nature. Additional manpower/labour (skilled and unskilled), overtime, extra shift of working, deploying sub-contractor for carrying out certain activities etc. More equipment might be diverted from other project or to hire some equipments for temporary period. Fabricator can be asked to work overtime for which extra cost is involved.

10.9 Crash Programme

In CPM network system expediting of an activity is often called "crashing". A graph is shown in Fig.10.1. As per this figure the minimum time to which an activity can be realistically reduced is called its "crash time" i.e. from 15 to 12 days and the corresponding direct cost is called its "crash cost". The intersection point i.e. vertical crash time from horizontal axis point "12" (days) with topmost point of graph line as shown on the aforesaid graph is called its "crash point". Generally extra cost (viz., Direct cost) is involved in such crash programme.

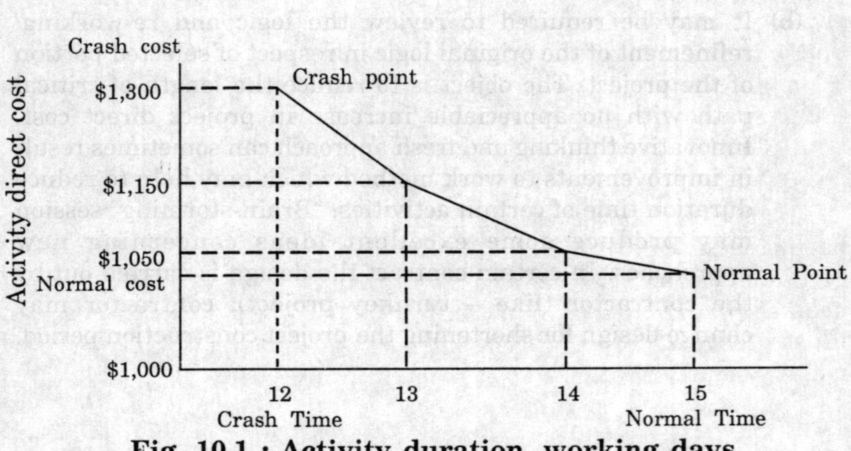

Fig. 10.1 : Activity duration, working days

It is required to maintain the effect of the expediting action on the floats of other activities. Not only the importance of float values but new critical activities formed must be identified.

10.10 Limitation of Shortening

There are numbers of restrictions in the process of shortening a critical path. These are summarized as under :

(1) Crash limit of individual activity. Some activities cannot be shortened at all.

(2) Shortening is limited by float. Shortening in duration of most activities reduces the float of other activities. This may cause another chain of activities to become critical.

(3) Shortening is limited by parallel paths. If one loop of parallel critical paths is to be decreased in length then a commensurate decrease is also to be made in other loop.

(4) Shortening is limited by crash critical path.

10.11 Milestone and Interface Events

Reduction of overall project duration has been dealt under foregoing paragraphs. It is sometimes necessary to advance the expected dates of milestone and interface events. If the T_E (earliest time) value of such an event does not satisfy an established time requirement then some action is to be taken by the contractor for shortening certain activities. The procedure is the same as that of shortening project duration except that the project critical path may or may not be involved. The early time for any event is determined by the longest path from project start to that event and effort is to be made to advance the event time by shortening the longest path. This longest path is to be studied thoroughly. If required, it may be done by least cost expediting method.

10.12 Variation of Total Project Cost with Time

There are three parts of Project cost as under

(i) Direct Cost

(ii) Overhead Cost and

(iii) Total Cost

It has been stated earlier in this chapter that direct cost may be increased due to shortening of project duration or advancing certain event and overhead cost of project is decreased. These have been shown on three graphs (Fig.10.2). The total project cost curve illustrates an interesting phenomenon. On some projects, there are usually some activities whose expediting costs on per day basis are less than the daily overhead expenses. The result of this is that at least some project expediting is possible at no extra cost or even a small gain to the contractor. This behavioral pattern of total cost might serve to motivate contractors to expedite their projects with the objective of minimizing the project's cost and commensurately increasing profit. This is shown on the aforesaid graph.

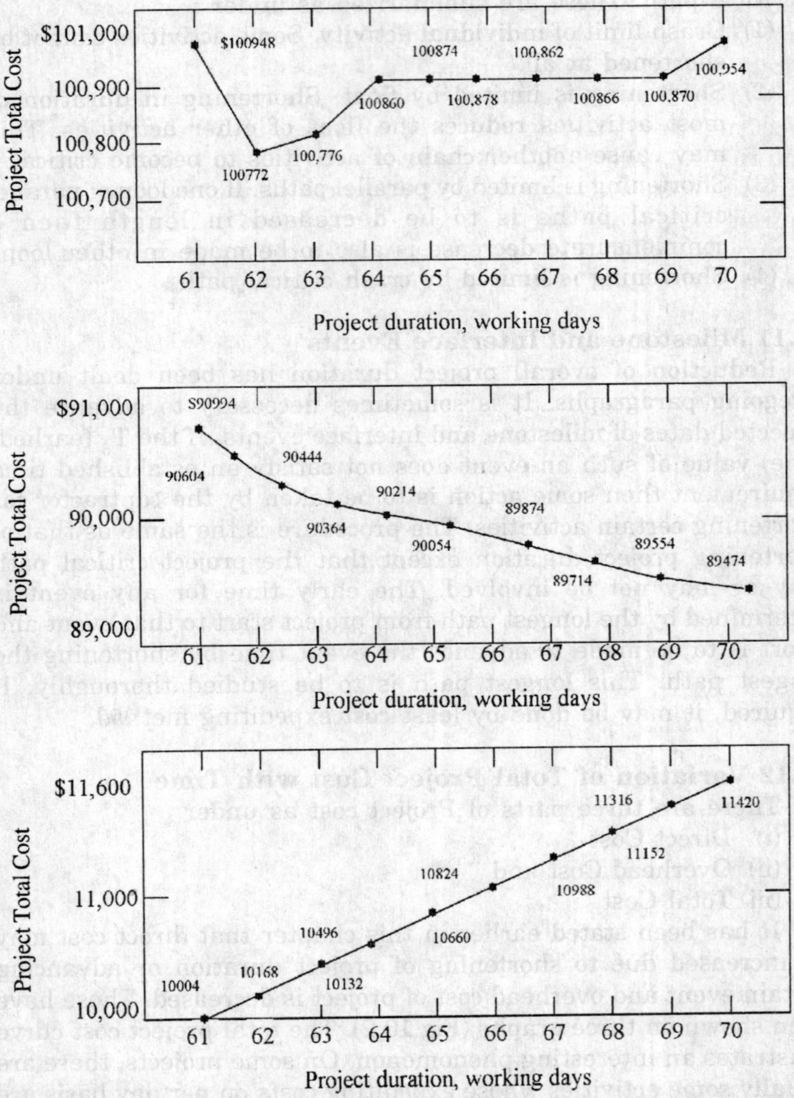

Fig. 10.2 : Example Project, Variation of Project Costs with Time

10.13 Project Lengthening/Extension

Contractor may like to lengthen/extend the work duration period in certain situation. As stated in one paragraph of this chapter the contractor prepare detailed working CPM network programme immediately after awarding the work by the owner. Contractor considered certain logic, number of manpower, equipments for preparing CPM network. As per this initial CPM programme, work completion period is quite less than the period stipulated in the contract agreement. After some initial period of construction necessity arises to shift some resources to his other newly awarded priority project; or the contractor reviewed his original CPM network and found that he may earn more profit (by reducing cost of work) if the completion period is extended/lengthened. Less resources will be required by lengthening the project completion period (which will remain with completion period stipulated in the contract agreement). In such a situation certain critical activities can also be relaxed. The effect of this is to lengthen the critical path and hence to increase the floats of all non-critical activities.

11

Construction Equipment and Machinery Management

11.1 Introduction

(a) With the improvement in technology and rapid increase in the labour rates, there is an increasing trend of mechanization in construction industry. Equipment/ machinery is important for time bound projects and it improves quality of works/better productivity. Certain types of works have to be done through plant/machine (like, hot bituminous works, large earthwork/rock cutting in highways, mountainous roads, rail track embankments, dams, canals, etc.).

(b) When the owner's organization take up the project construction and engage contractors, it is required for the owner's organisation to stipulate provision and particulars of plant/machinery/equipments in the NIT (tender documents). After acceptance of tender the contractor is required to make detailed planning, programme (based on CPM network) for arranging suitable type, size, numbers of machinery/equipments at appropriate time. Effective management of construction machinery/equipment by the contractor's organization is very important. In case of departmental works (i.e. works executed by the Government Department by procuring construction equipments etc.), detailed planning, deployment of machinery, labour, materials, operation, management, etc. are done by the engineers of the Department. However, it may be mentioned here that in Government organization most of the projects are got executed through the contractors. Equipments, machinery, pumps etc. which are permanent parts of the building/structure will not be dealt under this Chapter. Hence, details of management in respect of construction equipments/machinery to be carried out by the contractor and department/owner of the project will be dealt here.

11.2 General Planning

11.2.1 Immediately after acceptance of tender the contractor prepares detailed working planning and programming for the project construction through CPM network method. Thereafter, requirement of equipments/machinery for various activities is worked out. Calendar dates for requirement of construction equipments are also indicated on the schedule of equipments. Contractor is required to take special care for equipment planning and procurement of the same at right time. Manpower and materials are planned on the basis of same CPM network. Equipments can be used/operated with manpower; it handles the materials effectively. Production rate will be increased; quality of works will be improved. Necessary planning is also to be made for arranging energy (viz., diesel, petrol, electricity, light diesel oil, etc.) at project site for operating machinery. One of the largest single element of investment of the contractor would be on owning and operating of plant and equipment. The capital investment on the purchase and/or rentals/lease and operation of the plant and equipment being generally high; it has to be managed so as to ensure minimum operating, maintenance and repair cost, maximum productivity and returns on the investment made. The role of machinery and equipments in Civil Engineering works is very important. But necessary care and importance are not given on this part of project. This may be due to poor knowledge or aptitude on this subject. For completing large projects in time, it is essential to make proper planning. Necessary stipulation in NIT for executing the project with suitable machinery and equipment is required for completing in short/reduced period or on crash programme basis for time bound project. It is also required to use certain equipments and machinery for achieving good quality of works. As an example, it may be stated here that "Slip Form Shuttering" equipments are required to construct RCC chimneys, Lift Cores, Circular Tower for elevated water reservoir/tank etc. in short period. Risk is also involved for erecting very high scaffolding if conventional manual method is adopted for constructing aforesaid RCC structures. Practical knowledge of such equipment is essential for Civil Engineer for accurate operation during construction, otherwise serious problem will be faced due to variation in verticality beyond permissible limit. It may not be possible for proper planning and stipulating about correct type of machinery, equipments in the NIT/tender documents without proper knowledge. Stipulation is also made in the NIT for making advance payment to the contractor for arranging machinery and equipments at project site in time. Hence, Civil Engineer should have knowledge about application of various machinery and equipments. Functions of machinery/equipments are also required to be indicated. Number, type of machinery/equipments required with dates/period as per CPM network are to be stated on the schedule so that advance planning can be made by the construction agency. Men, material and money for machinery are also to be planned and arranged by the construction agency. Proper management of equipments and machinery during project construction is essential for successful completion of project.

11.2.2 Regarding role of various authorities/agencies on the management of construction equipments, machineries and instruments, it is stated that as per existing system, following are three categories of agencies for construction projects :

(a) Government Departments (viz., PWD, CPWD, Railways, MES, Irrigation Department, etc.)
(b) Government Undertakings (as owner of projects and as construction agency/contractor) and Societies, Trustees, Corporation, private organizations as owners of projects.
(c) Private Construction Companies (as agency/Contractor).

11.2.3 All the above stated three categories have role on the matter of construction machinery and equipments. Role related to planning (i.e. pre-construction stage) is mainly carried out by the Government Departments/owners of the projects. Construction agency/Contractor may be private companies or joint ventures of Government authority and private companies. Construction agency is mainly concerned for detailed planning and procurement of equipments, machinery and its effective operations. Certain Government organizations (like, Border Roads Organisation, Government of India) execute large construction projects and maintenance works departmentally (i.e. without engaging contractor). Such organizations purchase construction equipments and machineries themselves and the same are operated departmentally.

11.2.4 Civil Engineers generally do not seem to have aptitude and knowledge in respect of basic principles, functional operations, suitability of particular type of equipments, machinery in civil engineering projects. Without these aspects of knowledge and experience, it will not be possible for Civil Engineers to prepare proper specifications, stipulations of machinery and equipments in the NIT/tender documents. In the absence of proper stipulation of the specifications and use of machinery and equipments in the contract documents of project, problem will be faced during constructions. Quality of works and speed of construction are hampered due to non-use of necessary equipments/machinery. During construction stage, contractor is required to select correct type, size and number of machinery, equipments for completing certain items of work. Management of equipments and machinery during construction period is also to be done by Civil Engineers. Civil Engineers think that above stated functions relating to machinery and equipments should be done by Mechanical Engineers. As Mechanical Engineer is not generally available at site it has been found at project sites that Civil Engineers are required to plan and select types, sizes, number of various equipments, machinery and also to procure the same by purchasing or on hire basis in time; even management and operation of machinery are done at project site by the Civil Engineers. In view of above stated hard reality, Civil Engineers should have some basic knowledge about the construction equipments and machinery. There is another aspect also. As already emphasized, upgradation/ modernization of technology is required for various project constructions. Proper planning and construction methods/techniques by using suitable equipments and machinery are essential.

11.2.5 Cost of equipment in construction project varies from 8% to 25% depending upon the extent of mechanization. Good planning, appropriate selection, sizes (capacity), procurement, installation, efficient operation, maintenance/repairing etc. have important role in equipment management for successful completion of project. Civil Engineers should be familiar with the application and maintenance of various types of construction equipments. It is expected from engineers to increase output and ensure good quality construction at minimum cost. Rational planning and judicious deployment of equipments as per actual site conditions are important. Hence, it is essential to check actual site condition, magnitude of works, technology involved in construction, completion period of projects, availability of skilled manpower for operation and repair/maintenance of equipments etc. At planning stage, it is required to decide the extent of mechanization after checking locally available skilled labour.

11.3 Factors for Deciding Extent of Mechanization

(i) ***Manpower*** – Types and quantities of main items of project constructions are to be identified. Completion period is also relevant. As an example, for earthwork in excavation and filling for constructing hilly/mountainous road, high embankment constructions for rail track and highways, substantial numbers of unskilled labourers are required. If completion period is less, then it will be required to deploy heavy earthmoving equipments/machineries. Economy will also be achieved. Cost of labour and equipments can be roughly worked out for knowing economy. In case labour cost is more, then it is better to deploy equipments.

(ii) ***Type of works*** – Certain type of works cannot be done manually or it is not permissible. For example, construction of bituminous layers on road surface cannot be done manually; relevant equipments/machineries (viz., Hot Mix Plant, Paver Finisher, Road Roller, Tipper Trucks, etc.) are essentially to be deployed/installed for such works. Deployment of certain equipments also depends on weather/ climatic conditions, topography (viz., snow bound high altitude mountain), toughness.

(iii) ***Completion Period*** – This is an important factor for deciding to execute certain items manually or mechanically. If completion period is less then there will be mechanization.

(iv) ***Availability of Fund*** – Capital cost for purchasing machinery is quite high. Heavy investment is required at the beginning stage. Hence, extent of mechanization depends on availability of fund with the construction contractor.

(v) ***Availability of indigenous machinery*** – It is advisable to procure indigenous equipments and machinery because spare parts are available and the local mechanics will be able to repair and maintain the equipments. Imported equipments will be costlier, difficult to repair and spare parts may not be available readily.

(vi) ***Quality and Performance*** – It is possible to construct certain major items of works (viz., concrete/RCC works, bituminous works, etc.) in good quality/standard by only using appropriate equipments and machinery. As an example, dense bituminous macadam layer and bituminous concrete wearing course are to be laid as road pavements. This has to be executed through Hot Mix Plant, Paver Finisher, Road Roller, Weigh Bridge, Wheel mounted Loader, Tipper

Trucks, etc. Number and size (capacity) of these equipments/ machinery are to be decided as per quantum of work and the stipulated period. There is another important aspect i.e. very good quality and smooth finished road surface. Quality of mix with various ingredients (i.e. bitumen content, temperature control in respect of aggregates, bitumen and final mix at the time of laying at site) is to be ascertained as per laid down specifications. Proportion of various materials is to be ensured by weight (not in volume). This should be done batch-wise. For ensuring above stated quality of works, it is essential to use suitable type of Hot Mix Plant (viz., Batch type machine, electronic control for accurate quantity of various materials, controlling temperature for heating materials and of final mix product at Hot Mix Plant and laying site, etc.). Similarly, automatic electronic control Paver Finisher with high level of accuracy is to be arranged. Paver Finisher's purpose is to ensure laying of layer with smooth surface (i.e. surface accuracy is to be passed after checking with Bump Integrator/Roughometer). Accuracy standard is to be stated in the specifications of works. Hence, role of equipments/machinery is very important to achieve quality of such type of works and description of these equipments/machinery are required to be made properly in the NITs.

(vii) ***Social Object*** - In case the project has been taken for providing employment opportunities to people, then machineries, equipments should not be deployed or very less equipment may be deployed, wherever necessary.

11.4 Matching Equipments

For certain items of work, a group of equipments are deployed at the same time. As an example, for laying of bituminous layers on road surface, it is required to operate following equipments, machinery in one group (i.e. as matching units) :

(i) Hot Mix Plant (to produce bituminous mix)
(ii) Tipper Trucks (for carrying mix)
(iii) Paver Finisher (to spread mix in hot condition on road surface)
(iv) Road Rollers (different types) (for compaction of layer)

Capacity and number of above stated equipments should match so that all the activities/entire operations go smoothly and continuously. In case operation of one unit is held up or slowed down, then operation of other units will be affected or stopped.

11.5 Detailed Planning at Construction Stage

(a) Detailed requirements of equipments are to be indicated on the schedule of equipments (based on CPM network) where number, size of equipments are indicated alongwith date of start and completion of the relevant items of work. Equipment planning is to be done accordingly.

(b) Manpower planning is done for civil engineering construction. In a similar way, manpower planning is to be done for operation and maintenance/ repairing equipments. Arrangement of a field workshop alongwith adequate spare parts is most important. It has been observed at site that equipments generally do not work smoothly due to minor/ major fault or breakdown. For this purpose, skilled/ experienced operators, mechanics are essential. Training is to be given to them in advance.

(c) It is essential to check site conditions of project. Equipment selection and operation planning are to be done as per actual site conditions. If earthwork in excavation is to be done on hard soil/soft rock or soil mixed with boulders, then "Crawler" mounted Bulldozer will be required.

(d) Planning should be done scientifically. Success of project construction depends on good equipment planning at pre-construction and construction stages.

(e) Selection of equipments – Selection of specific equipment/ machinery and its size/capacity is to be done after considering job requirements, site condition, quantum of work, availability of electric power/other energy and skilled workers (operators), economical aspect, stipulated period of completion of job, climatic condition, altitude of place etc. There are some advantages and disadvantages in respect of electrical motor driven machinery over diesel machinery. These are to be analysed during selection of machinery. Important decision is to select the most suitable plant/ equipment which can pay for itself and determine the number and type. Lot of information on the specifications, production capacities, capital cost, cost of operation and maintenance are available in the manufacturer's manuals. But past experience in this regard is the most valuable. A well experienced Mechanical Engineer comes handy in this regard.

(f) Detailed specifications of various items/activities of construction project are prepared and NIT/tender documents are framed accordingly. From this stage of planning, stipulation for using equipments for certain major items/ activities of works is made in the NIT (tender/quotation documents) by the owner's organisation. After considering quantities of works, site conditions, available period of construction, type of works, economy etc. requirement of different equipments is decided.

(g) Contractor will make detailed planning in respect of equipments which are to be purchased or to be arranged on hire basis or equipments already available with the construction agency in working condition. Number of shifts is also to be considered. It is economical to deploy equipments in two shifts. However, it may be mentioned here that output of work will not necessarily double in two shift workings; efficiency is reduced in second shift (during night). It is also required to work out the number and size of equipments. Details of construction works of relevant items/ activities are to be worked out indicating time schedule from CPM/Bar Charts and Bill of Quantities of project work. Available working days are also to be worked out for aforesaid purpose. Climate for constructing relevant items is also to be checked from calender scaled CPM diagram and planning is to be done accordingly. Bituminous works, deep excavation works should not be done during rainy season.

11.6 Purchase of Equipment and Machinery

(a) This is another problematic matter, particularly in many Government Departments. In most Departments, there is no proper mechanical branch or competent and experienced Mechanical Engineers for project construction. Civil Engineers are mainly concerned for the project construction; but it is difficult for them to select correct type and size of equipments. They cannot prepare detailed specifications and NIT (tender documents) for purchasing machine/equipments. Technical and financial evaluation of tenders/quotations and to issue supply order may also be problematic for them. It is also not possible for Civil Engineers to inspect equipments for acceptance after checking specifications and performance of equipments. Even if one Mechanical Engineer (not automobile engineer) is deployed, still problem is faced as the sole engineer is unable to prepare/finalise NIT/Quotation documents, to finalise tender case and to complete the purchase process. Civil Engineers consider that such duties are supposed to be done by Mechanical Engineer. In the absence of competent experienced senior Mechanical Engineer, there is abnormal delay in procurement of correct type and size of equipments, sometimes it has not been possible to purchase certain costly equipments, machinery. Purchasing of equipments, machinery is entirely different from construction activities.

(b) It is required to check the lead time (i.e. supply period) for delivering the machinery and the time required to install/ commission the machinery. Fast moving/maintenance spare parts are also to be purchased alongwith new equipments. Suitable stipulation is to be made in the NIT (tender document) and the same are to be included in the supply order of equipments. After-sales service is another important aspect and the same is to be checked before issuing supply order. As first stage, technical bids are to be invited and after detailed scrutiny and pre-bid meetings, the pre-qualification of competent manufacturers/suppliers is to be finalized. Thereafter, modified bid documents are to be prepared on common basis for obtaining financial bids from the pre-qualified suppliers/manufacturers.

(c) It is essential to sort out the problem relating to responsibility of engineers for purchasing equipments and machinery, as already mentioned in the foregoing pages. In view of reluctance of Civil Engineers to accept the onus due to lack of knowledge and aptitude, Mechanical Engineers should do the job. However, it has been observed that even if there is a separate Mechanical Engineering Branch, they do not consider purchase of machinery and equipment relating to construction works as their domain. It is, therefore, suggested that a competent and experienced senior Mechanical Engineer on this field (i.e. construction equipments and machinery) should be placed under a senior Civil Engineer who will co-ordinate, give detailed requirements of equipments (particularly performance, facilities available with the equipment) and monitor various activities viz., preparation of tender documents, processing and finalization of tenders/quotations, inspection of equipments and taking delivery.

(d) If quantum of relevant items of works is not much and there is shortage of fund (as capital cost) for purchasing heavy equipments/ machinery, then these can be arranged on hire/lease basis. There is another method; a blend of both. Initially the equipment is arranged on hire basis. After operating it for some period, the same is retained/purchased from the owner of equipment by adjusting the amount already paid as hire charges. However, suitable percentage of amounts paid as hire charges is reduced during adjustment of accounts.

(e) *Factor affecting output and performance of equipments* – In order to plan the equipments for a particular job, it is necessary to know the output of the equipment for that type of particular job. Actual job condition is to be ascertained. Manufacturer gives information considering ideal condition which is different than the actual particular job condition. Hence, their production figure has to be modified.
The factors affecting performance of equipment on any particular job can be divided into two groups, viz., (i) Job Factors and (ii) Management Factors. Job factor depends on physical conditions, type of the job, etc. Management factor relates to arrangement of plants at site and managing the same in an effective manner. Operator's efficiency depends on the training and experience of operator, proper matching of equipment, sizes and numbers. Time is required for on-the-job servicing and maintenance. It also depends on availability of well equipped workshop, maintenance facilities, spare parts stock etc.; these are very important. Working days/hours for the equipments in a year is also an important factor.

(f) *Buy / Lease / Rent* – An important decision is whether to buy or lease or rent the equipment. It may be beneficial from the financial angle, to lease or rent the equipment. Problem of obsolescence sometimes counter-balances inflation and rising cost and the residual cost of equipment is even more than their purchase price.

(g) Once the equipment has been acquired it is important that the same is utilized to its full advantage as the idle equipment not only can add to losses but it is also likely to be rendered unserviceable. This is one where the optimum utilization is, therefore, of paramount importance and calls for the best of skills in technology and management.

(h) *Financial Impact* – High mechanization entails heavy initial investment on the procurement of construction equipments. However, in the long run, if the equipment is operated, maintained and managed efficiently, it would more than pay for the cost and in the ultimate analysis may pay better dividend to the contractor. Cost of equipment could be divided into "owning cost" and "operating cost". The owning cost should include interest on investment, taxes, insurance, depreciation, storage and overheads, overhaul and repair. The operating cost should include wages of the operators, cost of fuel and lubricants, minor repairs, and loading, transporting, unloading, erection at construction site and dismantling etc. For calculating depreciation amount generally "straight line" method is applied.

11.7 Operation/Maintenance during Construction

(a) Efficiency as well as life of equipments depends on skilled, trained and experienced operator. It also depends on efficient repairing and maintenance of equipments. For this purpose, a good "workshop" is essential. Regular servicing including cleaning, lubrication and check up is very crucial. Field repair arrangement is required to be made at project site. Inspection of equipments, machinery is to be carried out regularly so that preventive maintenance can be done. For large project or organization, a central workshop is also set up. Another important matter is arranging adequate number of various spare parts in the workshop. It has been found at ground that many costly equipments lie idle due to non-availability of spare parts and huge loss is incurred on this account. Hence, special attention should be given for spare parts and efficient functioning of workshop.

(b) Broadly, there are two types of maintenance :
(i) Breakdown maintenance
(ii) Preventive maintenance

(c) Certain maintenance tasks are of minor type and the same are carried out quickly at worksite. Top priority and importance should be given to maintenance of equipments. Success of project construction depends on good maintenance of equipment and efficient operation. Breakdown of equipment can occur due to unpredictable failure of components. This can be reduced/ eliminated by proper preventive maintenance. Proper planning/scheduling is done on a chart for regular inspection. Source of trouble is found out and needful is done before breakdown. Like medical field, "prevention is better than cure". Overloading and misuse of equipments and vehicles should not be done. It is required to keep the equipments/machines in good condition so that "quality" of the product is achieved and safety of workers is ensured. Monitoring should be done for achieving maximum production. Proper records are to be kept for various activities of procuring equipments, installation of plant/machinery, minor/major breakdown/repair works and preventive maintenance, operation hours, production records, etc. These should be analyzed for taking appropriate action. Spare parts register, inspection register, Log books, defect register, etc. should be maintained and the same should be checked monthly by senior engineer. If preventive maintenance is good, then downtime as well as overtime payment would be reduced. Expenditure on repairs would also be reduced and the breakdowns would be minimal.

(d) ***Services, maintenance and storage of Equipments*** – The life and performance of the equipment depends on preventive maintenance, scheduled servicing and proper storage of the equipment. There should be proper parking place under shed for all equipments/machinery in working/non-working condition. There should be optimum utilization of equipments. Action should be taken daily (morning and evening). Check list is to be prepared and displayed at site office for checking various parts/components/fuel/water/lubricants/battery/tyre pressure etc. and compliance operation in the morning. Similarly check list is also to be prepared for checking various points of equipment at the time of closing the work at evening/night. Inspection and servicing of equipments are to be done as per schedule and checklist containing various items.

(e) ***Records of equipments*** – Accurate and complete records of equipments and machinery are to be maintained both from the control point of view as also from the accounting requirements. Log book of each equipment is to be maintained on daily basis including idle period, repairing period, transportation period. Particulars of servicing, repair works, overhauling work, etc. are also to be recorded on the log book. Quantity of fuel used is also to be recorded in the log book. Periodical reports on prescribed form should be sent to all managerial concerned persons.

11.8 Training of Operators

Economic and efficient operation of the equipment are of vital importance for the success of a mechanized construction project. There is a dire need of operating and maintenance skills for operators, mechanics, helpers, foreman/supervisor.

11.9 Safety

Safety in handling i.e. transportation, setting, erection at project site, dismantling for shifting to another site and operation is very important. Instructions given by the manufacturers in this regard are to be strictly followed. Lubrication, repair works should be done by authorized skilled/trained person. Machinery should not be left unattended. Wire ropes of cranes, hoists etc. should be checked regularly. All site workers (skilled and unskilled) must wear helmets. There are many other instructions on safety matter.

Worksite, store, field workshop, fuel store, etc. should be cleaned on daily basis. Adequate area (open and covered) should be available for parking, repairing/ maintenance works.

11.10 Some Construction Equipments and Machinery including Pumps for De-watering

As per enclosed Annexure – I.

ANNEXURE – I

Statement showing some construction equipments and machinery including pumps for de-watering

Sl.No.	*Name of Equipment*	*Main use of equipments*
1 (a)	Tractor	Two or four-wheeled or crawler or wheel mounted type. The tractor is a versatile earthmoving machine. These are also used for pulling or pushing of various other equipments.
1 (b)	Bulldozer	Tractor is a basic machine which can be used for various applications like dozing, towing. The Bulldozer is an attachment mounted on a tractor for carrying out earthwork. Crawler type is used in rough/rocky/hilly areas. When attachment is with teeth, it is known as "Ripper" (for loosening hard earth and other material)

Sl.No.	*Name of Equipment*	*Main use of equipments*
1(c)	Excavator and Front-end Loader	Mainly for excavation work. Different types of front-end attachment can be used. The machine can be made with front-end loader and excavator attachment on rear side which is quite popular.
1(d)	Scrapers	For functioning of loader, hauling and discharging earth/material with the help of other equipments.
2	Crane, Grab Dredging Crane, Tower Crane	Lifting and placing of materials at different locations (by moving horizontally and vertically)
3	Hoist	For carrying materials vertically and loading/unloading at required level.
4	Compressor	These are available as stationary, portable, rotary types and used for drilling rocks, pile driving, guniting, pneumatic vibration, etc.
5	Belt Conveyors	These can be portable or fixed installation and used for transporting materials (used in Crusher Plant, Hot Mix Plant, Concrete Mix Plant, various other plants etc.)
6	Crusher Plant	These are used to crush stone boulder for obtaining stone aggregates, stone dust.
7	Batching type cement concrete mixing plant and Ready Mix Concrete Plant	Cement concrete mix is produced.
8	Hot Mix Plant (consists of bitumen storage tanks, aggregate heating unit, mixing unit, bitumen heating unit, Loader for loading aggregates, weigh bridge, control/operating panel room).	For producing hot bituminous mix (for road, pavement and Runway works)
9	Pavement/Concrete Breaker / Rock Drilling Machine	Worked on the basis of compressed air. Rock drilling is done for blasting operation.

Sl.No.	*Name of Equipment*	*Main use of equipments*
10	Guniting Equipments	Cement concrete (with fine aggregates/ sand, cement and water) is placed on existing concrete surface under heavy pressure.
11(a)	Road Roller (Two wheeler/Three wheeler/Tandem types)	For compaction of subgrade and various courses of road, runway, pavement.
11(b)	Sheep Foot Roller	For compacting earth embankment.
12	"Acrow" type centering, shuttering and scaffolding	Several system of steel pipes are used for building and other works.
13	Various types of Pumps and Well Points system (Ring type and multi-stage type)	Well-Points system is installed temporarily at deep excavation site (viz., Basement Construction) for dewatering underground water.
14	Slip Form Shutter	This is mechanical type shutter used for casting RCC Lift core, chimney, Circular Tower for Overhead Reservoir.
15	Needle Vibrator, Plate Vibrator, Table Vibrator	These are used during casting RCC/ CC works for compaction.
16	Wet Stone Aggregate Mixing Plant	This is used for wet mix macadam in road work.
17	Paver Finisher for bituminous courses, cement concrete pavement/road, wet mix macadam	Required type of Paver Finisher is used for spreading hot bituminous mix on road/concrete mix/wet mix aggregates for road works.
18	Rock Drilling Machine (Rig with Compressor)	This is used for drilling bore for installing deep tubewell in rocky strata.
19	Rig with compressor for Bore Pile	For making bores to construct bore piles.
20	Transit Mixer (Special Type Truck)	Used for transporting cement concrete from Ready Mix Plant. Conical shape large cylinder body which rotates around its axis.
21	Concrete Pump	Cement concrete mix is pumped for laying at casting location.

12

Financial Management

12.1 Introduction

Financial management is a basic necessity for an effective overall management and control in any industry for better end result. The planning/management related to various resources (viz., manpower, materials, construction equipments etc.) have been dealt under foregoing Chapters. But the same entirely depends on finance/ money/cash flow. Hence, finance is one of the most important resources in construction industry.

12.2 Owner's Organisation

Project construction is initiated by the owner. Owner's organization may be Government Department/Government Undertaking or private organization. Contractor's organization may be Government Undertaking or private company. Government Department/Government Undertaking locate source of fund (under Plan or non-Plan sector or loan from Government source) and budgetary provision is made year-wise and thereafter the decision is taken to take up the project/work. Finance Department of the organization is responsible to locate fund and allocate necessary budget provision year-wise for entire project. Necessary budgetary amount (year-wise) is proposed by the Chief Project Engineer/ Manager. In case these forecast budgeting amount is required to be modified (on lower or upper side) during the year the revised budget estimate is submitted by the Chief Project Engineer three or four months before the closing of financial year which is 31st March. After approval of these budgetary provisions (i.e. budget estimates) necessary fund is made available for making expenditure related to works. This is treated as cash "outflow" from the owner's organization for making payment to the Department's personnel, cost of materials (like cement, steel) to be purchased by the Department for issuing to the contractor, for making payment to contractors and others related to works/projects. In case owner's organization is a private one (viz., private company, group housing society, etc.) then they are required to plan first for arranging necessary fund (year-wise) for entire project and budget estimate (year-wise) is also prepared accordingly.

12.3 Contractor's Organisation

(a) After awarding the construction work to the contractor, the detailed planning and cash flow statement (month-wise) are to be prepared by the contractor. "Cash inflow" and "cash outflow" (month-wise) to the contractor are to be planned. Main source of cash inflow to the contractor is owner's organization (i.e. cash outflow of the owner's organization); these cash inflows are received by the contractor (generally every month) against the running account bills submitted by the contractor to the owner's organization. It is important to ensure that owner's organization make payment every month to the contractor. If there is delay in making these payments then progress of work may be hampered. Hence, special care and monitoring should be done by the Project Managers/ Engineers of owner's organization and contractor every month during project monitoring site meetings.

(b) When detailed planning and cash flow statements (month-wise) in respect of inflow and outflow are prepared by the contractor it will be seen that there are some gaps i.e. requirement of extra fund during certain period. This extra fund is to be arranged by the contractor from some other sources like, taking loan from bank/financial institute. Contractor is required to cost of this loan money i.e. to pay interest to the bank/financial institute. Hence, contractor is required to make plan for refunding the loan amount as early as possible.

12.4 Finance as a Resource

It is generally in short supply. Hence, the proper and efficient utilization of this resource is of vital importance to a construction organization. Financial management is an important aid to the overall management strategy. Financial accountability must be built into the performance and appraisal system. There are two types of capital requirements – (i) long term and (ii) short term. It is required to ensure its proper utilization through budgetary control. It is required to draw out a proper cash flow plan/statement to maximize use of this resource at the lowest possible cost. Care has to be taken to see that the interest element (i.e. cost of capital) is kept to the barest minimum and to make maximum effort towards the realization of advances and dues from the clients in good time.

12.5 Financial Control

(a) Important functions of the head office are arrangement for finance, accounting and internal audit and control. This will be clear from the organization structure as per chart (figure 12.1). Long term planning is done by the head office and its day-to-day function is done by its regional office/units. Financial concurrence should be a pre-requisite for deciding on any major financial commitment. A good knowledge of accounting and income tax will help the contracting company in transacting its business economically and efficiently.

ORGANISATION CHART FOR FINANCE DIVISION

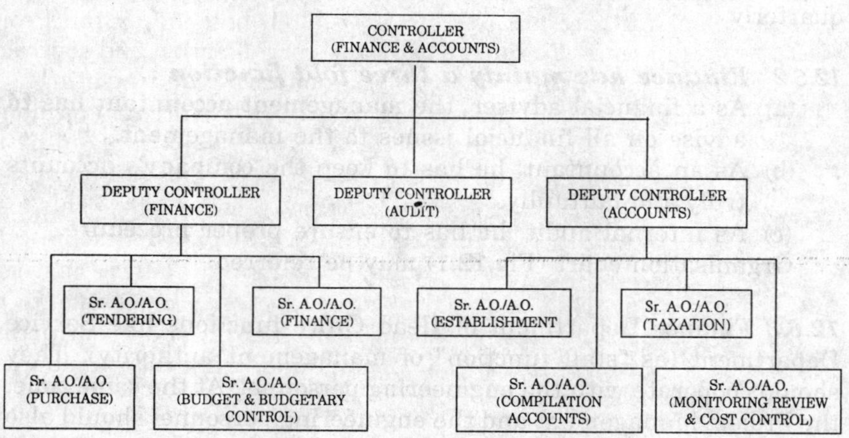

Figure 12.1 : Organisation Chart for Finance Division

(b) In case of construction industry it is practically difficult for carrying out effective budgetary control because it depends on terms and conditions of contract, site position and nature of construction work/project. The Head Office finance will exercise control on the field units/zones through the "cost ceilings". As soon as the work is awarded, the ceilings for various items including overheads should be issued by an independent authority at the Head Office level. This type of cost control and other safeguards will help to complete the job in the contracted period economically. However, head office should obtain recommendations from the field unit/ regional management.

12.6 Field Unit Working

12.6.1 Actual utilization of fund is done by regional/field unit. As such, their role is important. Finance Manager of field unit are to gather, process and report to head office on the financial information. Unit finance will review expenditure with reference to ceiling amounts. The unit accountant is required to keep a close track on the budget provisions, cash flow (inflow and outflow) positions and the collection of dues at unit level. He must keep a track of all items of work and keep proper records of construction materials, regular and deviated/extra items etc. so that due claims are sent to the client at correct time. Certain reports/returns on finance and accounts matters are to be submitted to management authorities quarterly.

12.6.2 ***Finance has mainly a three fold function :***

(a) As a financial adviser, the management accountant has to advise on all financial issues to the management.

(b) As an accountant, he has to keep the company's accounts truly and faithfully.

(c) As internal audit, he has to ensure proper procedure.

Organisation chart (Fig.12.1) may be referred.

12.6.3 Finance Department of Head Office functions like Service Department (as "staff function" of management authority). They should cooperate with the engineering personnel. At the same time, the Project Management and the engineering personnel should also consider that finance is an aid to them and not a bottleneck.

12.7 Cash Flow Statement

12.7.1 The cash flow statement of contractor's organization should be submitted to management authorities quarterly. A proforma in this regard is as per Table-VI. The cash flow indicates generation of resources, collection of dues and extent of investment made and needed in future.

Table No. VI
Construction Organisation
Forecast of Cash Flow for the Period Ending :
Name of Work/Unit :

Sr. No.	Estimated turnover during the quarter	Opening balance of cash in hand and at bank	Estimated cash receipts during the quarter		Total cash available	Likely cash expenditure during the quarter on works	Payments likely to be made for temporary hutments centring, shuttering, scaffolding (CSS) small tools & stores (STS)
			From clients	From other sources			

Payments likely to be made during the quarter in respect of past liabilities	Other payments	Total estimated cash payments	Net surplus/ deficit

Purpose : The cash flow indicates generation of resources, collection of dues and extent of investment made and needed in future.

Frequency : Quarterly

To :
S.A.O. from unit
Controller (F & A) from Regions.

From :
A.O. Unit Level
A.O. Region Level

12.7.2 Statement showing position of outstanding dues of contractor's organization should also be submitted to management authorities quarterly. A proforma in this regard is as per Table No. VII. Realisation of bills/dues from client is an important aspect in financial management as initial cash flow projections/plan go out of control if payments are delayed unduly from the side of client. The information helps management to take remedial/alternative actions in time.

Table No. VII
Construction Organisation
Position of Outstanding Dues
Name Of Work/Unit :

Figures in Rupees in million

Sr. No.	Value of Work	Less deduction Mobilisation	Security deposit	Any other deduction	Net payment due	Already received till date	Net outstanding dues for work	Due on account of P&M advance material advance	Any other amount with details	Total dues outstanding

Purpose : Realisation of bills/dues from client from important aspect in financial management as initial cash flow projections go topsy turvy if payments are delayed unduly. The information helps management to take remedial/alternative actions in time.

Frequency : Quarterly.

To :
Account Officer Manager C. E.

From :
Accounts Officer/Senior Accounts Officer

12.7.3 Senior Finance Officer should submit consolidated/abstract of financial review report of the contractor's organization monthly. At senior management level, the information gives complete financial position at a glance. Position of original plan/ forecast (unit-wise) in respect of investment made by the contractor's organization, inventory amount, outstanding dues, performance (i.e. anticipated profit and actual profit/loss), advances (i.e. recoverable by clients) etc.

12.7.4 Contractor prepares initially (i.e. immediately after receiving the letter of tender acceptance) the plan/forecast of cash inflow and outflow month-wise. Major part of cash inflow amounts is from the source of client (i.e. owner of the project). Mobilisation advance (general and in large project on plants and machinery), secured advances against materials brought at site by the contractor are to be paid by the owner as per stipulation made in the contractor agreement. The contractor prepares construction programme using CPM network; target gross values of construction to be executed month-wise are worked out on the basis of aforesaid CPM network/ programme. These values are shown as forecast cash inflow (i.e. to be paid by client). But on reality there may be a lot of deviation due to many practical problems. Contractors do not submit running account bills correctly and in time. One important matter is recording measurements of work done jointly i.e. engineers of contractor and owner's organization. It has been found that up-to-date measurements are not recorded on regular basis and signed jointly by both the parties on the Measurement Books (M.B.). Senior engineers/ engineers of owner's organizations are also required to carry out test checking of measurements on M.B. But these are not done in time. Due to these and other reasons lot of delay occur on actual receiving the payment by the contractor. Some times considerably reduced payment is received by the contractor. Hence, cash flow position is changed considerably and contractors face problems. Progress of work is hampered. Contractor is required to take appropriate action for solving such problem. He may discuss with various authorities of the owner's organization so that payment is expedited. He may also arrange some amounts from banks/financial institutions on loan. Cost of money (i.e. interest) is involved for such cash inflow. Contractor may try to purchase some materials on credit basis. It may be mentioned here that contractor is concerned with net amount of running account bills duly checked. Various recoveries are made from running account bills, such as :- installments against various advances made earlier (entire advance amounts are to be recovered when 80% work is completed), values of materials issued by the client at certain issue rates, income tax etc. Contractors make payment to material suppliers, labourers, construction equipments and on account of overheads. These are cash out-flows of the contractor. For purpose of planning (i.e. for fixing target/forecast) manpower cost may be considered as 25% of project cost, material cost may be considered as 45%, cost on account of equipments may be considered as 10%, 10% is on account of overhead and remaining 10% may be profit.

12.8 Measurements and Payments

12.8.1 Measurements of materials are to be recorded on the measurement books (M.B.) by the engineers of owner/owner's organization (Government/private) during purchasing materials. Measurements of work done at construction site are to be recorded on the M.B. jointly by the representative engineers of the owner's organization and contractor. They are required to sign on the M.B. This is very important permanent record as payment is made to suppliers/contractors on the basis of these measurements. These

M.B.s are permanent records and to be kept under proper custody for long period (at least 20 years in Government Department). It may be required to produce before Arbitrators/Court if any dispute crops up in future. Measurements are to be recorded accurately (degree of accuracy i.e. decimal digit, will be as per laid down rules). Problem of accuracy is faced for recording measurements and method of computation in case of earthwork in embankment of road/highway and for railway track. Measurements for such work are done by taking levels, original ground level i.e. before starting work and levels of finished embankment work. It depends on accuracy and type of level instruments and also error of taking the levels due to unskilled/inexperienced engineer. Accuracy of instrument is to be checked carefully. There are different methods of computations of earthwork in embankment and correct method (as approved by competent engineer) is to be adopted.

12.8.2 There is another example of special type of mode of measurement for item of bituminous mix. Bituminous mix courses are laid on road pavement. Mix is produced in hot mix plant and the same is carried to laying site through tipper trucks. Weight of mix is done on weighbridge at hot mix plant site (in metric tones). Measurement of such items can be done in following three methods:-

(i) By weight (in Metric Tonnes) :- by weighing on weighbridge at plant site.

(ii) Measurements of surface area of pavement are measured and thickness of layer is measured accurately at suitable intervals. Then volume is worked out. Weight can be worked out theoretically by density method.

(iii) By taking levels at suitable intervals before starting laying operation and after completion of the layer/course.

It is recommended to adopt the first method. This should be stipulated in the NIT (tender documents) accordingly.

12.8.3 Methods for various items of works are laid down in the manuals, codes published by various authorities viz., BSI, CPWD, MES, Railways, State PWD, National Building Code, etc. Mistakes are made by inexperienced engineers on method of measurements in respect of certain items; a few examples are stated below :

(i) Area of floor of rooms is to be measured from wall (and not from skirting surface).

(ii) Measurements of steel reinforcements are to be done as per actual length after placing and binding on the shuttering or in the columns, beams, etc. Supports (viz., chairs, pedestals etc. made with bars) are also to be measured and recorded on M.B.

(iii) Actual total quantity of wood used in door/window frame (chowkats) is to be measured i.e. wooden frame embedded in the floor, brickwalls are also to be measured.

12.8.4 As per laid down procedure 100% test checking of steel reinforcements in RCC works is to be carried out before starting casting work by one rank higher i.e. in Government Departments all measurements are recorded on M.B. by the Junior Engineer and test checked by Asstt. Engineer. Ten per cent test checking is to be done by Executive Engineer on the M.B.

12.8.5 Nomenclature of items are to be recorded on the M.B. as per schedule of quantities/B.O.Q. attached with the contract agreement. If any item has been executed not as per agreement or as extra item then the same (i.e. as per actual) is to be recorded on the M.B. Rate for such type of items are to be worked out (i.e. analysis of rate) as per laid down procedure.

12.8.6 All pages of M.B. are machine numbered. Necessary certificate is to be recorded on new M.B.s by Executive Engineer at the time of issuing to Site Engineers. Locations of items executed at site are to be indicated (if necessary with sketches) on the M.B. at the time of recording measurements.

12.8.7 Standard M.B. (SMB):– In case of repeated type of works (like, construction of same type of residential flats) measurements are carried out in respect of one flat or one block. SMB is also prepared in respect of certain items (viz., plaster, floor, doors, windows etc.) of completed building for future maintenance works (viz., painting, etc.).

12.8.8 Quantities are calculated in relevant units against all measurements recorded on the MBs. 100% calculation checking of quantities are to be done by a responsible person and to be signed on the MB (stating calculation is checked). In case certain element could not be completed at the time of preparation of running account bill then part/ provisional rate (P.R.) is paid; when this element is completed later on then full payment is made in future bill. P.R. is also paid on R.A. bill in respect of extra item till it is approved by the competent authority. In case there are some defects in certain items then contractor is instructed to rectify the same through "site order book". Finally, in case certain minor defect could not be rectified then "deduction item" is prepared (subject to approval of competent authority).

12.8.9 In practice, bills are prepared/checked by the field engineers of the owner and submitted to accounts branch. Necessary recoveries are made from the bills on account of instalments of mobilizations advance, materials issued by the owner's organization, security deposit, secured advance against materials brought at site, tax etc. Theoretical consumption statement of certain materials (viz., cement, steel, etc.) issued by the owner's organization is to be prepared for checking/control of these basic materials; variation should be within permissible limits.

12.8.10 After detailed checking the bill is passed by the accounts branch and payment is made to the contractor, supplier. This payment is the inflow of contractor's cash flow statement. The contractor's inflow amounts mainly depend on these running account payment. In case there is delay for these payments then progress of work may suffer. Hence, it is required to monitor the above stated matters (i.e. recording measurements on MBs, preparation of bills, its checking and payment) for ensuring timely payment. Running Account Bill consists of following documents :-

- Forwarding letter
- Abstract of bill on prescribed form and also on M. B.
- MB containing measurements, abstract of quantities
- Extra Item Statement/ Substitute Item Statement
- Statement of Recoveries
- Position of various advances
- Theoretical Consumption Statement of Materials (cement, steel) issued by the owner.
- Part Rate Statement (with reasons for P.R.).

13

Adopting Appropriate Technology and New Materials

13.1 Selection of Technology

13.1.1 The type of technology to be adopted in the construction industry depends on the nature and volume of the job, the time available for completion of the job and the sophistication required to be adopted. It relates to optimum utilization of available resources in the country. Wholesale transfer of technology from advanced countries to our country may create problems, it may distort our socio-economic structure.

13.1.2 There are two general approaches. First approach advocates the use of "appropriate technology" which is labour intensive techniques, involving low rate energy use. The second approach advocates the use of the most advanced and capital intensive technologies and is linked with the policy of expanding the energy necessary. The first strategy is better in our country at this stage. However, in long term planning, we should proceed for above stated second approach i.e. gradually adopting certain advanced technology for certain project, like trench-less technology for sewerage works in city (particularly in mega city). The third approach advocates upgradation of technology which is being used presently for certain types of work. This is mixed with labour force and machinery.

13.1.3 Examples of certain projects may be given here. In case of important international/national level project it is essential to complete the intricate/specialized construction project within short period on crash programme basis. It is also essential to achieve high standard quality of work. For such project it is essential to make suitable architectural and structural design using pre-fabricated and structural steel materials so that project can be completed in short time in good quality. Various construction equipments/ machinery should be used after appropriate selection. It will take considerable period in case conventional RCC frame structure is adopted. Ready Mix Concrete and pre-cast RCC units (slabs for floors) can be adopted. Appropriate type and size/capacity of cranes should be used. Slip form shuttering should be used for constructing circular/other shaped tower/cores for overhead reservoir, lift cores, chimney, etc. Large span roof can be designed with steel trusses, steel columns, steel girders; suitable interior design works can be done for achieving good aesthetical look.

13.1.4 However, in rural areas, the labour intensive techniques should be adopted particularly in agricultural sector. Some more examples of using modern technology are given here. Pre-cast pre-stressed girder design should be adopted in construction of flyover in city. It is required to adopt modern technology i.e. trench-less technology (micro tunneling method) for laying sewer line across railway tracks, below city roads in congested area where heavy traffic exists. Heavy earthmoving machinery is essential for constructing roads in mountainous areas. Special modern equipments are also required for rock cutting during road construction in mountainous areas. Special types of equipments are required to detect leakages and its remedial measures in public water supply system in city (there may be heavy leakages/wastage of 45% to 50%). Special techniques are required to be adopted to construct leakproof underground structures (basements, underground car parking, etc.).

13.1.5 Recycling of certain materials (viz., municipal solid wastes, scraps, waste papers, etc.) is another field where modern technology should be applied. Compost (manure) should be produced through compost plant by using municipal solid wastes. Energy (electricity, gas) can be produced by using municipal solid wastes. Papers are produced through re-cycling technology.

13.1.6 New technology increases productivity and the same increases standards of living. Increased productivity and economy of the country is improved. There is another theory for adopting modern technology. This is to maximize the rate of development which in turn leads to maximize the rate of increase of "labour power" or productivity. As per above theory, the "strategy for appropriate technology" is not one which encourages development. Its application is for increasing labour productivity. There is another aspect of capital intensive development; it is necessary to create appropriate financing, provisions for very long term, low interest loans and credit arrangement.

13.1.7 In our country, the factors which contribute to the selection of technology are the abundance of labour, scarcity of capital and difficulties in securing right type and quantity of managerial skills. Hence, it may be required to adopt labour-absorbing and capital-saving devices. In advanced countries which are under-populated, modern technology has to adopt labour-saving and capital intensive devices. In India, it is required to examine carefully about degree of mechanization in building industry keeping in mind about local needs.

13.2 Animal Energy

Animal energy is very useful in rural areas. The socio-economic aspects of the increased utilization of this source of energy for rural development besides agriculture, are important, because of the small load and short distance involved in rural areas. Increased use of animals in India will continue at least for some decades. There is some calculations of equivalent energy of thousands of megawatts of electric power from millions of work animal in India at a much cheaper price than conventional hydro, thermal or diesel power. However, there are some pitfalls in animal energy. In long term, it is required to use gradually right technology in stages in rural areas also.

13.3 Present Need

(a) In our country we should resort to selective changes and adjustments of greater innovative efforts rather than lavish imitation of advanced technology. Upgradation of existing technology should be done step by step. It is required to adopt modern technology in certain types of specialized/ intricate projects. To sort out certain problematic works it is required to adopt modern technology. Some example projects have been given above.

(b) "Technology transfer" and collaboration for transfer of know-how should be adopted in certain types of projects. We have achieved upgradation of technology in civil engineering projects successfully. It is required to make the best use of the available resources in abundance particularly labour and local materials of constructions and limited finances. It is required to improve the standard of working for labourers and to provide various facilities and to increase wages. Necessary steps are to be taken for improving productivity of labourers. Increasing trend in mechanization, pre-fabrication and adoption of industrialized methods of construction is unavoidable.

13.4 Pre-fabrication Technology

13.4.1 There are some disadvantages in cast in-situ RCC works. Due to certain practical constraints it is difficult to achieve good or very good finishing on exposed RCC works. Progress is slow in cast in-situ works due to various reasons like, curing time, period required for removing centring and shuttering from floor/roof slabs, rectification of surface of structural members. Time is required to carry/lift concrete and to place the same at required locations. Above stated negative factors/problems can be solved by adopting pre-cast technology. Unfortunately, use of this technology is very less in India as compared to many other countries. Pre-cast units can be made in the yard/ factory with effective quality control and speed. Degree of accuracy will be more in pre-cast work. It is possible to cure much better way than cast in-situ work. Strength of concrete will be better in pre-cast technology. It is possible to carry out parallel activities at project construction site. One important advantage is speed of construction by adopting pre-cast/pre-fabrication method. As workers wage rises, the pre-fabrication and industrialized method of construction is bound to be cheaper compared with conventional on-site construction. In advanced foreign countries, there is a lot of encouragement in pre-fabrication technology. Pre-fabrication can contribute to the standardization in planning and designing and rationalization of building techniques and erection work which can speed up the construction. It is also easier to maintain the quality of work in this system. Proper shuttering/ moulds with plastic lining can be made easily and very good exposed concrete finishing can be achieved in this pre-cast system.

13.4.2 Certain structures (viz., factory, workshops, etc.) should be made with steel members. Pre-fabrication of such structural members (viz., trusses, girders, etc.) are done at construction site in short time as parallel activity of project construction. These pre-fabricated members are erected in short time. Speed of work is achieved in this system. Accuracy in pre-fabrication is also achieved. In case of pre-cast technology of RCC works it is required to mix concrete and pour in the moulds and then curing at pre-cast yard/factory. But in case of steel pre-fabrication considerable time is reduced as compared to pre-cast system because there is no mixing of concrete, curing, making moulds etc. In our country steel frame work for multi-storey/high-rise building is not done due to high cost as compared to RCC frame structure.

13.4.3 As an example, one multi-storey building (two blocks of 18 and 14 storeys) has been constructed by adopting pre-cast technology. In this project, RCC columns and beams were constructed as cast in-situ. Simultaneously, the work of pre-cast RCC slabs for floors were being made at casting yard of project site after making good standard moulds with plastic lining. Sizes of floor slabs are quite large as per drawings. Immediately after completion of RCC columns and beams at various floors, the pre-cast RCC slabs were lifted with crane and placed on the beams of building structure. In this method of construction progress of work was increased considerably and good exposed concrete finishing was achieved. There was also economy in construction. Quality of concrete is very good as compared to cast in-situ method. The combination of pre-fabrication at various members of the houses in factories, coupled with "system approach" to drastically cut down the time taken to erect the houses at site. There are instances in several countries wherein hundreds of flats have been raised within weeks and townships are habitable within a few months. There is evidence that the completion time of the building projects could be reduced by 50% by adopting pre-fabrication building method. Conventional time consuming construction of buildings can create the blockage on capital. Due to longer completion period, the overall costs can be substantially higher.

13.4.4 There is another field for upgradation of technology in respect of manufacturing new materials. Several new building material industries are also being established to produce products for pre-fabrication and other constructions. A few examples are given on this matter. Fly-ash bricks are being produced at cheaper rate; this also solves the environmental problem due to non-disposal of huge quantities of fly-ash. Substitute of solid wood are being manufactured; this will be economical in building construction. Cellular concrete plant using cement/lime and sand/fly-ash, the establishment of a plant for manufacturing light weight building products, pre-stressed hollow cored slabs etc. are examples in development of new technology in the field of building materials.

13.4.5 There is no doubt that where speed and quality control and industrialized method of construction is bound to be cheaper and faster in the long run. New equipments are being manufactured for "material handling" and erection devices; this also accelerate speed of construction works. This has been dealt briefly under Chapter of equipment management. Certain modern techniques, construction methods, management have been dealt under Chapter of non-conventional structures.

13.5 Selecting Appropriate Technology

Project Manager/Engineer of owner department is required to examine various alternative technology and materials at preliminary and detailed planning and design stage of project. The object of management is to get the project planned, designed and execute proper way so that various aspects/objectives are achived like : –

Cost effective/economical, structurally safe, to complete the project in shortest possible period, to achieve best quality (though Total Quality management as dealt in a separate chapter), easy/less expenditure in maintenance of the project during its full life, to protect the project/structure from advance environmental effects (like aggressive sea costal effects, harmful/toxic emissions from factory or sewerage system etc.), availability of skilled worker/ operator for operation of construction equipments/machinery etc. Special care is to be taken for architectural & structural design including selection of materials. Many new materials are being developed/manufactured which are economical and easy to maintain during life period of project/structure. For this purpose necessary particulars/data of suitable materials are to be obtained from market. Economy is to be checked after considering total life of project and its maintenance. Life of project may be decreased due to wrong selection of materials and faulty design. After finalising aforesaid matters appropriate technology and materials are to be specified in the detailed estimate, NIT (Tender documents) and drawings. Hence role of management in selecting appropriate technology and materials is very important at planning and design stage of project. Some examples on above stated matters have been given under foregoing paras. Some more examples on technology are stated under subsequent paras.

13.6 Ready Mix Concrete (RMC)

(a) Rational utilisation of cement, conservation of materials as well as ensuring quality in construction are the objectives. RMC is one such technology which offers substantial scope for the conservation of cement through proper quality control in production of concrete. There may be saving of 15% of cement in large project construction. More important, RMC relieves the consumer from the tasks of procurement of materials viz. cement, aggregates and, their storage, production of concrete at site with adequate quality control; these are taken care by RMC supplier. Thus RMC is not only a "material" but also a "service". Hence lot of space at congested site is saved for other materials and site activities. For small construction job (where RMC is not used), the cost of quality control is viewed to be high. In view of shortage of cement in the country, the need for conservation of cement is paramount. Use of RMC is also useful from environmental improvement as well as marketing. Other technological improvements like, use of admixtures, bulk supply of cement in silos, pumping of concrete for laying at site etc. go hand in hand with RMC.

(b) Nearly two-thirds or more of the concrete produced in most of the developed countries are RMC. Presently use of RMC in India is far less. RMC is produced through plant which is installed in a factory type premises by a company who is known as supplier. Considerable cost is added on account of transformation flect of agitator/truck mixers/transit mixer trucks. It is essential to arrange machinery infrastructure for proper handling, placing and compaction of large quantities of RMC at the site of placement and pouring RMC. Project site engineers is to coordinator with RMC supplier effectively and to keep ready all site arrangement/ facilities for receiving RMC supply on specific date and time for specific quantity of RMC. Generally 3 cubic metres of RMC per trip may be considered as minimum at economic rate; site engineer may face problem receiving and placing the same within specified time (i.e. before initial setting of concrete). Due to this reason it may be difficult to use RMC in smaller construction works. It is required to work out quantity of concrete per day on specific date as per CPM calendar diagram. When quantity is more (more than 3 cubic metre per day/trip) then it may be possible to buy RMC at economical rate. Schedule of RMC supply indicating date, quality and time is to be prepared. Mix design of RMC and addition of specific admixture/plasticiser are to be prepared by the project engineer on the basis of structural design and the same are to be stated in the supply order of RMC. Necessary stipulation for the same is also to be made in the NIT (tender documents) of the project construction.

(c) ***Advantages of RMC***

Offers quality assurance, conserve materials, reduces

construction time, Reduces wastage of material, guarantee satisfactory performance and service life, enable "design mix" concrete instead of nominal concrete. It saves 15% cement consumption. It ensures uniformity of concrete. RMC technology is made accessible to various types of works and scattered construction jobs.

13.7 Jointing Reinforcing Bars

(a) In India reinforcing bars are usually spliced by laps. Mechanical or welding joint methods are rarely used. It has been found in some cases of structural failure that due to congestion of many bars bond between concrete and bars could not be developed. Spliced by laps of bars aggravate aforesaid problems particularly at the junctions of two beams or at the junctions of RCC column and beams. This problem can be solved if both ends of two bars are jointed with welding (Butt joint). There are following two distinct advantages of jointing bars by welding method –

   (i) Extra quantity steel (may be about 5%) can be saved.

   (ii) Congestion of reinforcement bars which prevents effective placing and compaction of concrete and which may seriously impair the performance of the structure. This problem can be solved by jointing bars with welding.

(b) There is upgraded technology on this matter. Reinforcement bars are welded through the method of "Gas Pressure Welding". This is basically "hot – forging process" which involves heating of the bar-ends to be jointed by "Oxy-acetylene flame" and applying hydraulic pressure. This method is common in Japan since last four decades. Skilled operator of simple equipments is required. There are three types : – (i) Manual process; (ii) Semi-automatic process; (iii) Automatic process. This method of jointing confirm IS : 1786 - 1985, IS : 9417. Higher degree of quality control can be achieved. Jointing process is fast.

13.8 Fibre – Reinforced Plastic (F.R.P.)

13.8.1 Presently plastics enjoy lot of use for a variety of applications in a variety of forms. It is possible to use plastics for structural purpose. Polyvinyl chloride (PVC), Acrylics are example of thermoplastics that can be used for light structurual applications.

Polyester, silicone resins are example of thermo – setting plastics. Polyester resins are most widely used.

13.8.2 Fibre - Reinforcements

Common reinforcing materials used in plastics are fibres (natural or artificial). Fibres possess appreciable tensile strength. Strength proportion depend largely on "resin & fibre ratio". Fibre content varies from 25 to 50%. Following types of fibres are used more : –

(i) Glass Fibre Reinforcements
(ii) Jute Fibres.
(iii) Asbestos Fibres.

13.8.3 Selection of particular type of F.R.P. materials is to be done as per necessity of items. Thermosetting plastics are good for structural use. Jute fibre reinforced plastic materials are light in weight and they are more flexible than glass fibre reinforced polyester.

13.8.4 Following data are to be assessed before designing on FRP structure –

(i) Life of Structure
(ii) Deflection Limit
(iii) Temperature of Operation
(iv) Details of environment such as weathering,expected alkali, acid attack.
(v) Method of fabrication

FRP material possess low modules of elasticity which makes them unsuitable for use in conventional structural forms such as beam and slab construction. Hence it is necessary to utilise those structural forms such as "folded plates", "Shells" etc. which give adequate structural stiffness by viture of their "Shape". The best suited polymer resins and fibre should be selected. The external surface of the FRP structure shall be protected against weathering by a gel coat.

13.8.5 PVC pipes of different sizes (12 to 150 mm dia) are very good for use of underground water supply, drainage line because PVC pipes are corrosion resistant. G.I. Pipes are used largely in India but it is damaged due to corrosion in underground water supply lines in short time (may be after 8 to 12 years), water is contaminated due to aforesaid reason and the same is dangerous for drinteing purpose. But difficulty is faced in India because proper jointing with various fittings can not be done with available average quality of glue/adhesive material. PVC pipes are very useful as "duct" for laying various utility/service lines.

Drinking water tanks of 300 to 1000 litres capacity are very popular for water supply system in houses and other buildings; these tanks are placed on ground surface/roof of buildings. Certain toilets fixures, fittings (viz, Flushing Cistern, commode seat, covered type shelves etc.) are also made with plastic material and the same is not damaged due to water and can be cleaned easily. Door shutters for toilets, partition system in commercial/office buildings are also constructed with plastic material, these are easy to maintain. Plastic corrugated sheets are used as roofing sheets. This material is much cheaper than wood which is a fire hazard material.

13.9 Fibre Reinforced Cement Concrete (FRCC) and Ferro Cement

(a) FRCC is a composite material of concrete mix with evenly distributed and closely spaced discrete fibres of steel, asbestos, glass etc. An important role of the fibres is to arrest an advancing cracks. The structural properties viz. tensile, compressive, shear and flexural strengths are considerable improved due to such additions. Cube compressive strength of concrete increases significant with the addition of fibres. By adding 1% (by volume) steel fibres, the shear strength is increased by about 50%. This material is very useful for certain items like, Man Hole covers, Slab Covers on storm water drains etc. If cast iron M.H. covers are put on the sewer/drain man-holes, the same are stolen for selling as scraps. Open man-hole on road/public place is dangerous as fatal accidents may occur. These items are made as precast in casting yard.

(b) Materials may be used as per following proportion : –
- (i) Mix – 1:1:2(1 cement : 1 coarse sand : 2 aggregate)
- (ii) Percentage fibre – 1% (by volume) of mix
- (iii) Aspect ratio (i.e. ratio of length of fibre and lateral dimension) – 70
- (iv) Water cement ratio – 0.6

(c) It has been found that the cracks are much less in the case of FRCC slabs than in RCC slabs. Due to using steel fibres the stiffness is increased.

13.10 Appropriate Technology in Environment, Public Health and Solid Waste Management related Projects

Adopting appropriate and modern technology in projects related to environments, public health engineering, municipal solid waste management will be dealt under separate chapters 20, 21 and 22.

13.11 Precast Prestressed Girder (Module from) of Delhi Metro Project

This is one unique example of adopting appropriate technology for achieving quick progress with very good quality in construction of large span girder of elevated metro rail track. Two photographs of the same (under construction) are as per Fig. No. 13.1 (a), (b). RCC pillars are constructed as cast-in-situ. The span varies from 20 to 31 metres. Precast prestressed small units (2 to 2.5 metres length and about 3 metres width) are made at casting yard. If the span is 31 metres then number of units required for one span/ girder may be 12 to 15. Necessary holes along these units are kept for placing steel wires during post-tensioning; holes are also kept for lifting. Suitable shape (grooves and projections) for joints with glue are also made on the precast unit during casting. High grade industrial glue is applied on the joints during fixing these units. Launching arrangement with Steel Structure is made from pillar to pillar for lifting and placing precast units, its jointing, placing steel wires through all the precast units and post tensioning. Anchoring of post-tensioning is done at pillars.

Fig. 13.1(a) : Pre-cast Prestressed Girder Installation (Delhi Metro)

Fig. 13.1 (b) : Precast Prestressed Girder Installation (Delhi Metro)

14

Management of Non-Conventional Buildings (PART A – HIGH RISE BUILDINGS)

14.1 Introduction

Various aspects of management in respect of conventional works have been dealt in foregoing Chapters. Now management in respect of special aspects of certain non-conventional projects/buildings will be dealt in this Chapter. Certain problems are faced during pre-construction, construction and post-construction period; this type of problems is not faced in conventional works/projects. Management to tackle some of these special problems are dealt here; examples of a few projects are as under :-

High-rise buildings, Cultural/Auditorium Complex, Sports Stadium/Swimming Pool Complex, Water Tower Reservoir with recreation facilities, Factory Complex, Airport, Underground multi-level Parking Complex, etc.

Projects related to Highway, Roads, Environment (viz., sewerage/ drainage, water supply systems, solid wastes) will be dealt under separate Chapters.

14.2 High Rise Buildings

It is a typical example of advancement in building technology. Important aspects of construction, economy and speed, have to be predominantly taken care of in the management of construction programme of a tall building for the success of the project. Paramount importance is to be given to "safety" aspect in architectural and structural designs. Most important is fire safety in high rise buildings; detailed designing of fire suppression and safety is to be done by competent fire consultant. Socio-economic studies should be carried out at conceptual/preliminary planning stage of the project. An effort should be made by sociologists, ecologists and psychologists to knit a cordial relation between nature, man, society, shelter and utilities as a comprehensive unit of environment. There is a lot of advancement in technology in method of construction and design of highrise buildings, skyscrapers (more than 400 storey/1455 feet height) in USA and European countries. There should be interaction and studies with international technical authorities on such technologies.

14.3 Practical Problems of Planning, Design and Construction

Tall structures have got their own problems of conception, planning, design and construction methods/activities at site. The comprehensive approach to the planning has to be directed towards achieving a good environmental condition and proper provisions for facilities from users' points of view. Care should be taken on the aspects of aesthetic, sociological and economic solution to urban land problems. Another important aspect is about continuous/uninterrupted energy (electricity) during use of building. This problem actually exists in India. Serious problem will be faced by the occupants of the high rise buildings if there is interruption in electric supply. The role of Electrical Engineer on this vital matter is very important. Fire safety and operation of lifts can not be ensured without uninterrupted power supply. Architectural design should be done for better natural ventilation and lighting. Another paramount important is structural design of high rise buildings. Latest method of structural design should be applied. Safety and economy are to be taken care while preparing structural design. The normal approach is the "systems approach" which embodies the principles of the most efficient techniques for ensuring a rational plan for the design and execution of works. High skill of management is required for effective planning and co-ordination in respect of various disciplines which are as under :-

Water supply and sewerage/drainage systems, internal electrical installations, lift installations, air-conditioning system, pumping arrangement, fire suppression/fire detection system, electronic and telephone services, security surveillance.

14.4 Co-ordination

14.4.1 The entire programme involves a high level of co-ordination between Architects, Engineers, Structural Designers, Electrical Engineers, Managers, Finance Manager, Accounts Manager, Managers of various services/utilities, Contractors/Sub-contractors, Suppliers, etc. During pre-construction/planning stage it is also required to compare/check all the drawings of architects, structural design/drawings, internal electrical installation drawings, water supply/sanitary installation drawings, fire suppression/fire fighting installation drawings, AC/Lift installation drawings, etc. Co-ordinated drawings of these services/utilities are required. Certain service lines may obstruct or create problem for other service lines. Hence, it may be required to change alignment of certain service lines. It is essential for all concerned Civil and Electrical Engineers, Architects to discuss together alongwith their drawings for necessary co-ordination. In advanced countries, the modern trend is to associate even the economist, political, administrative authorities, social scientists and psychologists at preliminary planning stage of the project. Environmental impact and other environmental aspects are also to be considered at this stage of planning.

14.4.2 Close co-ordination is of paramount importance in such type of building project as many disciplines, agencies, Architects, Structural Designer, Government authorities, Finance/Accounts Managers, Contractors, Suppliers, Various Consultants, Expert Associates etc. are involved. The Contractor must not only have good knowledge of the aspects of design but also must be fully aware of the modern techniques of construction of such high rise structures. Upgraded/modern technology should be adopted in such projects. Sequence of construction of various components of structure is also important. As an example, where deep basements (2 or 3 levels) is involved, it is required to construct retaining RCC piles before starting earthwork in excavation. After completion of sub-structure (i.e. basement) priority is given to construct all the lift cores upto full height; however, relevant structural horizontal connecting members are also to be constructed side by side. Certain new/additional stresses are developed temporarily during construction stage (on account of certain sequence of construction) and this aspect is also to be taken by Site Engineer in consultation with Structural Designer. Another example may be given. In case of 2 or 3 basements (with deep excavation of 14 or 15 metres) there will be heavy upthrust on the bottom of RCC raft and vertical RCC walls; this upthrust is due to buoyancy of underground water. As per structural requirement it is required to keep water level one metre below from the bottom of RCC raft of lowest basement so that there is no upthrust force on the raft. This is done by continuous pumping out of underground water during day and night for a period i.e. till upper basements/upper floors are constructed for equivalent downward force (i.e. weight of upper basements/floors). De-watering should be done by installing "Well Point System". The well consist of tubes from 40 to 50 mm in diameter and 5 to 7 metre long, called risers which are fitted with the screens and the ball valves at the lower tip. This is connected to a large diameter header pipe. The header is connected in turn to a vacuum pump, operated electrically or diesel driven. Depending on the rate of flow of water and area of ground/raft area, a large number of wells are to be installed at suitable intervals. Above stated pipes and wells are installed along ring shape. If the depth of bottom of raft is very deep then it may be required to follow multi-stage well point system. Well point slowly sinks to the desired depth (i.e. about one metre below the bottom of raft slab). In certain cases it may be required to seal anular space at the top and provision of filter media may be required to create desired vacuum. The well point system has helped in speeding up construction project.

14.5 Practical Problems and Solutions in Construction of High Rise Buildings

14.5.1 We face many practical problems mainly during construction and maintenance stages of High rise Buildings. This is a vast subject, where certain main problems, its causes and solutions will be dealt briefly. Causes of many problems (may be 40% to 50%) are due to faulty and inadequate planning/designing of various items. Hence, importance is to be given to planning/designing of the project. Roles of all concerned authorities, Engineers, Architects, Contractors, Experts/Consultants, etc. are important. Planning stages are in following three parts :-

(i) Preliminary/Conceptual Stage

(ii) Detailed planning stage (architectural and structural design/ drawings, specifications, etc.)

(iii) Tendering and award of works to Contractors.

14.5.2 Wind Tunnel Test

Serious problems were faced in one high rise building at Delhi as wind tunnel tests were not carried out at conceptual stage of building plans. Conceptual building plans were approved by all concerned authorities (viz., Development Authority, Urban Arts Commission and Local Body) without wind tunnel test report. But wind tunnel tests were carried out after finalization of detailed architectural and structural drawings and the proposed building plans were found as fail as per wind tunnel tests. The shapes of two towers were reverse curves like ")(". The velocity of wind was increased by six times at the shortest gap of two blocks. Besides delay for preparing revised building plans, extra cost on account of manpower/ extra fees of the Consultants were also involved. Hence, it is essential to make suitable models of the building and to carry out wind tunnel tests at conceptual stage itself to know adverse effect.

14.5.3 Proof Consultant for Checking Structural Designs

If the height of the building is too tall (viz., more than 15 stories) or the building is of intricate type (due to special shape or other reasons) then planning should be done for engaging a competent, experienced Structural Designer as "Proof Consultant". Practically, it has been found that due to long process of planning the pressure is put on Engineers by higher administrative authorities for inviting tender urgently. To avoid such difficult situations, tenders are invited without checking by Proof Consultant. Subsequently (i.e. after starting the construction work/executing foundation work), the structural Proof Consultant was appointed and major fault in structural design was found. Thus, serious problems/crisis cropped up. It takes considerable time to appoint a competent Structural Designer as Proof Consultant who requires sufficient time to check the structural design. It may be required for him to make a fresh design independently and then compare with the original design. Hence, it is required to decide about appointing Proof Consultant for structural design checking at preliminary planning stage.

14.5.4 Problem relating to Protection Measure for Deep Basement Works

High rise building is constructed in mega city. Generally adjacent buildings (very old or new tall buildings) are very near to the proposed high rise building. In large high rise buildings it may be required to construct 3 basements to fulfill the approved norms of parking and other service equipments (electrical sub-station, central AC, pump rooms, water tanks, etc.). It may be required to excavate upto 14 metres below ground level. It is not possible to provide side slope or stepping/benching on excavated basement area due to shortage of space. This safety and working problem is generally solved by constructing RCC retaining round piles at suitable interval as the same is economical. But there is some risk of collapsing RCC retaining piles. Slides of excavated area may occur during heavy rain. Such incident has happened at certain project site during construction of RCC raft and created havoc. Another safe method is by constructing diaphgram wall along entire length of periphery of deep basement area. But this is very costly. Cost of diaphgram walls may be 4 to 5 times of RCC round retaining bore piles. Final decision on method of above stated safety and working arrangement should be taken by a group of senior technical experts/engineers as conscious decision after considering various facts, site conditions, soil conditions etc. For large deep basement areas (three floors with extended areas) cost of diaphgram walls is very high (may be to the tune of Rs.6 to 9 crores). This important part of structural design is to be finalized at the stage of detailed designing and provision is to be made in the detailed estimate accordingly. If any adjacent property is damaged due to construction activities it will be required to pay damage/compensation (may be huge amount); accident can also occur.

14.5.5 Completion Period of Construction

Completion period of construction is stipulated in the NIT (tender documents). As per CPWD manual guidelines the period for constructing high rise buildings becomes very long. On the other hand administrative authority desires for too short period. Both position creates problem. If it is too short period then actually the same is not practically feasible. It is required to grant extension of times frequently. The CPM construction programme is prepared on the basis of stipulated period. It is required to revise the network several times. Several problems are faced during construction period for reporting to higher authorities, payment on account of escalation, too many changes about deployment of machinery, equipments, adverse effects on other related contracts (viz., electrical works of lift, AC, internal installations, sub-stations, etc.); Contractors of above stated services claim extra amounts due to delay of main civil contractor. Thus, due to unrealistic stipulated period many problems crop up during construction period. Hence, realistic completion period should be stipulated in the NIT in consultation with experienced senior engineer.

14.5.6 Clearance of Project from Various Statutory Authorities

Considerable period is required to obtain clearances from various statutory authorities at planning stage.

(i) Principle approval of the proposed scheme considering budgetary aspects, availability of land etc. by the top administrative authority of the organization.

(ii) Approval of conceptual drawings/scheme from statutory authorities (viz., Development Authority, Urban Authority,

Local Body, etc.).

(iii) Height clearance from Civil Aviation Department. Clearance from Archeological Department.

(iv) Approval of Fire safety scheme from Fire Services Authority.

Considerable period is required for above stated approvals. Clearances of Archeological and Civil Aviation Departments were not obtained in certain project; the work was held up due to aforesaid violation. Hence, it is required to monitor these matters at higher levels. It is required to meet and discuss with various concerned authorities.

14.6 Detailed Requirements of User Department

It is essential to obtain detailed requirements (like sizes of rooms, broad specifications, electricity points, AC, etc.) from various user Departments. Architectural drawings and details of various service facilities are to be prepared and the same is to be discussed with the users. Clearance is also to be obtained from the user Departments in writing at "preliminary planning stage". Problems are faced during and after construction from user Departments for making many addition/alteration. If necessary, it may be required to visit some existing similar buildings alongwith user Department representatives at preliminary planning stage.

14.7 Making Approved Samples as part of NIT

Provision of huge quantities of granite stone, marble stone, ceramic/vitreous tiles is made in the NIT. Problem of white marble stone slab item is faced during execution; the cost of white Makrana marble (first quality) is about double of the white Rajnagar marble. Dispute is cropped up during execution on account of quality. Hence, it is essential to make approved sample pieces of stone slab as part of NIT. Regarding ceramic/vitreous tiles it is better to stipulate names of 2 or 3 reputed manufacturers in the NIT.

14.8 CPM Programme

The CPM programme/network of entire project (i.e. including various services, electrical, lift, AC works and other sub-heads) should be prepared by planning branch at detailed planning stage and sent to all concerned branches/Departments of the organization. Various milestones of the project are to be shown in this network at planning stage. Major activities of pre-construction stage should also be prepared. It will be possible to monitor planning activity and co-ordinate on the basis of this CPM network. It is required to prepare one master CPM network of all sub-heads/disciplines of the project (indicating milestones) as integrated programme. This will also be useful for effective co-ordinations amongst all concerned Departments, branches, contractors, sub-contractors, consultants, etc.

14.9 Earthwork and De-watering of Underground Water

It is required to excavate earth for a depth of 13 to 14 metres (depending on 2 or 3 floor basements) on entire area of basements. Huge quantity of earth of good quality is available for which lot of demand exist in the market. As per normal method huge payment for earthwork in excavation and cartage of earth to dumping ground are made as per relevant items of CPWD Schedule of Rates. Trucks are not allowed to ply in the city during day time. Problem is faced for locating dumping ground for such huge quantity of earth. It may be required to make payment by the Department/owner for using dumping ground. It is also difficult to measure the carting of earth including lead during night (after 10 P.M.). There is another problem about measuring and payment of de-watering of underground water (to keep water table below bottom of raft i.e. 14 to 16 metre below ground level). It may be required to carry out pumping throughout the day and night for very long period (may be 6 to 18 months). As per requirement of structural designer above pumping is required till 3 or 4 floors are constructed to create sufficient downward load/force so that upthrust on the bottom of foundation/raft is counter-balanced. As per normal method the measurements are carried out on the basis of relevant item (i.e. in the unit of kilolitre per hour) of CPWD schedule of rates. It is not practically possible to measure correctly throughout the day and night for long period. Huge amount is involved on this item. Following method has been found beneficial to the Department/owner:-

> "Additional/special condition be stipulated for abovestated three items in the NIT (tender documents) i.e. earthwork in excavation, disposal/carting of excavated earth and de-watering of underground water. These items will be executed by the contractor. Entire quantity of earth (except certain quantity required for back filling) will be property of contractor. No separate payment will be made to the contractor for executing abovestated three items. Contractor is required to quote rates of other various items of NIT considering above position".

14.10 Upgradation of Technology

(a) For achieving good quality and expediting progress it is required to upgrade technology in respect of certain items of works. One most important item is RCC works particularly in basement floors (raft, outer vertical/retaining walls), exposed concrete finishing work etc. Serious problems are faced on account of seepage, leakage in the basements, lift pits. It is not only on account of faulty specifications of waterproof treatment but also on account of average/fair quality of concrete and its laying operation without using good equipments. Hence, "Ready Mix Concrete" should be used after finalizing proper mix design of appropriate grade of concrete. Appropriate water-cement ratio, slum are to be achieved for correct workability. It is required to use suitable plasticizer in the mix. In case quality of concrete mix and its laying operation are very good then RCC raft, wall should be water tight. Leakage/seepage is also due to defective construction joints in the RCC walls and raft slab. These joints should be done as per details given in the Code.

(b) There is another example of upgradation of technology. All floor slabs of a high rise building have constructed by providing "pre-cast slabs". RCC columns and beams are constructed as cast in-situ. Floor slabs of large sizes are cast on ground at project site after making special type shuttering (using plastic fiberglass lining on strong wooden frame). Good exposed concrete finish on the bottom of slab can be achieved. Then the pre-cast slabs are placed on the beams by lifting with crane. Considerable time is saved and good exposed concrete finishing is also achieved. Ready-mix concrete should be used.

(c) ***Slip Form Shuttering*** – As per sequence of construction it is required to construct the lift cores independently at superstructure. For expediting progress of lift-core construction, "Slip Form Shuttering" technique should be used. The form is lifted mechanically upward after casting of one metre height of lift core. This is a specialized work. Special agency with experienced competent operator has to be engaged. High operational skill is essential to maintain accurate verticality of lift core. Verticality is to be checked carefully at every one metre height and also at every 5 to 10 metres interval. Accuracy of verticality checking also depends on skill, correct type/ size of plumbob and string.

Installation of the lift in the shaft/core is directly linked up with correctness of verticality of shaft. There is possibility of large deviation of verticality at upper part of lift core. If deviation is too much, then it may also affect the structural stability adversely. Fabrication of lift cage/car and other parts are got done by electrical department through the lift contractor. In one project aforesaid fabrication was done on the basis of original drawings before construction of lift cores. But subsequently, considerable error of verticality was found. After rectification of verticality error the size of core becomes less and due to this reason it was not possible to install the fabricated lift cage/car. Serious problem was faced. Many contractual were disputes cropped up and project was delayed on this account. Fabrication of lift cage/cars and other parts is to be done only after measuring completed rectified lift cores (not on the basis of original drawings).

14.11 Finishing on External Walls

(a) ***Exposed concrete Finishing with pattern*** : Problem is faced if exposed RCC wall surface (external walls) is done with special type of pattern like vertical strips/beadings (2.5 cm width and 2.5 cm projection in round shape) and with space of 4 cm between two strips/beadings. Practically it is not feasible to achieving good finishing with shuttering and concrete. It is also difficult to maintain straight vertical lines of such type of beading/strips on external walls of 100 metres or more height. Skilled workers for such work may not be available locally. Ready mix concrete with special type of vibrator equipment should be used concrete should be laid through pump. There is another problem to maintain uniform colour of exposed concrete finishing on entire external wall surface because it is not possible to procure same colour of coarse sand and cement during long period (3 to 4 years). It may be required to apply "silicon" coat on exposed concrete surface to protect from rain/weather effects. It is recommended for not adopting above type of special pattern. Civil engineers should examine such proposal of architect before stipulating in the NIT. Suitable specifications, details of pattern, ready mix concrete, special type of shuttering with lining etc. should be stipulated in the NIT. It is better to select suitable type finishing material so that it is practically feasible to achieve good quality and to maintain easily.

(b) Practical problems have also been seen for finishing on external walls with stone slabs (fixed with metal hooks/ clamps). Some slabs fall down and fatal accident/ injury may occur to persons who walk near the building.

Problem is also faced to repair/maintain such type of finishing on tall building. Hence, special care should be taken for deciding specifications of external walls.

14.12 Problem related to Unsatisfactory Services of Architect-Consultant

In case Architect-Consultant is engaged for various services (viz., architectural drawings, structural design, electrical, sanitary, water supply works, etc.) problems are faced due to their unsatisfactory services. Due to lack of proper provision/conditions made in their agreement it becomes difficult to control the consultants. Proper standard agreement with effective terms and conditions for such consultancy services has not been evolved by organizations. Considerable delay occurs for various services and project works suffer badly. Even sometimes the Consultants stop their services and the work is held up. Generally, Architect-Consultant do not deploy competent experienced Civil Engineers for effective supervision. Special care should be taken to pre-qualify reputed competent Architect-Consultants and to prepare agreement with effective terms and conditions.

14.13 Awarding Work to Contractor

(a) Many problems are faced from incompetent contractor. Success of the project completion also depends on the contractor. To achieve the same, the stage of "pre-qualification of contractors" is most important. Hence, pre-qualification of documents (like, NIT) are to be prepared by experienced engineer with special care. Documents submitted by contractors are to be scrutinized carefully including cross-checking from relevant authorities. Inspections of their completed works should also be carried out.

(b) It has been found that sometimes tenders are rejected as the lowest rates are quite higher than the detailed justification of rates. But in certain cases it has been found that huge amounts were not considered as financial effect of additional/special conditions stipulated in the NIT/ agreement (as examples – earthwork in excavation, carting of excavated earth, de-watering as stated under para 14.9 of Earthwork and De-watering).

14.14 Some Other Problems

(a) Problem is faced during earthwork in deep excavation at basement area and during construction of RCC raft/lowest basement floor due to filling of entire area with rainwater. Heavy rain can create havoc at site. It is essential to make peripheral large size drains around the project site so that rain water does not get entry in the project plot. For this purpose levels should be taken on entire area (around project site) and to draw contour for ascertaining direction of surface flow of rainwater. This is like catch-water drain along mountainous road.

(b) ***Fire Prevention/Fire Suppression Works*** – Main problem is faced for obtaining NOC from Fire Services Authority after completion of building. Hence, it is essential to design and prepare details, specifications, building plans, etc. as per building bye-laws/codes. Compliance is to be done in respect of "12 point" of fire suppression system as laid down in the Municipal Bye-laws. It is essential to get clearance of Fire Service authority in respect of preliminary and detailed drawings, specifications of fire safety works before starting main building work. Proper approach roads for heavy fire fighting vehicles, equipments (viz., snorkel, long ladder, tender etc.) are to be constructed. Sufficient pavement space is required around the building for movement of aforesaid heavy equipments and vehicles. Fire Service Authority also requires certificate from the competent engineer about structural strength of certain load for movement of snorkel vehicle. In one building, the plaza/pavement around the building is constructed on roof of the extended basement, but structural strength of aforesaid roof was less than the requirements of Fire Services Authority. Due to this reason, problem was faced; issue of NOC was held up for long period.

(c) Lot of problems is faced for smooth progress of work and effective co-ordination amongst various departments if two Chief Engineers (Civil and Electrical) are in-charge of a project. There should be one Chief Project Engineer (who is Civil Engineer) for the entire project i.e. in respect of all sub-heads, disciplines including electrical/lifts/AC works, fire fighting works, etc.). Many disputes crop up if above stated project organizational structure is not followed. Adequate/ effective delegation of powers (technical, financial and administrative) are to be given to the Chief Project Engineer/ Manager for proper project management. Planning process should also be done under same Chief Project Engineer.

(d) ***Completion Plans/NOC*** – In most of the building construction projects, problems are faced for approval of completion building plans, various services completion plans, viz., water supply system, sanitary system, "Fire Suppression/ Fire Prevention System", Electrical installations, AC system, Lift installation, etc. Problems are faced to get clearance of electrical installation, Lift installation, Pump installation, etc. from electrical inspection authorities of Government Department. In case Architect-Consultant is engaged for the project then finalization of abovestated matters is to be done by the Consultant in a time-bound manner. Approval for connections of water, sewer and electrical is also to be obtained by the Consultants in time. Most of these important matters remain pending for very long period (may be from 9 to 24 months). It may be stated here that in case completion plans are not got approved from local authorities then it will not be possible to occupy the completed building. There will be considerable loss on account of rent of the building. For avoiding above problem, the building construction should be done as per various rules/regulations/codes/bye-laws/Acts/ approved plans.

14.15 Verticality of Lift Shaft/Lift Core

14.15.1 Author has lot of practical experience on this problem relating to verticality of lift shaft/lift core. Structural work is involved in construction of lift shaft/lift core in multi-storey/high rise buildings. It is required to ensure proper verticality with high accuracy. Installation of lift in the shaft is directly linked with correctness of verticality of shaft. Guide rails of steel sections are fixed on the wall of lift shaft with steel bracket and clamp. These steel guide rails have to be fixed vertically with high accuracy because lift cage/ car is operated through electrical and mechanical components which are also installed with high accuracy. In case there is error in size of lift shaft/core then size of lift cage/car is to be changed. If the lift shaft is not vertical with required accuracy then it may not be possible to install the lift equipments keeping the originally designed/ specified size of lift car/cage. Method/construction technique of shuttering for lift shaft is important. In case equipment/method of "Slip Form Shuttering" is adopted for accelerating the progress then it is essential to deploy experienced skilled workers; otherwise it is difficult to achieve accuracy in verticality. Purchase and installation of lift is done by a separate agency/Department (i.e.

Electrical Engineering Department and their lift contractor). Lift shaft/core is constructed by Civil Engineering Department and civil contractor. Electrical Engineering Department generally award the lift work to lift contractor side by side. Thereafter Electrical Department/contractor starts fabrication/ manufacturing of the lift components in their workshop on the basis of "tender drawings" and sizes as stipulated in the lift installation contract. In case the lift shaft is not constructed strictly as per original drawing then serious problem is faced for installing various parts/equipments of lift. If the error of verticality of shaft is quite high then it may not be possible to install the lift parts already fabricated by the lift contractor as per tender drawing. Error in verticality of lift shaft can be rectified upto certain extent with great difficulties. If error is beyond certain limit then it will not be possible to install the lift parts. Hence, fabrication of lift components must be started only after construction of lift shaft and after checking measurements as per actual site position and not as per original/tender drawings.

14.15.2 Remedial Measures for Verticality Problem of Lift Shaft/Core

(a) *From the side of Civil Engineers*

(i) During construction of lift shaft, it is essential to check verticality of lift shaft/core at every one metre height with mason's plumb accurately by the mason/contractor's engineer and department's/owner's site engineer. Overall verticality of lift shaft is also to be checked at each 6 metres' height and at final/full height. Verticality checking at full height should also be checked by other method (viz., with Theodolite).

(ii) In case any error/non-verticality is found then the same is to be marked on the lift shaft (inside and outside) with paint and error is to be measured (i.e. how much millimeter is out of plumb). This is also to be plotted/ marked on the drawing (vertical section of shaft/core). There may be bulging out (i.e. convex) of concrete wall of shaft and concave on surface of wall at certain locations. Thickness of shaft wall is also to be checked and shown on the drawing (vertical section).

(iii) Abovestated defects/error are to be rectified suitably. It may be required to chip off the bulged portion and to fill up the concave portion by micro-concreting (if necessary by adding reinforcement). Error of verticality is permissible upto certain extent as per requirement of structural design engineer and Lift manufacturer.

(iv) Abovestated steps are to be taken side by side during construction of lift shaft/core and not after completion of entire height of lift shaft. It will be very difficult to check/rectify error later on because very high scaffolding is to be re-erected again at considerable cost and time. Safe scaffolding (inside and outside the lift shaft) is essential for checking verticality properly by the site engineers. It has been found that checking of verticality and other defects are not done by junior and senior engineers due to non-availability of proper stable/safe scaffolding. If any major error/defect is found during construction at every metre height then the relevant portion of lift shaft should be dismantled before proceeding further. If this is not done then it may be required to dismantle longer length of shaft later on.

(v) One aspect of checking verticality of shaft is from the view of minimum requirement of Lift manufacturer. There is another angle of checking of verticality of lift shaft, that is "structural stability" which is to be examined by the structural designer on the basis of site checking data/error as plotted on vertical section of drawing. Structural designer is to give solution for structural stability and remedial measures are to be taken as per recommendation of structural designer. Lift agency should fabricate the lift equipments/parts/car after construction of shaft and as per actual site measurements. If necessary, it may be feasible to install lift equipments after reducing size of the lift car/ cage; or by increasing length of brackets for fixing guide channels. Adjustment upto certain extent can be done by the lift agency.

(vi) Slip from Shuttering — As stated under para 14.10

(b) *For Electrical Department / Lift Agency*

Electrical Department/Lift agency is required to start fabrication/manufacture after completion of lift core/shaft. Actual site measurements of completed shaft are to be taken and fabrication of lift equipments/parts/car (cage) are to be done as per actual site measurements. Fabrication/ manufacturing of lift equipments/car should not be done on the basis of original/tender drawing before construction of lift core. If above system/ stages of construction/ fabrication is not adopted, then serious problem will be faced if there is difference of measurements/sizes of constructed core/shaft than original drawing. Hence, the work of lift installation should be awarded to lift agency at appropriate stage of civil engineering construction work of lift core.

14.16 Building Technique

(a) In most of the advanced countries, industrialized methods of building techniques have paid good dividends. The important techniques adopted by them are modular coordination, pre-fabrication and industrialized methods of construction. The basis is the "system approach" which is based on the most efficient principles of planning, design and execution. With the system approach based on the use of on-site pre-casting unit, considerable savings, both in time and expense, is possible because of repetitive use of components. One example in this regard has been given under para 14.10 (b). In certain project, productivity may be 50% more and site labour may be 50% less.

(b) In the management of a highrise building construction programme, it is possible to adopt the techniques of industrialized methods of construction, as labour intensive methods are not ideally suited for highly sophisticated construction of this type. For any efficient construction programme, after the completion of foundation, floor to floor time of construction assumes importance, as the time of completion of project will depend upon the rate of construction of intermediate floors. In case casting of RCC frame works and RCC floor slabs are done in conventional method (i.e. cast in-situ) then it will depend on number of centering and shuttering sets for each floor; quality will also depend on quality of shuttering.

14.17 Construction Layout – Layout and Benchmarks

(a) A good construction layout can pay good dividends in a construction programme. Before starting the construction work the contractor is required to prepare a site layout plan showing location and space for following construction facilities :-

Proper circulation/approach roads within the plot, store for cement (or silo), open space/yard for stacking construction materials (viz., sand, stone aggregates, steel bars), store shed for certain materials (fittings, utility items, doors, windows, boards, timber, tiles, etc.), Carpentry workshop, space for construction equipments on level hard ground and pavement as repair shop, site office, toilet, canteen, correct location for crane, hoist for materials lifting, hoist for persons' lifting, pre-casting yard, bar binding/steel fabrication yard, construction and drinking water arrangements, security guard room, field laboratory, plant for CC, temporary electrical installations for illumination on entire site, crèche, etc. One sample site layout plan showing abovestated arrangement is as per Fig.14.1. Locations for various arrangement should be decided so that the same do not obstruct to carryout other activities and double handling is avoided.

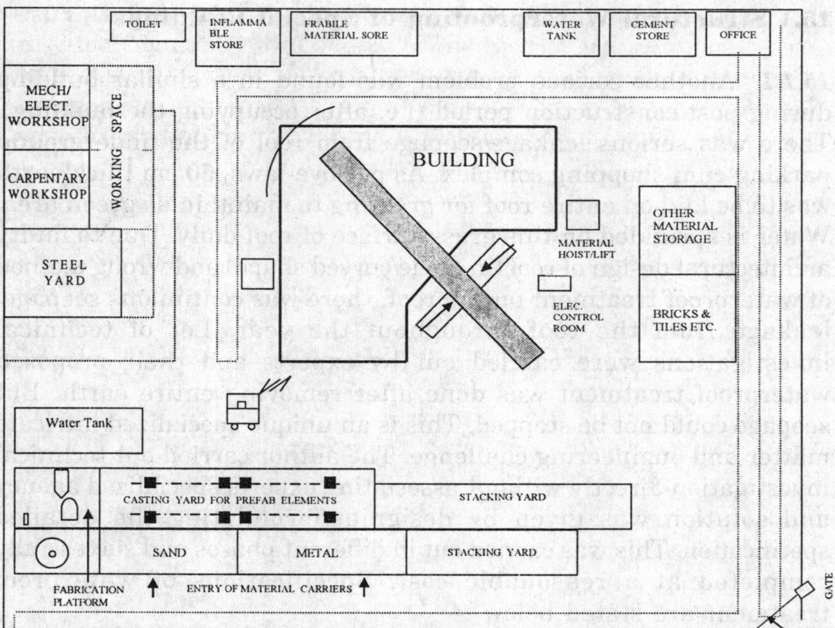

Fig. 14.1 : Construction Layout for a Multistoreyed Building

(b) It is important to check and re-check the benchmark for level control with a high standard level instrument. The other most important thing is to layout the reference lines and the centre lines of building on ground, with necessary well constructed pillars which are to be maintained till the construction work is in progress.

15

Mangement of Non-Conventional Buildings (PART B – OTHER THAN HIGH RISE BUILDINGS)

15.1 Structural Waterproofing of Special Buildings

15.1.1 Another serious problem was found in a similar building during post-construction period (i.e. after occupying the building). There was serious leakage/seepage from roof of the underground parking-cum-shopping complex. As per bye-laws, 50 cm thick earth was to be laid on entire roof for grassing to maintain as green area. Water is sprinkled on this grass surface of roof daily. Due to faulty architectural design of roof (in dome/curved shape) and wrong method of waterproof treatment on the roof, there was continuous seepage/ leakage from the roof throughout the year. Lot of technical investigations were carried out by experts and their proposed waterproof treatment was done after removing entire earth. But seepage could not be stopped. This is an unique specialized/intricate matter and engineering challenge. The author carried out technical investigation directly without associating experts/specialized agency and solution was given by designing/formulating the detailed specification. This was carried out in different phases and successfully completed at a reasonable cost. Specifications of waterproof treatment are stated below :-

15.1.2 There are many structural joints on the RCC members (roof slab, beam, walls). These are another source of leakage/seepage. This is more problematic. Special type of joint treatment is to be done so that structural joints function for expansion/contraction. Technical details of water proofing have been dealt separately. Waterproofing treatment for such special problem (i.e. structural waterproofing) is required to be done in combination of various types of layers with heavy duty rubberized/aluminium foil sheet laid with bituminous waterproof compound. It may also be required to treat porous/honeycombed RCC slab by impregnation/injection method. In this method, 10 to 12 mm diameter hole is made in the

RCC slab (half thickness of slab) and then cement slurry mixed with certain chemical (viz., Tape Crete/Pidi-crete) is injected in these holes (which are at 20 cm interval on both directions) under suitable pressure through handpump.

15.1.3 Architect is required to give suitable shape of roof (preferably flat/slope) and not in irregular shape with many deep notches. There will be stagnation of water on deep notches without outlet which is covered with 50 cm thick earth for grass. Hence, it is essential to check architectural drawings from engineering and waterproofing angle before approval. Generally such type of checking of architectural drawings is not done by the civil engineer and even if some observations are made, yet architect does not agree. Under such circumstances, the decision of seniormost engineer (viz., Chief Engineer) should be final.

15.1.4 Another serious problem was faced by the users of the complex. As lowest floor level/foundation level is about 10 metres below ground level and highest level of ground water table is quite high (one metre below GL), considerable up-thrust of water exists on the bottom of RCC raft and on external surface of walls. Waterproof treatment done was not able to resist such water pressure and considerable leakage was found. In spite of carrying out some rectification works, there is always stagnation of water upto certain level at lowest floor. Pumping is required to be done daily on account of the leaking water. It is not feasible now to excavate for making effective waterproof treatment on external surface of walls. Waterproof treatment done from inside of the building was not found effective.

15.1.5 Hence, it is very important to prepare detailed specifications of multi-layers of different types waterproof treatment before inviting tender. Samples of such treatment should be done before finalizing specifications and tendering. It is better to carry out one sample work of small area and then (i.e. after observation) the work should be done in phases. Specifications can be modified after observing sample work. Normal/ conventional type of one particular waterproof treatment will not solve the problem. Structural design of RCC raft and walls is also important for resisting water leakage/ seepage. Mix of concrete should be richer and correct proportion of ingredients (stone grits, cement Grade 53, coarse sand) is also important. Compactness and density of concrete are also important so that

RCC raft, walls and roof act as water barrier. If proper concrete is used with proper compaction, then such RCC raft and walls are expected to be leakproof/waterproof. It is better to use "ready mix concrete" mixed with suitable admixture (like, plasticizer).

15.2 Swimming Pool

15.2.1 Swimming Pools are constructed in different sizes. In India most of the swimming pools are open to sky. Some swimming pools are indoor type. Hot and normal water are available in indoor pool. Indoor swimming pool is used throughout the year.

15.2.2 One important problem is faced in construction and maintenance of swimming pool. This is about coming out of tiles from floor and walls during use and maintenance of swimming pool. Main reason for the same relates to structural design of pool and specification of finishing items for floor and walls particularly about fixing/binding material. Soil investigation is to be carried out correctly particularly to ascertain highest water level of underground water during past some years. If soil investigation is carried out during summer season then this will not be relevant for structural design. Underground water level is to be checked during and immediately after rainy season. There will be upthrust on the bottom/floor of swimming pool due to buoyancy effect of underground water. The RCC floor and walls are to be designed considering aforesaid upthrust and side pressure of underground water during worst condition. Some allowance should be kept for rise of water table over the life of the structure. Mix of concrete is to be designed suitably so that it also works as waterproof media. Gradation of stone aggregates and coarse sand is to be done correctly. Water- cement ratio is to be made correctly. Compaction of concrete is to be done properly during laying. Making construction joints correctly is very important. This is one main source of access of underground water inside the pool. Proper waterproof treatment is to be made on "external and internal" surface of floor and walls. It is better to use "ready mix concrete" with suitable admixture in RCC works to get waterproof structure.

15.2.3 Generally ceramic/vitrified tiles are fixed on the floor and walls of pool. Suitable adhesive material is to be used. Normal method of fixing the tiles with normal cement mortar may not be satisfactory. Extra precaution should be taken for proper binding/

fixing strength. It may be desirable to put a coat of certain binding and waterproof chemical on the RCC surface before fixing tiles.

Provision of above stated design and specifications is to be made in the detailed estimate and tender documents.

15.2.4 In case faulty/inadequate structural design is done and problem on account of high upthrust (due to buoyancy of underground water) is found, then there is no other alternative but to strengthen the floor slab by casting another RCC slab on the original floor slab. However, depth of pool will be reduced on account of above proposal.

15.2.5 Another problem is maintaining clear and unpolluted pool water. Lot of dust is carried with wind and pool water, is polluted by this dust. Water is also polluted due to use of the pool by overcrowded public. Chlorination equipment is installed for disinfecting water of the pool. The water is circulated by pump continuously through chlorination equipment. Hence function of chlorination equipment and pump are very important. These equipments should be procured from good manufacturers and operated efficiently. Pool should also be cleaned for draining out the water and re-filled with fresh water after cleaning the dust etc. periodically.

Specifications of floor finishing around the pool at top level should be as per requirement of user (i.e. Senior Swimmer/Coach). This space is used for movement of swimmers for going inside the pool.

15.3 Auditorium

15.3.1 Large capacity multi-purpose auditorium is intricate project. Many problems are faced during design, construction and operation stages. Some problems are due to defective planning and design. Hence, pre-construction stage is very important. Many matters are of specialized nature. Number of experts are to be associated at different stages as under.

15.3.2 Special Stage Lighting

Special stage lighting is to be dealt by stage lighting expert who fall under group of "theatre field". General electrical engineer cannot design such lighting system. Control for such special lighting system is done through operation of "Dimmer Control Panel". It is specially designed as per number of various stage lighting fittings manufactured by a few firms in India. It may be required to import certain stage light fittings to give special lighting effects for various types of cultural programmes. One famous international company on stage light fittings is M/s Rank Strand, London. It is also essential to obtain detailed requirement of stage lightings from users (viz., Managers of Play Troupe, Musicians, Dance/Ballet Troupes of Western and Indian styles, Songs, Concerts, Conferences, etc.). It is required to design in a balanced way for various types of performances. Interactions with artists and Managers of their groups/ troupes are important.

15.3.3 Acoustic Treatment

Another most important specialized subject is "acoustic system" of the auditorium. Sound system is also a part of this subject. Role of acoustic expert is very important. Acoustic treatment is to be designed by an "acoustic expert" having practical experience. Acoustic treatment relates to sound absorbing materials of auditorium chairs, ceiling, wall cladding/paneling with special type of sound absorbing materials, carpet specifications for floor of auditorium, special wooden floor of stage, wings system, stage curtain, etc. Various materials as stated above including a few complete auditorium chairs are to be got tested as per direction of acoustic expert. Sound system is also to be designed by expert on this subject in coordination with acoustic expert. Echo effect on any part of auditorium will be a serious hazard. In case roof of central portion of auditorium is constructed with steel truss and covered with CGI sheet, then it will be required to provide a sound proof barrier (packing with cotton type material) just below the CGI sheet of roof. This is due to lot of sound from CGI sheet roof during rains. Above barrier will also reduce external heat. If any deficiency is found in acoustic treatment and sound system later on (i.e. after completion of auditorium) it is mainly on account of faulty design and it will be very difficult to rectify in piecemeal. Auditorium may not be popular on account of abovestated acoustic deficiency. Acoustic panels for fixing on walls are made with cotton placed on wooden board and covered with thick cloth and fixed the same by nailing wooden beading/strips at suitable intervals. Good sound absorbing boards are fixed on cotton base in some auditoriums.

15.3.4 Main Stage

Design and specifications of "stage" are important part of auditorium. Size, shape, specifications of stage materials (floors, curtains, wings, cyclorama/back screen etc.) are to be decided carefully and samples of these materials are to be approved by the users i.e. artistes/managers of cultural programmes/troupes. Requirement with size, floor level, etc. for "Orchestra Pit" in front of stage is to be obtained from theatre expert. For international cultural programmes (particularly Ballet groups/Play groups) it is required to obtain their requirements through ICCR, Government of India. The wooden floor of stage is designed with 3 or 4 layers of wooden planks/scantling, strips (10 to 15 cm wide with tongue and groove joints). Neoprine bearings (rubber based) are fixed on cement concrete base and thereafter wooden members in 3 or 4 layers are fixed. Wood samples (for topmost layer) should be shown to users. There has to be proper resilience effect on the stage floor during performance. Necessary slope is to be given on the floor as laid down in the design book/drawing of auditorium. Sample of curtain cloth is to be approved by the theatre expert. Mechanical system of various curtains (horizontal/ vertical) is to be designed properly in consultation with theatre expert so that operation and maintenance can be done easily. Skilled workers on such work are to be engaged.

15.3.5 Auditorium Balcony for Sitting

Problem is found in balcony of auditorium. Sometimes huge crowd reaches the balcony. Cantilever projection of balcony is also quite large. If structural design is not done properly, then there will be "deflection" beyond permissible limit. Balcony may not collapse but there may be vibration/slight vertical movement. If the structural design is faulty or construction is defective then the balcony may collapse when huge crowd stands on the balcony. Hence, special care should be taken at the stage of structural design and construction so that problem is not faced during use of auditorium.

15.3.6 Cyclorama (Back screen)

It is better to provide permanent white back screen on wooden frame in parabolic shape and finished with plaster of paris material. Size, shape, finishing of such screen is to be approved by the theatre expert and cinema operator. The auditorium can also be used for cinema programme/projection system.

15.3.7 Auditorium Chairs

Design of auditorium chairs is very important. Type of chair is to be decided i.e. with or without pushback. Material and design of handle and cushion (rexin or leatherite fabric) are to be decided carefully. Comfort of the chair is also important. Various types of sample chairs are to be made for examination by experts and users and then final sample is to be made and approved. Aesthetical aspect and acoustic effect also depend on design and material of chairs.

15.3.8 False Ceiling

Design and construction of false ceiling are quite difficult. It is designed in such a way so that various stage light fittings (at upper level) are fixed in the false ceiling. The fittings should not be visible to audience but light will focus on the stage from left and right side angles. Number of rows of fittings on horizontal direction are fixed in the ceiling. Frame for ceiling is made in step form. Lot of space is available (2.0 to 4.5 metre height) inside the ceiling (i.e. between false ceiling and bottom of roof). Proper arrangement with steel frames is made for laying many electric cables and ducts and for movement of workers/operators at horizontal and vertical directions. Catwalk is made with steel frame. This area (i.e. inside false ceiling) is fire hazard due to existence of huge electric cables/wires, many stage light fittings, general lighting installation, etc. Special protection arrangement is to be made in this space. False ceiling materials must be fire resistant type. In small auditorium, stage light fittings are installed below ceiling and the same are visible. This is less intricate and fire hazard is also less.

15.3.9 Fire Safety Measures

The source of fire is from intricate electrical installation, many special stage lighting fittings, dimmer control panels and huge electrical cables/wires. Electric transformer, generator, motors, etc. are also used here. Detailed requirements of fire safety measures have been laid down by Fire Services Department. There are 12 fire safety points which are stated under para 15.5.4. Electric transformer, generator set, etc. should be installed outside the auditorium building and not in the basement of auditorium. This is also required to give good acoustic effect in the auditorium. Vibration from transformer, generator disturbs the acoustic level of auditorium. If fire accident occurs in the auditorium during programme then this will be disastrous; many people may die (as happened in the fire accident of Uphaar Cinema Hall, New Delhi, when 69 persons died due to fire accident). Hence, it is essential to prepare design for fire safety measures and get it approved from

Fire Services Department before starting construction work. After completion of work, the NOC is to be obtained from Fire Services authority before using the auditorium. Fire resistant paint coating should be done on various exposed materials like, main stage, curtain, wings, wooden panels/doors, chairs, false ceiling, etc.

15.3.10 Auditorium/Hall for Cultural Programme/Cinema/ Conference Purposes in Schools, Colleges, Office Buildings, Institutes

Such project may be divided in two parts, viz.,

(i) Construction of building, installation of electrical/stage lighting, furniture, stage, etc.

(ii) Acoustic work

15.3.11 In case the aforesaid first part of work is executed properly but acoustic work is neither designed nor executed properly then the whole purpose for constructing the building may be defeated because of echo sound/noise problem. Audibility of programme would be spoiled. It may not be possible to enjoy/hear the programme clearly. In some cases such auditorium/hall without acoustic treatment is not used.

15.3.12 Large halls/small auditoriums are constructed in many senior secondary schools, colleges and various other institutions. Various types of programmes viz., cultural programmes, seminars, cinema/ film projections, conferences, debating programmes, annual functions, etc. are held in the hall/auditorium. Hence, it is also essential to design acoustic treatment and execute the same with proper materials. But unfortunately in many such halls/auditoriums neither acoustic treatment is done as per standard nor a partial treatment is done. Designing of acoustic treatment is a specialized field; an expert in this field should design and submit details of acoustic treatment i.e. specification of materials, details of fixing, wall paneling and false ceiling treatment, etc. Execution should also be done under his periodical supervision. Acoustic standard should also be tested. It has been found that such halls/auditoriums constructed in schools, colleges, institutions are not used because there is a lot of echo sound in these halls due to defective or no acoustic treatment. In such cases, importance or attention is not given by any authority for acoustic treatment. Such halls have been abandoned for purpose of cultural programmes, film projections and conferences. For reducing cost of acoustic treatment, it may be possible to bring the acoustic standard upto certain mid-level so that the same is functional for purpose of schools' functions. In case of public functions, the acoustic treatment in public auditorium/ cinema halls/important conference halls has to be of good standard; otherwise booking of defective auditoriums will be less, thereby affecting earnings. Acoustic treatment should be of high standard for international programmes.

15.3.13 To solve the abovestated problem, it is essential to get the acoustic treatment designed from an experienced acoustic expert on the basis of the approved building plans. Details/specifications of various acoustic materials are to be given by him. Details of wall paneling, false ceiling treatment, etc. are to be given with sizes and thickness of materials. Materials used in auditorium chairs are also important from acoustic point of view. Type/specifications of floor carpets (with base layer) are also important for acoustic purpose.

15.4 Industrial Projects

15.4.1 Civil engineering works form an important component of any industrial project. The construction component may be 30% to 50% of the overall project cost. Common management matters related to civil engineering works have already been dealt under foregoing chapters; the same which are applicable in civil construction of factory/ industrial project will not be repeated here. Certain special management aspects which are related to industrial/factory project will be dealt here.

15.4.2 In a typical industrial project whether metallurgical, fertilizer, petro-chemicals, power stations, or any other process/ production unit, the civil engineering components of the project will consist mainly of foundations of various machinery, plants (besides foundations of structures) and housing for various machinery/ plants and steel structures of extra height. Long column and large span beam/truss are involved. Due to this reason considerable structural steel fabrication, welding work, erection of steel girders/ beams, trusses and other roof components are involved. Skilled welders and fabricators having good experience are essential for such projects. Static and dynamic load on account of various components for plants, gantry are to be considered as extra for structural design of foundations related to factory structure and plants/machinery. Feasibility report for the proposed project is important. After approval of feasibility report the Detailed Project Report (DPR) is prepared. One part of DPR is civil engineering works. Various components of civil engineering works are designed by architect and structural designer. Special care should be taken on structural matter particularly for exact accurate location, size of holes for foundation bolts, base plates, levels, etc. These are to be designed after discussion with mechanical engineers of owner's organization and machine manufacturers. Owner of the project may engage architectural and structural design consultants who have good experience on such type of projects. If township is involved for large industrial project then Town Planner should also be consulted. Generally, detailed "shop drawings" for fabrications, components of structure, foundations for machinery are to be prepared by the contractor; necessary stipulation in this regard is to be made in the NIT (tender documents). Efficient running of the plant depends upon proper design and construction of the foundation work of machinery and the structures with sound and accuracy. Proper control and high degree of precision of aforesaid part of work is of paramount importance. It is also important for connected ancillary civil engineering works such as services required for efficient operation of the plant. Facilities for townships, ancillary buildings such as recreation, shopping, hospital and various services viz., water supply, sewerage, drainage, electrical works, solid waste management, environment management (pollution control), roads, horticulture/ landscaping works, etc. are to be planned properly in consultation with senior management authority of factory.

15.4.3 Management style may be little different in such type of projects. It is important for expediency in decision making as well as co-ordinated approach between the field and regional office on one hand and regional office and the head office on the other. Decentralisation (i.e., to give adequate delegation of authority and responsibility to a workable level of management) will expedite the progress of various activities. It is essential for architect, structural designer and Project Manager (civil construction) to obtain detailed requirement of users i.e. in-charges of various branches/departments. Requirements during operation/production of factory are also to be obtained. All concerned persons should discuss about these requirements during preliminary planning period. Architect is required to send detailed drawings to all in-charges of user branches/ departments for confirmation. In industrial project several disciplines are involved. During construction, site coordination meeting with all concerned persons is to be held weekly/fortnightly. Minutes of meetings should be issued by the Project Manager within 3 days.

15.4.4 Concrete Work in Foundations

(a) Lot of concrete work is done in foundation for steel columns (which are generally long columns). Base plates are fixed on the top of concrete foundation with heave type bolts. For this purpose, it is required to leave pocket holes of correct size, depth, locations as per shop drawings. It is also required to construct concrete foundation for many machinery of the plant. These machinery are required to install on the concrete foundation as per drawings. Pocket holes of required size are to be left in the concrete foundation correctly as per shop drawings which are supplied by the machinery supplier.

(b) It is required to be ensured that problem is not faced during erection/installation of machinery, its operation and maintenance of the machinery during production stage of the plant. Due importance must be given to this primary item from the very beginning and at every stage. This will avoid chipping, breaking, widening of pocket holes etc. at the time of erection of equipment; this will save unnecessary work and save time. Checking of reinforcements, dowels and form work before starting of concreting is of utmost importance. Correct lines, levels and pocket hole locations etc. can be achieved in the connecting structure only if the form work is maintained as per stipulated dimensions. It is absolutely necessary to check and re-check at different levels (viz., Asstt Engineer and Executive Engineer) in respect of the layout lines and levels of shuttering before the concreting starts; the correct fixing of "inserts" is also to be ensured as per drawings. It is also to be ensured for proper maintenance of these form works, dowels, inserts, lines, levels during concreting and after concreting and also during removal of form/shuttering materials.

(c) Concreting of machine foundations should be done continuously and as far as possible, without allowing any construction joint. This will help in eliminating weaker planes in the machine foundation, particularly the ones subjected to heavy static and dynamic forces such as high vibrations and reversal of loads.

(d) Lot of structural steel works are done in factory/industrial building. Hence lot of steel fabrication is involved in plant buildings, gantries, cranes, girders, trusses, supports for equipment, piping etc. Normal trend is to resort for wielding. It is important to select trained welders. All welding jobs have to be done strictly as per specifications laid down. Sophisticated testing of welding including x-ray testing is to be done. Precision and good quality can be achieved if steel fabrication is done in the workshop. However, contractor prefers to carry out fabrication job at project site. Necessary equipments are to be arranged (viz., proper tools and tackles) for erection of fabricated components. Proper lines and levels of these structural members are to be checked carefully. Special safety aspect is to be checked during erection of these structural members particularly when height is more.

15.5 Fire Safety Works

15.5.1 One most important requirement for high rise/multi-storey buildings is fire prevention/fire detection/fire suppression system. This is also important for certain other buildings like, auditoriums, cinema halls/PVR Complex, Shopping Complex. Special attention is required to be given by the management authority for this important work. This is a specialized multi-discipline work consisting of civil, electrical, mechanical engineering and Fire Officer having practical experience. Main problem for this work is to obtain NOC (i.e. approval of completed works) from Fire Services Authority. The building can not be occupied without the approval of fire safety works by the Fire Service Authority.

15.5.2 This work is of specialized nature. Various items under this discipline are executed during construction of building. Detailed design, drawings, planning, specifications of materials and installation works, are to be prepared before starting the building construction works. Complete scheme/drawings, specifications of materials/ installation works are prepared as per Fire Services Code and instructions / requirements of Fire Services Department. Aforesaid documents alongwith building plans are to be submitted to Fire Services Department for approval. After completion of fire service works, the same are to be approved/cleared by them. As per building bye-laws, the building cannot be occupied unless fire prevention/suppression /detection works as executed are approved by their Department. It is very difficult to obtain NOC (No Objection

Certificate i.e. approval of fire fighting works executed) from their Department. Due to this reason, some completed buildings could not be occupied for a long time. To avoid such serious problem, it is required to execute strictly as per approved drawings/specifications. Objections raised by Fire Services Department are to be complied with on priority basis.

15.5.3 Building Plans related to Fire Safety

Certain items/requirements related to architectural/building plan, are to be included like, – separate underground tank of required capacity for fire fighting works, emergency fire escape staircase, fire refuge area at different floors of multi-storey building, approach road of required width for operation of Fire Fighting vehicles/ Tender/ Snorkel vehicle, long ladder, water tanker of large capacity and other equipments. If there is plaza/roof of basement (open to sky) then certificate of structural engineer is required about certain minimum load carrying capacity of heavy vehicles/snorkel. This structural requirement of Fire Services Department is required to be given by the structural designer before starting building construction work. Serious problem is faced if this requirement is not tied up with structural designer bore starting construction works. If approach road to building is narrow and heavy fire fighting vehicles/ tender/snorkel and other heavy equipments cannot reach upto the building and also turning/movement space is not sufficient for easy exit then NOC is not issued by the Fire Services authority. Fire exit staircase is to be designed and constructed in such a way so that evacuation time of all occupants of the building is within the permissible limit. Architect is also required to prepare building plans considering proper "Compartmentation" as per fire safety requirements. All requirements of Fire Services authority must be complied with for getting the NOC. It may be seen from above that fire fighting work does not mean only installation work; building plans are also to be prepared by architect as per fire safety requirements.

15.5.4 As per Fire Service Department/Fire Service Act/Municipal Bye-laws, it is required to comply with the following "12 points" which are minimum standard :–

(1) Means of access
(2) Underground/Overhead water static tanks
(3) Automatic sprinklers system
(4) First-aid Hose Reels

(5) Fire extinguishers with ISI certification mark
(6) Compartmentation
(7) Automatic fire detection and alarm system/manually operated electrical fire alarm system
(8) Public Address System
(9) Illuminated exit way marking signs
(10) Alternate source of electric supply
(11) Fire Lift with fireman switch
(12) Wet Riser Down Comer system

15.5.5 The most important and a major part of fire fighting system is installing "Automatic Sprinkler System" in all Shopping Complex, basements and in various floors if height of the building is more than 15 metres. This is to be designed by a specialist/ expert designer in this field. Number of sprinkler head depends on floor area of each room/hall/ passage. The bulb (quartz) will burst at a certain temperature (i.e. 68°C in normal office building) and sprinkling of water like shower will start automatically. This system of installation is very effective to suppress the fire. Make/quality of sprinkler head should be of high standard. Periodical actual testing by putting heat at required temperature is to be carried out. Such random testing will be carried out by Fire Officer before issuing NOC to occupy the building.

15.5.6 There should be minimum use of wood in the building. If wood or wooden doors are provided, then the same should be fire-resistant type. Certain time is stipulated in the Fie Fighting Code or Circular issued by Fire Services Department about fire resistance capacity (i.e. Fire rating "X" hour); if automatic sprinkler system exists then fire rating for wooden door should not be more than one hour. Sample is put in a furnace of laboratory and checked. Such doors are manufactured by specialized firms. Certain special material/ chemical compound is put inside the door shutter/panel. If required certain part of shutter may be made with special type of fire resistant glass (reinforced with steel wires). Testing of such door is to be carried out in the laboratory and the same is to be given to Fire Services Department. It may be mentioned here that it is difficult to achieve architectural /aesthetic view due to above requirement. Hence, architect is required to select suitable type of material for partition panel/cabin work, wall paneling, false ceiling work, windows, doors, etc.

15.5.7 It is essential to check and maintain entire fire prevention and fire suppression system in perfect condition. Testing of various fire installation/equipments is to be carried out at site by operation. Sudden mock exercise of fire incident in the building is to be done periodically. Availability of sufficient water in the underground fire fighting tank is to be checked frequently. Working position of "Wet Riser System" is also to be checked periodically. This installation is very effective for suppressing fire at any floor.

15.6 Management of Private House Constructions

15.6.1 Individual owners construct their house. This is one most important sector in the society. In terms of number of houses constructed by individual house owners the same is highest in construction industry. This is related to population of country. Population of India is second highest in the world. In India these vast numbers of house owners face lot of problems for constructing their houses. Besides house constructions by individual house owners, the private houses are also constructed by group housing society, developers/builders etc. Here management aspect of houses constructed by individual owners will be dealt briefly. Though it is not a specialised or non-conventional building yet the management is different type. Individual private house owners face lot of problems for constructing their houses. Most of them construct their only one house with their life saving. Due to poor management they suffer loss of money, construct house of shorter life due to bad quantity of materials of workmanship, harassed/cheated by architect/ contractor. These houses may be divided in three groups viz.

(a) Low Income Group	Rs. 8 to 10 lacs
(b) Middle Income Group	Rs. 10 to 30 lacs
(c) High Income Group	Costing Rs. 30 lacs and more

15.6.2 These costs varies from place to palace (like, small towns, smaller city, metropolitan city). Cost of houses is increased in metropolitan cities, capitals. It is required to know certain major problems faced by the owners during new construction of houses, addition/alteration/renovation of houses; some of these problems are stated below.

15.6.3 Generally house owners depend on the Architect for preparation of building plans, structural design, architectural and engineering supervision, fixing contractor with terms and conditions, rates, payments etc. But practically engineering supervision of works including casting of RCC columns, beams, slab (roof/floor slab) is not carried out at all or not carried out properly. In most of the private individual owner's houses construction, there is no specific system/independent supervisory person for actual site checking of quantum of cement used in structural work; there is serious consequence on account of afore said lapse. It has been found in some works after completion of building that the half quantity of cement was used as compared to theoretical requirement. After carrying out certain reliable tests (viz. core testing of RCC columns, beams, roof/floor slabs and cement content test in brick work) it was found that strength of RCC works was 50% to 60% less than structural requirement. No civil engineer or architect supervised/ checked cement consumption during casting of RCC works; quantity of cement used in the cement mortar of brick work was also not checked. As per Architect's interpretation of contract document it is not their duty to supervise casting of RCC works; Architect also did not advice the owner for engaging civil engineer for such checking and supervision. Serious problem of structural stability of the house was faced. Due to certain disputes the contractor abandoned the construction works after receiving extra payment (i.e over payment). Contractor's agreement was prepared in consultation with the Architect such agreement is crucial for smooth progress, achieving good quality of works, control on release of payment to contract etc. Type of contract is another crucial factor. Mode of measurements and payment are not checked by any engineer. Many serious problems are due to faulty contract agreement.

15.6.4 Another serious problem is faced by the owner is due to faulty structural design. Architect's structural design engineer is generally not competent for satisfactory design. Due to this reason the structural failure (cracks in structures of building) is observed after compeletion of building. It is difficult to rectify such defects satisfactorily; it may be required to dismantle certain part of building and reconstruct the same.

15.6.5 After completion of house the house owner/occupant faces lot of practical problems in respect of faulty architectural design particularly from functional points. Maintenance problems are also faced. It is required to carry out some addition/alteration in the newly constructed house.

15.6.6 Following Problems are faced after Completion of the Construction Works

(1) Testing and commissioning of various service/utility lines (viz., water supply, sewer, electric, telephone, TV, pumps, etc.)- If there are blockages in pipelines / leakage from pipelines, then rectification is to be done.

(2) Locks, Tower Bolts, handles of doors, windows, cupboards/ almirahs are to be checked and defects (if any) are to be rectified. Flushing cisterns are important sanitary installations and generally it creates problem. Best quality of reputed manufacturer should be purchased. Electric and water connections of Geysers are to be tested. To check whether all the necessary light and power plug points have been installed at correct locations.

(3) Generally, it is found that adequate slope is not given on floor of toilets and it creates a lot of problems; some time water flows inside the bedroom. Similarly proper slope is not given on balcony floor. The slope is to be checked by pouring water on the floor. Proper slope is also not given on the roof.

(4) Normally Architect does not prepare "Completion plans" for approval of the statutory authority (i.e. Municipality/ Development Authority). In case this is not done, then it will not be possible to obtain approval of building plan in future for proposed addition/ alteration of building from municipal authority.

(5) Necessary approval for water and electric connections is not obtained by the Civil Engineer/Architect and get these connections done.

(6) Site clearance, inside cleaning of house is not done. Centring, shuttering, scaffolding materials brought on hire basis are to be returned to shop quickly.

(7) Contractor is supposed to attend maintenance complaints for a period of at least 6 months after the completion of the work. But some time the same are not attended.

In view of above, it is required to withheld suitable amount from the dues of the Contractors, Civil Engineer and the Architect.

15.6.7 From management point of view owner is required to make proper planning and programme for taking up the house construction. Selection and purchase of plot is most important, there are many aspects for the same. Details/Check list on this matter have been dealt in another book of author. Various aspects of management in owner's house construction principles are applicable in case of houses constructed by Low Income Group, Middle Income Group and High Income Group. However, the owner of low income/middle income groups may comply various requirements without engaging full time architect and engineer. He has to select architect and civil engineer (full time/part time) carefully and agreements are to be made as first step. The owner should finalise the agreement of architect with detailed scope of services, stage payments etc. in consultation with a Civil engineer. One Civil engineer (part time) is also to be arranged either through architect or an independent engineer (full time/part time) for large house construction. After complying this basic requirements and approval of building plans the owner is required to take very special care for selecting a reliable experienced contractor in consultation with Architect and Civil engineer. Detailed terms and conditions, type of contract etc. are to be got prepared through the architect and engineer for engaging a competent contractor. Apparently lower rates of contractor with faulty terms and conditions are of no use. Hence planning stage is most important for overall and effective management of house construction. Many problems relate to this stage.

15.6.8 Building Plans

(1) After taking physical possession of the land, it is required construct compound wall/fence to avoid the encroachments. Then it is required to select a good Architect for preparing building plans and to get it approved from the concerned authorities. Selection of proper Architect and his scope of services is very important. Terms & conditions, detailed scope of his services are to be finalised. Owner must consult/ check from a civil engineer at this stage.

(2) The owner will explain his detailed requirements of proposed house/additional construction on the existing house. Plan of vacant plot/ existing house is to be prepared. Architect will prepare a few sketches/ conceptual plans for obtaining suggestions/comments of the owner. The owner will offer his comments/ suggestions on the aforesaid sketch plans. Thereafter, the Architect will prepare one preliminary plan for the final comments/discussion with the owner. During above stated process following factors should be kept in mind.

(3) *Functional Aspect*

Functional aspect of building is most important It has been found in many houses that the owner faces lot of practical problems during stay in the house. A few examples are stated here. Function of kitchen is important. This is used for the maximum time. It is required to keep many articles, utensils, kitchen equipments, proper washing facilities, extra gas cylinders, cooking burner set with chimney, cooking range, refrigerator, etc. in the kitchen. However, it has been found in many houses that the size of the kitchen is too small but the size of the toilet is larger than the kitchen. This disproportionate size is not acceptable. Size/ floor area of the drawing room is generally kept quite large; but size of bedrooms is too small which is another important aspect. The occupants need proper space for keeping essential articles, furnishing for clothes, etc. More comfort is required in the bedroom. Some time the number of bedrooms is increased by reducing the sizes of all the bedrooms. It may be better to make at least one bedroom with sufficient floor area by dropping another bedroom/ reducing floor area of the large drawing room. It is important to make a balance in area distribution considering various functional aspects. Increasing the number of rooms by considering the commercial aspects, is not a good proposition. Balancing of the floor areas should be done as per the requirements of owner/ occupants.

(4) Size of toilet should not be too small or too large. Again balancing of area distribution is required for this important provision. Covered shelf (in plastic material) of appropriate size is essential in toilet/bathroom. As window is generally kept closed, it is better to make louver type shutter with frosted glass for good ventilation. Some time the internal layout of toilet is prepared in a faulty way; western type commode/ WC has been installed at the centre and the same is just opposite to the door. It may be mentioned here that it is very difficult to change such a faulty internal layout of toilet at a later stage after construction.

(5) Space for keeping shoe rack should be provided with specific indications. This is important for keeping the house clean and hygienic. Locations for refrigerator, cloth washing machine, crockery cabinets, TV/ Music System, room coolers/

AC, book shelf, box room/stores, dressing table, etc. should be indicated on the drawings.

(6) Another important aspect for preparing building plans is related to various service lines and its maintenance. As these service lines (viz., water supply, sewerage, electrical, drain pipeline, telephone line, TV etc.) are kept concealed in the walls, floor and columns, it becomes very difficult to maintain/ repair. Hence, size and location of the service shafts is important. Generally service shafts are found very small and it is not possible for worker to go inside for cleaning and repairing. In multi-storey residential flats, it may also be required to construct shafts for disposing the garbage and to make arrangement at ground level for lifting the garbage for disposal in the municipal garbage bins. Effort should be made to install service lines (particularly water supply and sewer pipes) on the surface of walls so that it is easier to maintain/ repair. It may be possible to prepare building plan in such a way so that exposed water supply/ sewer pipes are not prominently visible.

(7) It has been found that in some building plans sunshade/ chhajjas are not proposed on the external windows, doors. Problem is faced as the rain water enters into the rooms. It is difficult to construct proper sunshade after the construction of building.

(8) *Staircase*

If house is to be planned in duplex type, then care is to be taken for internal staircase to keep adequate head room and normal height of each riser. It is to be ensured that domestic furniture could be lifted to upper floor through the staircase without problem. In many duplex houses, it is not possible to carry doublebed, divan, almirah etc. through faulty designed staircase (having less head room, narrow width and sharp turning point). There is no use of such bedrooms where essential normal furniture cannot be taken through faulty designed staircase. This is highly objectionable. In case houses are constructed in two or more storeys, then it is to be designed suitably (with proper width, normal height of riser, etc.) so that difficulty is not faced for lifting furniture etc. Staircase must be provided for going to the roof of building for maintenance of overhead tanks and other maintenance works.

(9) Security and safety aspects are to be ensured during the preparation of building plans. If there is central courtyard and shaft open to the sky, then necessary grill should be provided. Railings should be in proper height and opening between vertical components should be less for the safety of children. Internal design of cupboard/almirah should be as per requirement of the owner. Overflow pipes from the overhead tanks, outlet pipes from the balconies should be connected with the drain at ground level through the pipes, so that overflowing water does not fall on the roof to avoid seepage on ceiling and walls.

If required, a model of proposed large building may be made for better understanding of the owner.

15.6.9 Method of Constructions/Type of Contract

(a) *First Type of Contract*

For the method of construction for buildings in Government Departments or large housing complex of private organizations (viz., Housing Society, Big company's housing projects etc.) item rate tenders are invited.

In case of construction of a house for individual owner, there may be some practical difficulty to follow the above stated lengthy procedure. However, in case an individual owner has proposed to construct a large house and he has the arrangement of Architect and Civil Engineer, then he may follow the above stated procedure (i.e. pre-qualification of the Contractors, item rate tendering etc.). During finalisation of detailed scope of services, terms & conditions, fees of Architect the owner should consult with a Civil Engineer.

(b) *Second and Third Type of Contract*

The second type of contract is fixing rates of different items (at least main items like, earthwork, brickwork, RCC works, flooring item, plaster items etc,) with materials and labour. The third type of contract is fixing labour rates for different items with labour rate Contractor. Materials are to be purchased by the owner and to be brought at site. The owner may avail part time/full time services of one Civil Engineer for estimating the quantities of various materials and to buy the same from the market. The construction work will also be supervised by his Civil Engineer including verification of bills of the Contractor and Architect. In case Civil Engineer is not engaged (part time or full time) then verification of contractor's running & final bills must be done by the Architect. In case of second type of contract, Civil Engineer will supervise construction works, measure various items of works, check the quality of materials & verify the bills of Contractors etc. Scope of services required from the Civil Engineer and his fees are to be decided by the owner after carrying out negotiations. Suitable Civil Engineer and Architect are to be selected as per their practical experience and track records. Special care is to be taken to prepare detailed contract document/agreement with the contractor. Architect may not be able to prepare proper agreement; this should be got done through an experienced Civil Engineer. Many private owners faced serious problems due to defective agreement.

(c) It is relevant to state here that in case owner has sufficient time and can take some physical strains, then the third method of contract is better. Under this method it will be easy to ensure the use of "good quality" of materials and to use full specified quantities of materials (viz., cement, steel of reputed company and grade). Quality and names of manufacturers, brand name, grade of certain materials like GI pipe, CI/ Wrought iron pipes, sanitary and other fittings, water proofing works, electrical wires and fittings, name of manufacturer and brand/grade of tiles for floor, walls, cladding, etc. are very important and it is easy to ensure quality under the third type of the contract. It is also possible to save the construction cost under the third method. It may be possible to achieve overall economy even after paying fees to the Civil Engineer. Owner will have lot of flexibility for changing or deciding type/brand/quality of materials without contractual disputes. Most important is to achieve the good quality of works; the life of building will be more and there will be less maintenance problem under this third type of the contract. However, owner has to ensure, procurement of the required quantities of specified materials in time at site. Security guard is to be deployed to stop pilferage of materials. The Civil Engineer is also to check theoretical consumption of various materials, materials actually procured and actually used in execution of various items of work. Due to faulty agreement entered with the contractor the owners face serious problem to sort out disputes. The owner suffers heavy financial loss; abnormal delay occurs in completion of work. It has been found in many agreements that the owner is required to release 80% payment to the contractor after actual completion of 45% or 50% of works. This is dangerous situation. Architect is mainly responsible for advising owner to sign such faulty agreement.

(d) *Fourth type of Contract*

It has been observed that some owners/Architects finalise agreements with the contractor for constructing private houses on the basis plinth/covered area rate for working out cost of house/contract amount. Stage payment/Running account payment is made to the contractor on the basis of certain percentage of aforesaid contract amount/cost of house at different stages of construction (viz. foundation level, lintel level, first floor level, roof level & wood work, floor, finishing works, etc.). In case scope of work, or specifications of works are changed and the total amount is reduced then there will be over payment of considerable amount to the contractor. It may not be possible to control payment and checking of various matters on the above stated method. Even detailed estimate on item rate basis is not prepared by the Architect/owner. This method is not acceptable. Cost of house may be worked out on plinth/ covered area basis for the purpose of budget or arranging fund or to know the approximate cost of the house. Serious problem is faced by the owner by entering agreement with the contractor as per above stated faulty method (i.e. cost of contract amount based on plinth/ covered area basis). Unfortunately Architect recommend to owner for above type agreement in many cases.

(e) *Fifth Type of Contract*

It may also be possible for adopting a mixed method i.e. (a) some major items (viz,, Brickwork, RCC and concrete works, earthwork, plaster items, etc.) be got executed by fixing a Contractor on item rate basis (materials and labour) and (b) the remaining items (viz., water proofing works, various service lines, flooring work, woodwork, various fittings of water supply, sanitary installations, electrical works, painting works, etc.) be got executed on labour rate basis (materials are to be purchased by the owner from the market). Cement and steel bars for RCC works should be procured by the owner.

N.B. It is required to prepare detailed estimate on the basis of item rate for various types of contracts so that necessary checking/control can be done during execution/payment.

15.6.10 Scope of Architect's Services

(i) To prepare conceptual/ preliminary building plans for discussing the detailed requirements of owner alongwith the alternative plans.

(ii) To modify/ prepare plans for final discussion with the owner.

(iii) To prepare final building plans etc. for submission to the concerned statutory authority (viz., Municipality/ Development Authority) and to comply with their observations and to obtain their approval.

(iv) Preparation of detailed working drawings/ structural design/ drawings and to hand over 5 sets of drawings to the owner,

(v) Preparation of specifications of various items.

(vi) Preparation of the detailed estimate for the construction works at item rate basis.

(vii) Preparation of the completion drawings and to obtain completion certificate/ NOC from statutory/ Municipal authority,

(viii) Inspection at site to sort out architectural and structural design problem, stipulating details of inspections, detailed scope of engineering supervision.

15.6.11 Scope of Civil Engineer's Services

(i) To check the architectural/ structural design and drawings for modification of the faults and to improve engineering/ functional aspects.

(ii) To pre-qualify the Contractors for executing certain items with materials and labour and certain items on labour rate basis.

(iii) To decide the method/ type of contract for executing the construction works.

(iv) To check the detailed estimate submitted by the Architect and to make necessary additions/alterations, if required.

(v) To obtain quotations from the Contractors and to verify the rates from the market trend and to negotiate with one suitable Contractor for finalizing the contract alongwith the terms and conditions (including arrangement of tools, equipments, centring and / shuttering materials, ladder, ropes, grinding machine, concrete vibrator, etc. arranging temporary electric connection & water for construction purpose, at site.

(vi) Checking and supervising various site works particularly RCC, brick works.

(vii) Purchasing of construction materials, its account, watch and ward and constructing the site godown. Working out the quantities of materials to be purchased alongwith grade / brand name / manufacturer's name, etc.

(viii) To check the quality of works, use of correct quantity of cement, steel reinforcements, testing cement concrete etc.

16

Management of Highway Project

PRE-CONSTRUCTION STAGE

16.1 General

Proper and effective management are required for Highway projects to achieve the basic principles and objectives. Different types of management and techniques are involved in respect of highway projects and other road works. Practical problems are faced during pre-construction, construction and post-construction stages and it is required to know such problems so that proper management techniques and planning as well as design are carried out in advance accordingly. System and procedure for highway projects are different than building works. Hence, various theoretical and practical problems will be stated here and solution for the same will be dealt.

16.1.1 There are different categories of roads like :-

(1) City/Town Road
(2) Major District Road
(3) State Highway
(4) National Highway
(5) Rural Road

Another classification of roads is as under :-

(1) Hilly/Mountainous Road
(2) Plain Road
(3) Road on Rolling Terrain
(4) Flexible Road (like Bituminous Pavement)
(5) Rigid Road (like, cement concrete/RCC Pavement, Stone Block Pavement)

16.2 Pre-construction Stage (preparation of DPR)

Pre-construction stage is divided in following three parts :-

(1) Preliminary Planning
(2) Detailed Planning and Designing (Detailed Project Report)
(3) Tender Stage

All public roads and highways are constructed and maintained by Government authority (who is the owner). As a first step, it is required to examine and decide about necessity for construction viz. widening/strengthening/re-construction/new construction of certain reaches/new construction of bye-passes alongwith new construction/re-construction of culverts and bridges. For this purpose, the demand of public and different authorities are to be checked. Preliminary data are to be collected about justification/necessity of the proposed highway project (State Highways/ National Highways). If it is required to construct highway along new alignment for connecting certain towns and villages then such project is to be dealt in a different way which will be dealt under separate chapter. Construction/re-construction/widening/strengthening of existing highway, construction of new bye-pass are being dealt in this chapter.

After collecting preliminary data about necessity of proposed work of existing highway of required stretches (indicating kilometre stone/distances and names of places/ towns and villages). Chief Engineer's office will examine and submit report regarding necessity of the construction work. The principle approval is to be accorded by the competent authority for carrying out Feasibility Study Report. Thereafter Feasibility Study is carried out through a Consultant who submit its F. S. Report.

16.3 Stages in Preparing DPR

If the proposed project is feasible, then further action is taken for preparing Detailed Project Report (DPR) for following stages :-

16.3.1 (1) Preliminary Project Report (PPR)
(2) Draft Detailed Project Report (Draft DPR)
(3) Design and Drawings
(4) Bridge Design
(5) Financial and Economic Analysis Report
(6) Final DPR

Sometimes Feasibility Study Report and other stages as stated above are taken up for preparing complete DPR under one contract with the consultant.

16.3.2 Engaging Consultant

Many problems are faced at tender stage, construction stage and post-construction stage due to faulty detailed field survey and levels, site investigations and wrong designing. Hence, special care is to be taken during above stages. Generally, preparation of complete DPR including feasibility study is got done by the Government Department through experienced competent Highway Consultant. As per necessity, the Consultant engages certain specialized experts viz., Bridge Designer, field survey agency, Geologist, Financial and

Economic Expert, etc. Appointment of reputed and experienced Highway Consultant is an important step. Quality and standard of DPR will depend on the Consultant. Hence, utmost importance is to be given for fixing proper criteria of eligibility of Consultants. Proper pre-qualification documents are to be prepared for inviting applications/technical and financial bids.

16.4 P. Q. of Consultants

16.4.1 There are some practical aspects and problems about pre-qualification of Consultants. It has been observed that some Engineers submit applications/bids in the name of some other established Consultants having experience in preparation of DPR. Both have understanding or internal agreement to carry out the assignment of DPR preparation. Names of Engineers and Experts are given in the applications; these engineers are engaged temporarily for this particular job. There may not be any staff, consulting engineers, experts, designers on regular/permanent basis. Names of many engineering staff/key persons given in the bids are changed after awarding the consultancy assignment. This is objectionable. Performance of such type of consultants will not be satisfactory. It may not be possible to get reliable/correct DPR from such consultants. Many problems will be faced during construction stage on account of faulty DPR. Actual CBR value of sub-grade (kilometre-wise/certain reaches) may not tally as stated in DPR. Actual traffic survey may not tally with data stated in the DPR. Alignment plan of existing highway (including location, radius of curves) and long sections at certain reaches of DPR may not tally as per actual site position. Due to above stated position/data of important nature, serious problems will be faced during construction period; contractual disputes also crop up due to faulty DPR and wrong design. It has also been seen that regular (i.e. in-house arrangement) key persons like Team Leader, Senior Highway Engineer, etc. are not available with middle level highway consultant. Such consultant hire/recruit all the senior and junior engineers for one specific DPR on temporary basis; some of these engineers leave the job suddenly during preparation of DPR due to various reasons. Due to above stated position, accuracy /standard of DPR may not be satisfactory. As DPR will be part of NIT (tender documents) and construction agreement, lot of problems are faced during execution of construction works. Contractual disputes with contractors crop up due to unsatisfactory DPR. Hence, it is required to select reputed and competent consultant; for this purpose, it is essential to prepare PQ (Pre-Qualification) documents properly and to verify records (i.e. Technical Bids) of consultants/bidders carefully. Success of project will depend on selection of competent consultant. This should be monitored and checked at higher levels (viz., GM/Chief Project Manager/Chief Engineer) of the Government Department.

16.4.2 P.Q./Technical Bid Documents

Many conditions/requirements are laid down in the PQ documents; some of these are stated below :-

Total value/turnover of the consultant's firm, list of similar consulting assignments (i.e. preparation of DPR of highway project) and year/period of consultancy assignments of highway project with amount of fees during last 5 years. List of key persons (i.e. at the rank of Team Leader, Highway Engineer, Design Engineers, Material Engineers, etc.) working with the consulting firm (indicating month and year), list of testing instruments owned by the consultants viz., for CBR values, BBD tests, etc. for field laboratory, mobile laboratory for carrying various tests, Auto levels required for DPR, permanent arrangement of various software's, printing of larger size drawings etc. It has been found that some consultants (middle level/lower level) are unable to make payment in time to their survey agency, bridge designer and other experts (viz., environmental expert, economic and financial experts) and due to this reason above stated associates are unable to carry out their jobs in time and standard of their services becomes poor/fair. Field investigation/condition survey of existing bridges and soil investigation by boring are not carried out satisfactorily. Necessary facility for carrying proper condition survey of bridges is not arranged by the main consultant. Suitable agency should be engaged by the main consultant for carrying out soil investigations by boring method including drilling in rocky strata for bridges. Suitable conditions and stipulations for above stated matters are to be made in the PQ/Technical Bid Documents for appointing consultant to prepare DPR. Main consultant should have at least one Asstt. Material Engineer as their permanent engineer who knows how to carry out various tests of materials viz., CBR values, BBD tests, field density test, operating Roughometre/ Bump Indicator, Plasticity Index Limit of moorum/binding material for WBM courses, various tests of bituminous road works, tests for stone aggregates, sand etc. Suitable condition is to be stipulated in the bid document for submitting different parts/volumes of DPR in time as per schedule given in the bid documents. Condition related to imposing compensation/penalty (with sufficient amount) on account of delay in submission of final DPR is to be stipulated in the bid document.

16.5 Bidding Procedure

(a) Bid notice is issued in the newspapers, internet and other places. Bid documents are sold to interested bidders. Bid documents can also be obtained from internet. However, necessary cost of same is to be deposited at the time of submitting bid. Bid is generally submitted in three envelopes as under :-

(1) Envelope No. I – Earnest Money

(2) Envelope No. II – Technical Bid as per requirement stated in the bid documents issued by the Department

(3) Envelope No. III – Financial bid on prescribed form. Only rate is to be quoted. Unit is generally in Rs."X" per kilometre length of road.

(b) Pre-bid meeting is arranged by the Department with all consultants who have applied or purchased PQ/Bid documents before receiving the bids (say about 10 days before). Points raised by the consultants are clarified during meeting. Final decisions on relevant points are informed to all the consultants by the Department through a letter which is treated as part of bid documents.

(c) Envelopes No. I and II are opened on specified date and time. Technical bids are examined in detail and marks are allotted by a Technical Committee of the Department. Bidders who get pass marks are pre-qualified. Then financial bids of pre-qualified bidders/consultants are opened. If rate of lowest bidder is higher than the market trend then negotiation is conducted. Thereafter, lowest bidder/consultant is appointed through letter of acceptance/Work Order. Then agreement is also signed by both the parties. Date of start is considered as per provision stated in the bid document/ contract agreement.

16.6 Bar Chart Programmes

It may be stated here that as per bid documents bar chart programmes are also to be submitted alongwith technical bids for various stages of DPR as indicated below.

(a) Bar Chart Programme for completing/submission of DPR in stages (indicating number of weeks) A sample Bar Chart is given in Figure 16.1.

WORK/ACTIVITY SCHEDULE SHOWING FIELD INVESTIGATIONS AND REPORTS ILLUSTRATED WITH BAR CHARTS

A. FIELD INVESTIGATION Sheet No. 1/2

Sl. No.	*Item of Work / Activities*	*MONTHS*									
		0.5	*1.0*	*1.5*	*2.0*	*2.5*	*3.0*	*3.5*	*4.0*	*4.5*	*5.0*
1.0	**Inception Report and QAP**										
1.1	Project Appreciation	■									
1.2	Work Programme	■									
1.3	Design Standard	■									
1.4	Key Plan	■									
1.5	Review Data and Documents	■									
1.6	Quality Assurance Plan	■									
2.0	**Traffic Survey**										
2.1	Volume Count		■	■	■	■					
2.2	O.D. and Commodity Movement Survey		■	■	■	■					
2.3	Axle Loading Characteristics				■						
2.4	Intersection Volume Count				■						
2.5	Speed Delay Characteristics				■						
2.6	Pedestrian/Animal cross, Traffic Count				■						
3.0	**Engineering Survey & Investigation**	■	■	■	■	■					
3.1	Reconnaissance Survey & Investigation	■	■	■	■	■					
3.2	Topographic Surveys	■	■	■	■	■					
3.3	Soil & Material Survey & Testing		■	■	■	■					
4.0	**Utility Services & Other Physical Feature Survey**			■	■	■					
5.0	**Road & Pavement Investigation**	■	■	■	■	■					
5.1	Road Inventory Survey		■	■	■	■					
5.2	Pavement Condition/ Investigation Survey		■	■	■	■					
6.0	**Investigation for Bridge, Culverts, Drainage & Structures**		■	■	■	■					
6.1	Inventory of Bridge/ Culverts and Structures		■	■							
6.2	Hydraulic and Hydrological Investigations		■	■							
6.3	Geo-technical Investigations and Sub-soil Explorations		■	■							
7.0	**Land Acquisition Survey & Plan**			■	■	■	■	■			
8.0	**Re-settlement/Rehab. Plan**			■	■	■	■	■			

Fig. 16.1 : Bar Chart Showing Programme of Consultancy Services for Preparing D.P.R.

(1) Inception Report
(2) Feasibility Study Report
(3) Preliminary Project Report alongwith layout plan of Highway and Long-section, condition surveys, preliminary project cost estimate
(4) Draft DPR alongwith various drawings including cross-sections of roads at 50 metre interval, environmental impact, rehabilitation/ resettlement and socio-economic report, land acquisition particulars, encroachment details, bye-pass construction, project cost estimate, culverts, etc.
(5) Bridge Design with General Arrangement Drawings, Soil investigation report for bridges, financial and economic analysis report, project cost estimate.
(6) To comply with observations of the Department on draft DPR and to submit final DPR alongwith tender documents (NIT) package-wise to the Department.

(b) Bar Chart showing deployment of key persons viz., Team Leader, Highway Engineer, Material Engineer, Resident Engineer, Geology Expert, Bridge Designer, Finance and Economic Expert, etc.
Period (in weeks) of each person (by name) is to be shown on the bar charts. A sample bar chart is given in Figure 16.2.

TOTAL MANPOWER DEPLOYMENT SCHEDULE

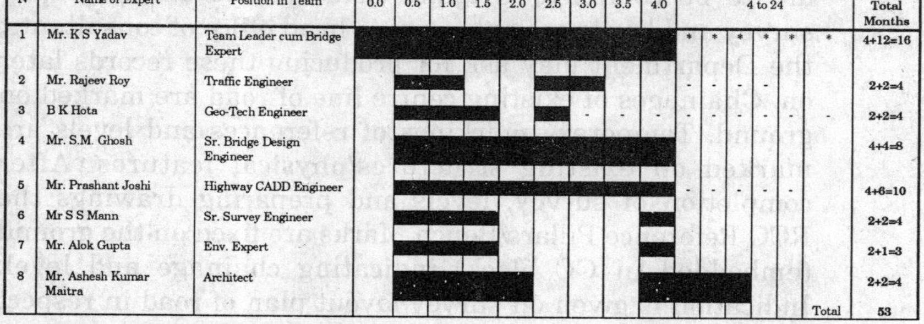

*** Denotes Intermittent Deployment

Fig. 16.2 : Manpower Deployment Schedule of Consultant for preparing D.P.R.

16.7 Certain Practical Aspects of DPR Preparation

(1) Start and completion dates of various activities related to preparation of DPR like, Mobilisation work, setting out office at site, arranging staff, vehicles – Jeep, motorcycles, survey agency, telephone, FAX facility, setting up of field laboratory and laboratory in-charge, etc. To carry out reconnaissance survey along entire length of project road.

(2) Submission of Inception Report alongwith two Bar Charts showing deployment of key persons (by name period-wise) and programme of various activities, submission of various parts/volumes of DPR (sample bar charts are as per Figures 16.1 & 16.2. If name of any key person actually deployed at site is changed from the original proposal (i.e. bar chart submitted alongwith technical bid) then permission is to be obtained from the Department. Hence, effort should be made for not changing any key person.

(3) To start the physical topography survey along the Highway from starting point with "Total Station" instrument and Level work with "Auto Level" by the survey agency engaged by the consultant. These are to be supervised/checked at site by the surveyor of the consultant side by side. Closing of level at the end of day by fly level on reverse direction should be done to know the error. Level books are to be checked by independent staff of the Department/consultant. Detailed requirements have been laid down in IRC Manual/Code (No.IRC:SP:19); various requirements as per this Code are to be complied. Original field records of topography survey and levels are to be sent to head office of consultants; the Department may ask for producing these records later on. Chainages of existing centre line of road are marked on ground. Temporary markings of references and levels are marked on existing structures/physical features. After completion of survey, levels and preparing drawings the RCC Reference Pillars/Bench Marks are fixed on the ground (embedded in CC block) indicating chainage and level. Indication is given on survey layout plan of road in respect of agricultural land, forest land, barren land, villages, towns, markets, encroachments, hilly area, existing Right of way, bridges, culverts, causeway, rivers, streams, nalla etc. Long-section and cross-section of rivers and streams at culverts/bridges are also to be surveyed and levels are to be taken for a length of about 100 metres on upstream and

downstream sides of bridges/ culverts. If any item as stated above is left out during survey work at field then it becomes difficult to do it later on because a lot of problem is faced (including extra expenditure) for re-arranging survey agency and consultant's staff, vehicles, etc .

As next step survey agency is to prepare and submit drawings i.e. layout plan and long sections along centre line of road (Kilometre wise). Appropriate scales are to be adopted for layout plan and long section (horizontal and vertical). Width of existing Right of way is to be obtained in writing from concerned authority (viz., PWD/NHAI) at personal level. This is essential to work out land acquisition details. The list of RCC reference pillars (indicating chainages and levels) and names of villages/towns is also to be made as part of DPR. Level of Bench Mark of Survey of India at nearby area is to be traced out from their maps and then to be checked at ground. If it is not possible to trace out their Bench Mark then B. M. of nearest railway station may be taken and the reference are also to be marked on the layout plan.

(4) One important aspect in highway design particularly from the point of safety is about connections of many roads emerging from many nearby villages and towns. Thus there are many junctions on the highway with abovestated village/ town roads. There are safety problems at these road junctions/intersections. Hence, it is essential to examine in details during survey and design of road and service roads, road junction design etc. It is required to carry out detailed survey in respect of abovestated village/town roads (at least 100 metre length) and area around these road junctions. Effort should be made during designing for proposing service road parallel to highway at certain vulnerable reaches and connecting with highway at reduced number of locations.

16.8 Feasibility Study Report

This report is very important because Government authority takes policy decision on certain issues viz., priority of project, beneficiaries, arranging funds for the project, necessity of new bye-pass roads at certain reaches (mainly in urban areas), rough requirement of acquisition of land (with type/category), environmental impact assessment (E.I.A.), socio-economic and re-settlement position, strategical aspect from security angle, etc. Rough cost estimate of the project is also to be indicated in this report.

Rate for bridges and culvert construction is taken on per metre length basis (as per past experience). On the basis of above position the Government authority takes decision whether the project is to be taken up or not, or to be deferred/dropped. After receiving observations on this feasibility report from the Departments, it is required to comply with same by the consultant before preparing PPR. It may not be practically possible to prepare proper E.I.A. report after carrying out environmental field screening survey in short time. However, it is essential to prepare Initial Environmental Evaluation Report for including in the Feasibility Study Report.

16.9 Traffic Survey

16.9.1 Field Work

Most crucial factor in designing highway is actual accurate detailed traffic survey along entire length of road (reach/section-wise). If BOT tender is invited, response will depend mainly on correct traffic survey. Design of highway including culverts and bridges depends mainly on designed traffic load. Projection of actual present traffic load is done on the basis of certain guidelines as stated in the IRC Code/Highway design book. Hence, special care, planning and arrangement is to be done for carrying out traffic survey by counting various types of vehicles like slow moving, light vehicles, commercial heavy vehicles with axle particulars, animal-driven cart, etc. Traffic survey is to be supervised closely by a Highway Engineer/Resident Engineer. Traffic count stations are to be selected properly so that traffic from different directions (up and down) are covered in counting. Pedestrians are also to be counted at major road intersections in urban areas so that design of intersections can be done correctly (with or without pedestrian subways). Traffic counting is to be done during 24 hours for a period of 7 days continuously. A number of persons including Asstt. Resident Engineer/Field Engineer are to be arranged in two shifts (i.e. 12 hours shift). 20% extra persons (enumerators) are to be considered. These enumerators may be arranged from local people and they are to be trained. Traffic counts are to be recorded on prescribed forms/register. Traffic (up and down directions) is to be counted on hourly basis. At road junctions vehicles going straight, right turn, left turn, are to be noted. For heavy vehicles (viz., truck, buses, trolley trucks), it is required to stop the same with the help of police and number of axles, axle load is to be checked at nearby "Dharam Kanta" (weighbridge). These axle/load of heavy vehicles are very important because design of crust thickness mainly depends on "axle particulars and axial load". Some vehicles are also to be stopped to know their original starting place and final destination place (i.e. "O-D" survey).

16.9.2 Proper planning is to be done for arranging required number of enumerators, and supervisors (viz., Field Engineer, ARE/RE). Accommodation is to be arranged near traffic counting stations and their food, vehicles etc. Sufficient money is to be obtained in advance from head office of consultant by the Team Leader/Resident Engineer for various expenditure like, Jeep, Diesel, food, wages for enumerators, accommodation and miscellaneous expenditure. All these traffic survey data are to be recorded on register in correct format as per IRC Code. Field Engineer and ARE/RE are required to sign on all pages with date. These field records are to be sent to head office of consultant. The Department may require these records later on. BOT contractors raise their observations on traffic survey report of consultant. Traffic data is crucial for BOT contractor as their loss/profit mainly depends on traffic data.

16.9.3 Processing for Design Work

After completing traffic survey at different count stations along highway the data is processed and analysed for design purpose. Number of various types of vehicles are converted in one common unit i.e. PCU (Passenger Car Unit). Various factors are available in the Code for each type of vehicle for converting in PCU. Vehicle count is multiplied by the factor for converting in the unit of PCU as equivalent. The statement of traffic load in PCU is prepared for different count stations. Total traffic load in PCU is stated on the statement.

16.9.4 Thereafter projected traffic volume in PCU is worked out for 15, 20, 25 and 30 years. Design period for the purpose of formation width should be taken comparatively for a higher period (viz. 20 to 30 years) depending on category of highway (i.e. national highway or State highway). Carriageway width depends on number of traffic lane :- single lane (3.5 metres), intermediate lane (5.5 m), two lane without kerb stone (7 m), multi-lane width per lane – 3.5 m. For the purpose of carriageway width and designing crust thickness designed period of 15 to 20 years may be considered.

16.9.5 It is also required to work out traffic volume in the unit of MSA (Million Standard Axle). Crust thickness of road depends on this value and CBR value of sub-grade. MSA is worked out as per formula given in the IRC Code (No. 37-2001). There is a complicated procedure for working out of annual growth of traffic. However, to simplify annual growth is taken as 7.5%. Construction period (3 to 4 years) is added to projection period. Correct lane distribution factor and vehicle damage factor are to be taken from IRC Code. Total pavement thickness is found out from the graph sheet available in the Code No. IRC:37-2001 on the basis of traffic volume in MSA and CBR value of sub-grade. Different courses of required thickness (WBM/WMM, Bituminous Macadam/Dense BM and Bituminous Concrete Wearing Course) are then worked out. There are two types of CBR value i.e. (i) Soaked in water for 4 days prior to testing the cube in the instrument and (ii) unsoaked CBR value. Generally soaked CBR value is used in the design.

16.10 Inventory and Condition Survey of Existing Road and Culverts and Bridges

(a) A team consisting of Field Engineer, Asstt. Resident Engineer and 4 labourers alongwith a jeep, pick axe etc. is to carry out detailed inventory and condition survey of road including culverts and bridges. For bridges the bridge engineer is required to check the entire bridge structure including condition of foundation. For this purpose, it is required to arrange necessary facilities (like, ladder/stage by erecting scaffolding below bridge, certain accessories). Existing cushion of RCC pipe is also to be checked as generally it is less than the permissible (i.e. 0.6 metre excluding thickness of road crust). Photographs are also to be taken.

(b) If existing road crosses the railway track (or another road) through under bridge then existing clear head, width between two abutments of the bridge are to be measured and to be examined in respect of availability of minimum clear headway as per Code and feasibility of widening of existing road etc. If problem exists then it may be required to construct bye-pass.

(c) It is very important to ascertain the high flood level at all bridges and culverts. This should be ascertained from PWD/ NHAI/Irrigation Department. Enquiry should also be done from local people. Due to unreliable information on above matter the embankment of highway, level of bridge, culverts were kept on lower side in one project and serious situation (i.e. flooding over the road pavement and bridges) was faced. Decision about raising of existing height of embankment of road and bridges will depend on correct data of high-flood level. Discharges under the bridges and culverts are also worked out at design stage on the basis of "Topo Maps" of Survey of India of appropriate scale and highest rainfall data which is available in the Meteriological Department. If any major fault is detected about HFL later on then considerable change in design of road embankment, bridge and culvert design will have to be made. Hence, special care is to be taken on abovestated matter/data collection on the basis of past history.

(d) It is essential to ascertain the CBR value of existing sub-grade and other technical details of road crust condition. Pits (size 1 metre x 1 metre x depth at 500 mm below sub-grade level) are to be excavated at edge of road pavement i.e. 0.5 metre on the pavement and 0.5 m on the shoulder. Material samples (viz., sub-grade earth, stone aggregate, bituminous layer) are collected in bags and slip containing location (in chainage) is to be put in the sample bag. Such pits are to be made at certain interval (one kilometre or less or more as per uniformity of road reaches). CBR tests and other test of these samples are carried out in the field laboratory set up by the consultant. It has been found that problem is faced due to non-availability of good Asstt. Material Engineer who knows how to carry out various tests; before starting the field work it is essential to arrange an engineer who possesses good knowledge for carrying out testing.

16.11 Benkelman Beam Deflection (BBD) Test

This test is carried out on flexible pavement (bituminous) to ascertain the requirement of strengthening the existing pavement/ crust. Pavement deflection is measured by the Benkelman Beam. To get correct result it is required to ensure that the beam is placed on a "hard level ground". Road reaches are to be identified on uniform condition considering classification of :-

(i) Good
(ii) Fair
(iii) Poor

These are recorded on the register of condition survey. Project road is divided in "homogenous sections". Design of road pavement/crust is made in respect of various sections/ reaches of highway. These are to be identified clearly and recorded in the DPR and layout plan indicating chainage. BBD test is mainly relevant for designing "overlay" with bituminous layer (i.e. where reconstruction of existing road is not required). Recommended minimum overlay thickness is 50 mm. Bituminous Macadam with an additional wearing course of 40 mm BC. Before doing overlay, the old surface shall be corrected and brought to "proper profile" by filling the cracks, pot holes, ruts, etc.

16.12 Preliminary Project Report

16.12.1 After carrying out field works as stated on foregoing paras the Preliminary Project Report is prepared. Layout Plan and Long section of entire length of highway project are made part of PPR. Existing Right of Way is also shown on the layout plan. Existing curves with radius etc. are also shown on the layout plan. Statements of various surveys (viz., condition surveys of road, culverts, bridges) and test reports of various materials, report on sources of construction materials (viz. stone aggregates, sand, moorum, select soil, RCC pipes with location of manufacturing factory, water, etc.) are made as part of PPR.

16.12.2 During carrying out condition survey of existing road alignment and bridges it is required to identify the following important deficiencies :-

(a) Certain bridges are submersible type. The deck level is quite lower. Due to this reason gradient of approach roads of the bridge is very steep which is not permissible as per Codal requirement. For important highway it may not be required to construct submersible bridge. In such a position it may be advisable to raise the deck level of existing bridge and to re-grade the existing steep gradient of approach roads. Such type of reaches of highway including bridges are to be designed suitably.

(b) ***Sight Distance*** – The sight distance is very important along Highways particularly at horizontal curves and vertical curves (i.e. crest) on hilly/ mountainous roads. Visibility is an important requirement for the safety of travel on highways. Existing structures, trees on curves, sharp curves create problems. Vertical curves (crest type) are designed suitably so as to provide sufficient sight distance for "safe stopping". It may be required to improve the alignment of road for better visibility and to increase sight distance. Sharp horizontal curves are to be improved by making appropriate larger radius. It may be required to check/carry out survey for improving certain small reaches (may be 2 or 3 or 5 kilometres) of existing road alignment and to change the alignment. Sharp curves/steep gradient and zig-zag alignment can be improved by changing alignment.

(c) ***Super-elevation***

During condition survey of road it is also required to identify the curves where super-elevation is very less than the requirement. The same is to be improved by designing properly.

16.13 Capacity/Roadway Width

Right of Way and Formation Width depend on traffic volume considering various types of vehicles in the unit of PCU. Design period is to be decided carefully. It is required to consider diverted traffic (plus or minus) due to development of other roads in future. Floating traffic as extra on account of special festivals is also to be added. This is to be assessed as per relevant IRC Code (IRC:73-1980). For deciding Right of Way and Formation Width (i.e. Roadway width) it is advisable to consider higher design period (viz., 25 to 30 years). Lesser design period (viz., 15 to 25 years) may be considered for deciding carriageway width (i.e. number of traffic lanes). "Design traffic" (i.e. projected traffic) is worked out as per IRC Code. Construction period is to be added with abovestated projected period to get design traffic.

16.14 Shoulder

From technical point of view the shoulder is very important. But generally, this is neglected during designing and deciding specifications. Width of each side shoulder is 2 m to 2.5 m. Edges of carriageway is protected by the shoulders. Run-off water often creates furrows in soft shoulders. Bullock carts damage the soft shoulders badly and the same is a hazard. Due to this reason, vehicle is parked on the carriageway during sudden breakdown. As per NHAI half width of shoulder should be constructed with WBM/WMM base course and one layer of bituminous layer. Edge of carriageway is to be marked with paint.

16.15 Gaps on Central Verge/Median

Necessary gaps (cuts) are kept on the central verge/median after proper study at site i.e. to know vehicle travel routes from adjacent villages and towns. Number of cuts/gaps should be less as far as possible (not less than a 2 km interval preferably it should be 3 or 4 km). But local people insist on lesser intervals.

16.16 Preliminary Project Report (PPR)

(a) On the basis of abovestated field works and completing certain items of DPR the Preliminary Project Report is prepared alongwith certain drawings, statements/lists of various matters like, material resource survey/quarry locations along road lay out plan in strip form, list of bus stops, parking locations, trees to be removed, preliminary project estimate, environmental impact assessment, rehabilitation/re-settlement matter, inventory and condition survey, CBR values of materials/earth and test reports of various materials, water, traffic survey data, BBD test report, identification of homogenous sections (reaches) of existing road, toll plaza, safety measures, etc. Summary report of PPR is important and to be prepared carefully by senior engineer. The PPR should be submitted in time as per bar chart to the client (Department).

(b) After receiving observation from the Department on the PPR, it is required to comply the same under supervision of senior management.

16.17 Design of Highway, Culverts and Bridges

As per abovestated position and relevant IRC Codes and Highway design book, various design works are carried out. A few practical aspects are stated below:

16.17.1 Final layout plan of highway, roadway width (formation width), carriageway width, widening of road, designing/improvement of existing sharp curves (horizontal and vertical) alongwith super-elevation, designing sight distance, improvement of existing steep gradient, road crust design, surface drainage system along and around the road including catch water drain in hilly areas, sub-soil/road crust drainage by laying drainage layer with granular material, identifying various sections (reaches) of highway for widening, overlay work, reconstruction of existing road, deciding alignment with other design matter for new bye-pass road, designing slushy reaches of road by special treatment (viz., CC/RCC pavement with proper drainage system), design of culverts and bridges, designing reconstruction and additional culverts, cause-ways and bridges etc.

16.17.2 Special care is to be taken to design bridges (reconstruction/ raising existing submersible/low-level bridge/new bridge) by engaging reputed experienced bridge designer. At this stage, General Arrangement Drawings are prepared. However, detailed working design and drawings of bridges are to be prepared by the contractor at construction stage and get it checked from the Department who will engage proof consultant at construction stage. Design calculation sheets, drawings in respect of above stated components of highway are to be finalized and the same will be part of DPR. Drawing folders are prepared. Final designed cross-sections of road are prepared at 50 metres interval which are a huge number of drawings. Layout plans and long-sections of road are prepared at appropriate scale kilometre-wise; there may be one or two kilometres length of road on one drawing sheet.

16.18 Land Acquisition

The matter of land acquisition is a very lengthy process. It involves a lot of field works; number of site inspections are required for sorting several issues. Land areas which are to be acquired are to be marked on the finally designed road layout plans and the same are to be coloured. Type/category of land/land use of the same are to be ascertained from concerned Departments (viz., Revenue Department, PWD, etc.). Revenue maps are to be checked. Plot numbers, khasra number, name of land owners, etc. are to be ascertained. Final layout plans of highway indication land to be acquired are to be superimposed on the revenue maps. There are 3 or 4 prescribed formats for preparing statements of land acquisition. If there are any structure, trees, crops, etc. then details of same are to be surveyed and marked on larger scaled drawings. Detailed report on the same is to be prepared. Matter of rehabilitation/re-settlement is to be examined in details and report in this regard is to be prepared. How to rehabilitate/re-settle affected persons, structures, owners of agricultural land? These problems are to be discussed with concerned persons including owners. To carry out abovestated job it is required to engage retired persons of Revenue

Department (viz., Tehsildar/Kanungo/Patwari). Proposed lands which are to be acquired are to be marked accurately on the updated revenue maps with the help of officials of Revenue Department/ aforesaid retired persons. Thereafter, the same are to be marked on ground. Actual land use as per site is to be checked and compared with land use as per records of Government Department. Categories of land like, Government land/forest land/barren land, agricultural land, green land, etc. are to be ascertained and to be stated. Area of land (category-wise) is to be worked out accurately. Necessary letters for land acquisition with supporting document to be issued by the client (Department) to concerned authorities are to be prepared with the help of consultant. Both the Client/Department and consultant are required to pursue the matter through letters and meeting/joint inspections. A lot of time is required for reconciliation of various issues with all concerned persons. Cost of land acquisition, re-settlement/rehabilitation, compensation is to be worked out and to be added in the cost estimate of project. Allotment letters for required lands are to be issued to the road development department by the Revenue Department. Orders are also to be issued by concerned authority in respect of compensation/cost of land to be paid to the owners (private/Government Department).

16.19 Preparation of Draft DPR

16.19.1 This is towards the end of finalization of DPR. Some practical aspects and problems of DPR preparation have been dealt from management points of view. There are many other aspects, items which are to be dealt for preparing DPR as per requirements of IRC Codes and books published by the Ministry of Surface Transport and Highways, Government of India. Guidelines issued by NHAI, Road Development Organisations of State Government are also to be followed. The draft DPR consists of a number of volumes (may be 5 or 6 in case of State Highways and 10 to 12 volumes in respect of National Highways) including drawings and tender documents. Compliance reports on the observations of Department given on the Feasibility Report and PPR are to be brought out clearly in the draft DPR.

16.19.2 Environmental Aspects

There is one important part of DPR; the same is on the environmental assessment impact and rehabilitation/re-settlement and proposed mitigation measures. DPR will not be approved by the competent authority without proper report on this important aspect. Environmental screenings (survey) along entire length are carried out and recorded in the register kilometer-wise. Strips of 500 metres to 1000 metres on both sides of centre line of road are screened (particularly on the Right of Way). All trees within proposed Right of Way are to be marked on the road lay out plan. Water bodies (pond, lake, rivers, streams, nullahs, wells, waterfall, etc.) are to be marked. Particulars of underground water (viz., level/depth during summer season and quality etc.) are to be ascertained. Impact during construction particularly due to dust nuisance is to be stated. Mitigation measures are also to be proposed. Rainfall data, temperature during different seasons are to be ascertained and to be stated in the DPR. Demographic details (population, literacy rate, various castes, etc.) are to be collected from district administrative offices. The "Socio-Economic Screenings" are also to be done and stated in the DPR.

16.19.3 Tender Documents for Construction Work

(a) Construction may be done on BOT Contract basis or on joint venture basis i.e. Department make part of payment (viz., 40% to 60%) as grant/subsidy to the contractor. If there is no response from the contractor on BOT or BOT concessionaire basis then item rate tender may be invited from the pre-qualified contractors as done for various works (which have already been dealt in earlier Chapters). Tender is to be invited "package" wise. If the quantum cost of project is high or length of highway is considerably long then the project length may be divided in two or three parts (i.e. packages). DPR is also to be prepared package-wise (indicating chainage) although most of the matters will be same in the DPR of various packages. Tender documents are also to be prepared package-wise. DPR is also part of tender document. Tender documents for inviting on BOT basis are of different/special type which will be dealt in another book. Financial and Economic analysis are to be got prepared through an experienced expert. This is a voluminous record prepared through particular software programme. This record is important particularly for BOT contractors for ascertaining financial/ economical viability.

(b) In case of BOT contract it is essential to state toll charges for different type of vehicles which are to be approved by the top administrative authority/Ministry of State

Government. Toll period is also to be stipulated in the tender documents and the same can be negotiated at appropriate stage. It is difficult to assess/calculate about reasonable toll period because traffic growth during toll period (15 to 20 years or more) may be more or less.

(c) Cost estimate of each package (if there is more than one package) is to be prepared. Generally, schedule of Rates of State PWD or Bridge Department/Highway/ Road Development organization is adopted. Decision in this regard is to be obtained from the concerned Chief Engineer of the Client/Department.

16.19.4 Crossing Rail Track and Congested Road of Urban Areas

Highway crosses existing rail tracks at number of locations. Some crossings may be at same or with higher level; at such position generally "level crossing" type is adopted and road is to be designed accordingly. In case rail track is on lower level and there is heavy road traffic then Road Over Bridge (ROB) may be proposed and design is to be done accordingly. In case rail track level is considerably higher than road level then road under bridge is constructed with necessary clear head for road traffic. For above stated proposed works it is required to obtain permission from railway authority. Railway authority may execute certain part of such works. In case highway passes through congested urban areas (large towns) then it may be required to construct "Flyover" for uninterrupted highway traffic and safety of pedestrians and slow moving vehicles of town. Necessary permission is to be obtained from PWD/ Municipal authority for constructing "Flyover". Installing traffic signal at the crossing with highway is generally not permissible. Detailed study/technical investigations are to be carried out for these major projects and design is to be got done through expert traffic engineer/bridge designer.

16.19.5 After considering above stated items/components the draft DPR is prepared in different volumes package wise and submitted to the Department. After receiving observations from the Department, the draft DPR is modified suitably and final DPR is submitted to the Department.

16.19.6 Monitoring and Reasons for Delay

(a) It has been found that delay occurs in completing the

preparation of DPR. It is essential to monitor the progress of various activities with reference to Bar Chart programme (vide figures 16.1 and 16.2). This should be done weekly in the head office of Consultant alongwith Team Leader and other staff. Monitoring meeting should also be done in the office of Chief Engineer/G.M. of the department monthly alongwith Team Leader of Consultant. Co-ordination with various agencies, authorities is also carried out during these meetings. Pending decisions are to be sorted out during these meetings. Quality of various activities is also to be reviewed/checked. Reasons for delay in respect of various activities are to be found out and the bottlenecks are to be removed on priority basis. If required modified Bar Chart programme is to be prepared. Actual progress (physical and financial) is to be marked on the Bar Chart and monitoring is done on the basis of the same.

(b) *Reasons for delay*

There are various reasons for delay in preparation of DPR; a few of these are stated here. Generally major part of delay from the side of consultant. Progress of field activities is hampered due to shortage of engineering staff and delayed in fixing of agencies for field survey, bridge works (viz. condition survey, decisions on proposed bridge works and its design including sub-soil investigation through boring). Field engineers are unable to engage vehicle (Jeep, Motor-cycles) and labourers due to shortage of money which is to be sent by head office of Consultant in time. Various activities are not carried out simultaneously in field agencies' offices and head office of consultant. Progress of testing of materials, BBD tests etc. is delayed because field laboratory is not equipped properly with instruments and experienced Asst. Material Engineer. Experienced and efficient engineers leave their jobs frequently due to bad service conditions and inadequate salary. Deployment of inhouse staff is very less; even Team Leader, Highway Engineer, R.E. are not on the regular roll of the consultanting firm in many cases. There is also delay from the side of Client (Department). Decisions on issues and intermediate reports (viz. Feasibility Study Report, PPR, draft DPR etc.) submitted by the consultants are not communicated by department in time. Delay occurs for making payment to the consultant.

CONSTRUCTION STAGE

16.20 Tender

16.20.1 Government Department is to take necessary action for obtaining administrative approval of project estimate from the competent authority on the basis of DPR. The DPR submitted by the consultant is to be technically sanctioned by the Chief Engineer. Then it is required to invite tender for project construction package wise. It is required to examine the possibility of awarding the construction work on BOT basis (without grant/subsidy). If there is no possibility then it is examined for awarding the work on BOT with subsidy/grant basis (i.e. joint venture). Tender documents (on BOT with subsidy/grant) be checked finally by the Accounts Branch and GM/SE of the Department. Tender notices be issued as per normal procedure on all India basis. Applications for pre-qualification of contractors (i.e. concessionaire) are to be received at first stage as per PQ documents. On receipt of applications/technical documents the same are scrutinized/verified by the Department. Then technical committee considers the applications, particulars and check the various criteria for pre-qualification. The contractors who fulfill various criteria are pre-qualified. Thereafter, tender documents are sold to the pre-qualified contractors. On receipt of tenders the same are processed/scrutinized by different branches and placed before the "Tender Committee" who considers the same. Various conditions given by the contractors are tabulated. Statement and comments of the Department are given against each conditions. The contractor who has submitted "best offer" to execute the project on BOT/BOT with subsidy/grant basis is identified. Thereafter negotiation is conducted with the "best offerer". Various terms and conditions are sorted out and final negotiated offer is obtained. Letter of acceptance/ Work Order is issued to the contractor/concessionaire and agreement is also signed by both the parties.

16.20.2 In case there is no response from BOT contractor (with or without subsidy/grant) then approval is to be obtained from the competent authority to execute work from Government fund and then to invite item rate tender from pre-qualified contractors. BOQ/ Schedule of Quantities of items of works as already prepared by the consultant with DPR is finally checked by Manager (EE)/ GM(SE). Entire tender documents for such type of tender are checked by EE/SE and Accounts Branch. Thereafter normal laid down procedure is to be followed and the work is awarded to the lowest contractor.

16.21 Engaging Supervision and Quality Control (SQC) Consultant

(a) *Scope of Services of Consultant*

Generally, Department (State Highway or National Highway) engages the SQC Consultant for management of project construction including supervision of construction works, its quality control. Measurement of works are also recorded on Measurement Books by the consultant jointly with the contractor/concessionaire. Bills submitted by the contractor/ concessionaire are checked by the consultant and submitted to the Department with recommendation. Consultant will ensure that work is executed as per specifications and proper quality. For this purpose effective system of collecting random samples independently and tested independently. Test reports and registers are to be maintained by the consultant. Testing of materials carried out by the contractor is also to be supervised by the consultant. Various field tests and quality of finished items are also to be carried out. Checking of undulation of road surface is most important; this is to be done thoroughly with string and Bump Integrator/Rugho Meter.

(b) Bids are invited from the consultants for the purpose of Supervision and Quality Control in three envelopes i.e. (i) Envelope I – Earnest Money, (ii) Envelope II – Technical Bid for pre-qualification to check eligibility and (iii) Envelope III – Financial Bid (generally in percentage of cost /actual cost). After receiving bids the same are opened as per laid down procedure and lowest bidder is engaged by issuing letter of acceptance/Work Order. Agreement is also signed.

(c) It may be stated here that although stipulated period (i.e. date of start and date of completion) is given in the agreement entered with the consultant yet practical problem is faced when contractor delays the construction work considerably. Consultants face problem to engage engineers and to maintain vehicles, office etc. without extra fees during long extended period of construction. Fees are paid on the total cost of work done i.e. as per final bill of contractor. Final bill of contractor is also to be checked and recommended by the consultant.

16.22 Construction Activities

16.22.1 After awarding the construction work to the contractor/ concessionaire and engaging SQC (Supervision and Quality Control Consultant), the construction activities at site start in full swing. Mobilisation activities start at site from the side of contractor and SQC Consultant. Considerable mobilization works are to be done by the contractor; due to this reason mobilization period of 2 to 3 months is stipulated in the agreement of contractor/concessionaire. Certain planning work is also to be done by the contractor. Generally Bar Chart programme is prepared. Details of planning to be done by the contractor has been dealt under Chapters 6, 7 and 9 CPM network planning is much better than bar chart programme. However, in practice, contractor adopts bar chart planning in highway/road works.

16.22.2 As first step of construction the mobilization and planning of construction activities are to be carried out by the contractor. Entire length of highway project as stipulated in the agreement is divided in 2 or 3 or 4 sections (reaches) as working arrangement. These sections have no link with sections indicated in the DPR. Contractor divides the entire length of road for deploying his resources and to carry out construction activities in different sections/ reaches simultaneously. Dividing in sections depends on certain working arrangement of contractor like, length of highway, quantum/ amount of work, conditions/ problems along certain reaches, availability of resources, etc. Activities of one section are repeated on other sections. Manpower, construction equipments/machinery/ plants, material collections, laboratory setting etc. are done section-wise. Specific numbers (like, Section 1, 2, 3 etc.) of sections are given for the purpose of carrying out planning, preparing bar chart/ CPM network, deploying manpower, equipments/machinery etc. Progress reports are also prepared section-wise and the same are compiled and submitted to higher authorities. Site offices with parking space, laboratory, Hot Mix Plant, Stores etc. are set out at each section. Communication facilities (viz., mobile phone/wireless system) for entire project road is essential. Quarry survey report stating sources of various materials (stone, sand, moorum, select soil, etc.) is available in the DPR; the same is checked by the contractor. Accordingly, suitable location/site is identified and crushers are set up for producing stone aggregates of different sizes and stone dust. Travel distances for bringing stone boulders and carrying stone aggregates, stone dust at different locations/ construction sites are considered during planning and installing crusher machinery for different sections. Similarly, locations for installing Hot Mix Plants for bituminous works are also selected and installed. Laboratories for different sections are installed. Travel distance for construction activities in respect of different sections is important for increasing efficiency of manpower, machinery and to achieve economy. Setting up of field workshops for repairing and maintenance of machinery, vehicles is most important so that the same are in working order and it works efficiently and economically. Planning is to be done by the contractor in such a way so that machinery, vehicles, road rollers, hot mix plants, crushers, etc. do not remain idle.

16.22.3 On the basis of CPM/Bar Chart programme (with calendar dates) the requirement of manpower viz. skilled and unskilled workers, efficient operators, mechanics, supervisors/ engineers, etc. and vehicles and equipments, machinery are worked out by the contractor. Proper levelling/balancing of resources are carried out by modifying certain activities. Contractor is required to arrange adequate fund (month-wise) as per works programme in advance so that work progress is not hampered due to inadequate "cash flow".

16.22.4 Sequence of Construction Activities

(a) It is essential for contractor to plan proper sequence of major activities. If there is provision in the contract agreement for preparing working detailed structural design and drawings of construction of culverts, bridges from the end of contractor then competent Bridge Designer (Consultant) is to be engaged on top priority basis. These design and drawings are to be submitted to the Department for approval. Proper coordination and regular monitoring for aforesaid important matter are essential at top management levels of Department, SQC Consultant and contractor. Planning is to be done to take up construction of culverts and bridges "at beginning of project construction" activities. But unfortunately contractor starts the work of formation and road crust first and achieve more progress on this part of project. Subsequently, technical faults are found after constructing culverts and bridges. Defective joints along culvert/bridge abutment walls and road crust are found and it is difficult to rectify the same properly.

(b) Another technical problem is faced. As per DPR it is required to construct RCC slab culverts at certain locations (where necessary cushion is not available for RCC Pipe culvert). But contractor is interested to construct RCC Pipe culverts without providing minimum cushion of 0.6 metre (excluding thickness of road crust). This is done because extra time is required for RCC slab culverts and contractor earns less profit. Contractor may provide aforesaid cushion over pipe by filling earth over top of pipe and making hump on road surface. This is not acceptable at all. Such type of culvert work is to be rejected. Minimum cushion is essential from structural point of view.

Sequences of various activities can be planned and shown effectively in the CPM network diagram; if this is followed then many technical problems and defects can be avoided.

16.22.5 Availability of DPR, various drawings and contract agreement in field offices and certain site problems

(a) As part of mobilization it is essential that abovestated documents must be available at all field offices (section-wise) of contractor and SQC Consultant. But unfortunately, these important records are not generally available in various field offices particularly section offices. How site engineers of contractor and SQC Consultant will ensure correctness of detailed construction, quality of works as per specifications, correct payment to the contractor without above stated important records ? Actually, due to above stated reason, serious mistake was done for laying bituminous courses in one project and huge overpayment was done to the contractor. It may be stated here again that DPR is part of contract documents and the same must be implemented. If any deviation/change is required then prior approval is to be obtained from C.E./C.P.M.

(b) It has been found that RCC pipes are procured from manufacturers where there is no arrangement for three-edge bearing test and water pressure test at factory. It is not possible for contractor to arrange these larger size testing equipments. Hence, RCC pipes are not to be procured from such manufacturers where tests can not be carried out by the Consultant/ Department. Test checking should be done even if the pipes are ISI Marked.

(c) It has been found that as per DPR it is required to re-construct certain unsafe old bridges but the concessionaire/BOT contractor does not re-construct the same and open the highway for traffic so that he can start collection of toll money early. Such a practice is not permissible.

(d) An example over payment of huge amount (in terms of crores for project value of Rs.200 crores) is given here. As per DPR provision a layer of extra sub-base is to be laid above the sub-grade and below granular drainage sub-base layer. This was required to raise the sub-grade level with selected soil of CBR value of 5% (minimum). This layer was laid with granular material (like moorum/GSB) because such granular material was readily available at nearby borrow area at cheaper rate. Contractor was paid at the rate of GSB item which is much higher than the item of selected soil of CBR value of 5%.

(e) ***Super-elevation***

It has been found at site that there is gross negligence in providing correct super-elevation at horizontal curves in many State Highways. Even templates for checking super-elevation are not available at sites. This is highly objectionable. Super-elevation is designed for all curves and written on the layout plans of highway. Generally, super-elevation is between 4% to 7%. Super-elevation is to be reduced gradually through transition curves on both ends of the horizontal curve. It is essential to construct sub-grade, GSB, Base Course (WBM/WMM) and Bituminous course with required quantum of super-elevation. If the same is not done at sub-grade/base course then work should be stopped.

(f) ***Undulation of road surface***

"Public is mainly concerned for good/smooth riding quality". It is essential to check the undulations on finished road surface on cross and longitudinal direction of carriageway. There should be no bump on road surface due to hump, joint at slab culverts, bridges, expansion joints of bridges, junction with road embankment and abutments of bridges/culverts etc. Finished bituminous surface is to be checked with straight edge and also with 100 m string on longitudinal direction during laying mix with Paver Finisher. Roughometer/Bump Indicator is also to be used for checking surface undulation on longitudinal direction lane-wise. Necessary correction is to be done during laying operation.

16.23 Road Safety Audit

16.23.1 Certain practical problems faced during field works, design works for preparing DPR and during construction works have been dealt in previous paras. There is another category of problems i.e. "safety for road users". Accidents occur frequently on the roads for various reasons. Pedestrians are most vulnerable road users. Number of deaths and injuries in respect of cycles and other two wheeler motorized vehicles are also high. Hence, it is essential to take special care for ensuring safety of these road users. For this purpose a new system, "Road Safety Audit" is being introduced in India. A lot of importance is being given in developed countries for safety of road users particularly vulnerable road users. Highways pass through both urban and rural areas. Problems in respect of urban areas (particularly in city/town) are more serious than rural areas. Road Safety Audit should be carried out by independent unit in respect of (i) existing road and (ii) also during various stages of field works/surveys, "design works of DPR". Tender should be called only after completion of Safety Audit process of DPR including sorting out recommendations/observations stated in the Safety Audit Report. Structural design, quality checking during construction works do not fall under scope of safety audit.

16.23.2 A few examples are stated here. In urban areas, the pedestrians face many problems about their safety on roads. There is a lot of conflict between pedestrians and vehicular traffic. Condition of footpaths/walkways/road berms/shoulders of highway is generally bad/broken. Pits, trenches are excavated very frequently for various utilities but are not restored properly. Many obstructions like malba, garbage, protruded sluice valve chambers, manholes, pillars of utilities, stagnated water, etc. are found on the footpath. Due to these reasons, pedestrians walk on the carriageway and the same is unsafe; accidents occur due to this reason. It is also difficult to drive vehicle due to such pedestrians. In congested areas railing should be provided along footpath and central verge for their safety and also for smooth driving of vehicle. It may be required to construct RCC toe wall type along central verge (keeping height of 40 cm to 70 cm) for ensuring that pedestrians do not cross the central verge and vehicles can not cross the central median by accident or by violating traffic rules. Accidents also occur due to lack of planning/designing of parking areas. Serious shortage of parking space is found in the urban areas and due to this reason vehicles are parked in haphazard manner and it becomes difficult for driving vehicle and walking/crossing pedestrians. Accidents also occur due to this reason.

16.23.3 Another important factor is visibility/sight distance for drivers. Many obstructions like sharp curves, blind corners, structures, trees and various poles (electric, sign boards, etc. hoardings) are found along the road and the same creates serious problems and accidents occur. Accidents also occur due to non-provision of proper super-elevation on the sharp curves. Poor lighting in urban areas also creates problems for pedestrians and cyclists.

16.23.4 Segregation of different category/Type of traffic viz., pedestrians, cycles, two wheeler motor vehicles, heavy vehicles, light vehicles, slow moving conveyances, hand carts, cycle rickshaws, etc. should be done as far as possible by road marking, kerb stone, railing, service road, etc. Service road can also be used as a cycle track and for pedestrians; it may also be possible for parking vehicles at certain locations. If speed breakers are not constructed as per standard dimensions then the same becomes cause of accidents/ injuries.

16.23.5 There is lot of importance of road shoulders of highways in rural areas. These shoulders are not constructed properly and also they are not maintained on regular basis. A lot of damages/erosions are found on the shoulders. Due to these reasons accidents occur. Pedestrians are not safe while walking along shoulders. Vehicles slip out from roads due to non-existence/heavy damage of shoulders. Vehicles can not be parked temporarily on the shoulder during emergency when it becomes off-road due to sudden mechanical defects. When vehicle becomes off-road/out of order suddenly the driver is forced to park on part of carriageway for emergency repair work. Problems are faced by the road users due to inadequate numbers of destination signboards. Serious problems are faced due to less sight distance. Visibility problem is faced due to various obstructions, blind corners, trees, projected rocky strata, sharp horizontal bends, hairpin bends, vertical curve (crest type); the same is also cause of many accidents.

16.23.6 Another safety measure on highway is stated here. Lay-byes and bus stops allow vehicles to stop safely and with the minimum adverse effect on other traffic. This is best done with segregated areas joined to the main road only at the entry point and exit point. Vehicles then can stop off the main carriageway without interfering with other traffic and with less risk to passengers getting on or off.

During audit process the remedial measures for various safety problems are taken.

16.24 Monthly Progress Report

One important requirement is to prepare the monthly progress report (MPR) of construction of Highway Project by the SQC Consultant engaged by the Department. Basically it is the responsibility of the Team Leader of consultant for preparing proper MPR and to submit by 3rd of every month to the Chief Engineer of the Department. On the basis of this report, progress and various problems/hindrances are reviewed by various officers/Managers of the Department. This is also brought to the notice of Chairman/MD of the Department/Secretary of the Ministry with brief note. This is also placed quarterly/half-yearly during meeting of the Board of the Government organisation/ Department. Hence, the MPR should contain correct data and progress report. The MPR may contain pages 45 to 60 alongwith Bar Charts, manpower deployment, machinery/equipments, abstract of test reports, physical and financial progress of major items of works, consultant's personnel and their activities etc. As for example, copy of Index/Contents of a MPR is enclosed vide Annexure II. Copies of Physical Progress (S-curve), Progress of major works (Histograms) and Physical Progress Chart/Bar Chart of road crust works are given vide figures 16.3, 16.4 and 16.5.

S-CURVE

Physical Progress (Up to 30th November 2002)

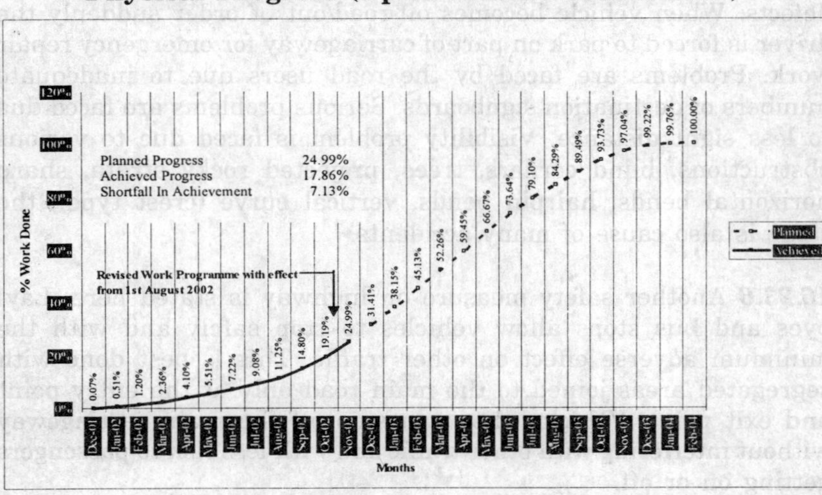

Fig. 16.3 : S-Curve Showing Physical Progress

Progress of Major Works : Histograms

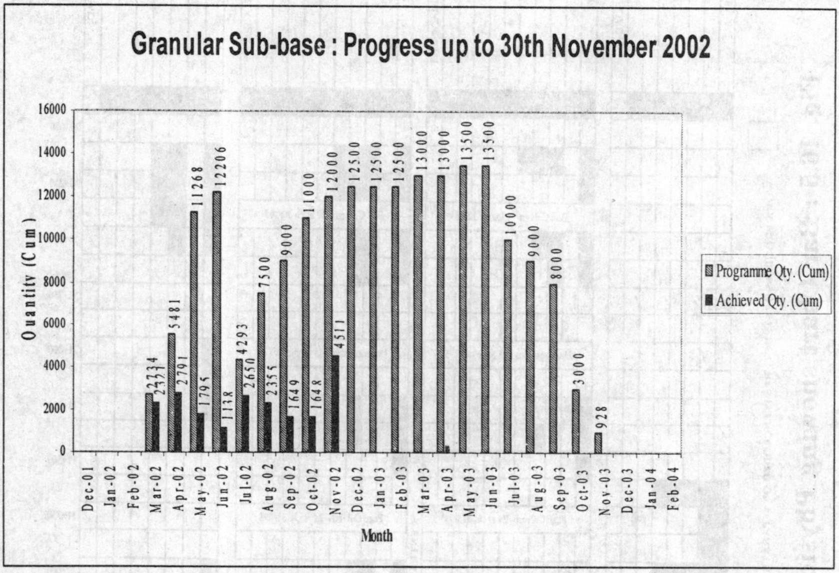

Fig. 16.4(a) : Histogram Showing Progress of Major Works (G.S.B.)

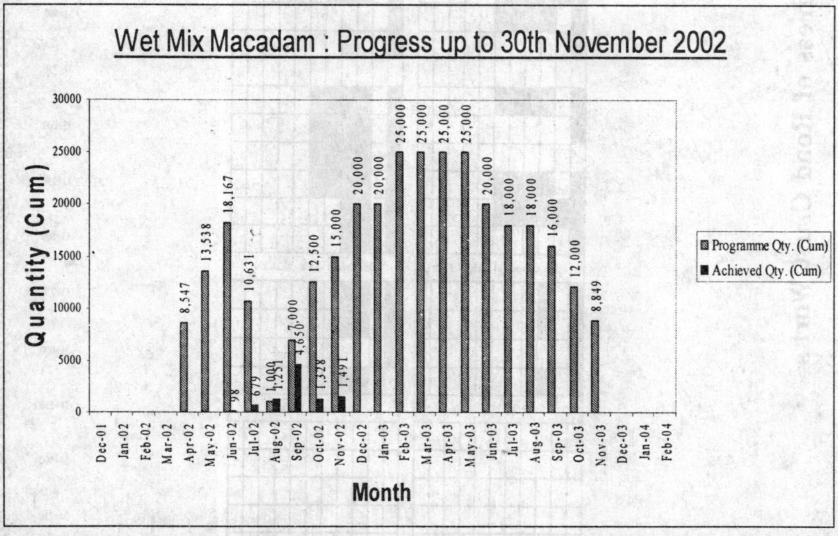

Fig. 16.4(b) : Histogram Showing Progress of Major Works (W.M.M.)

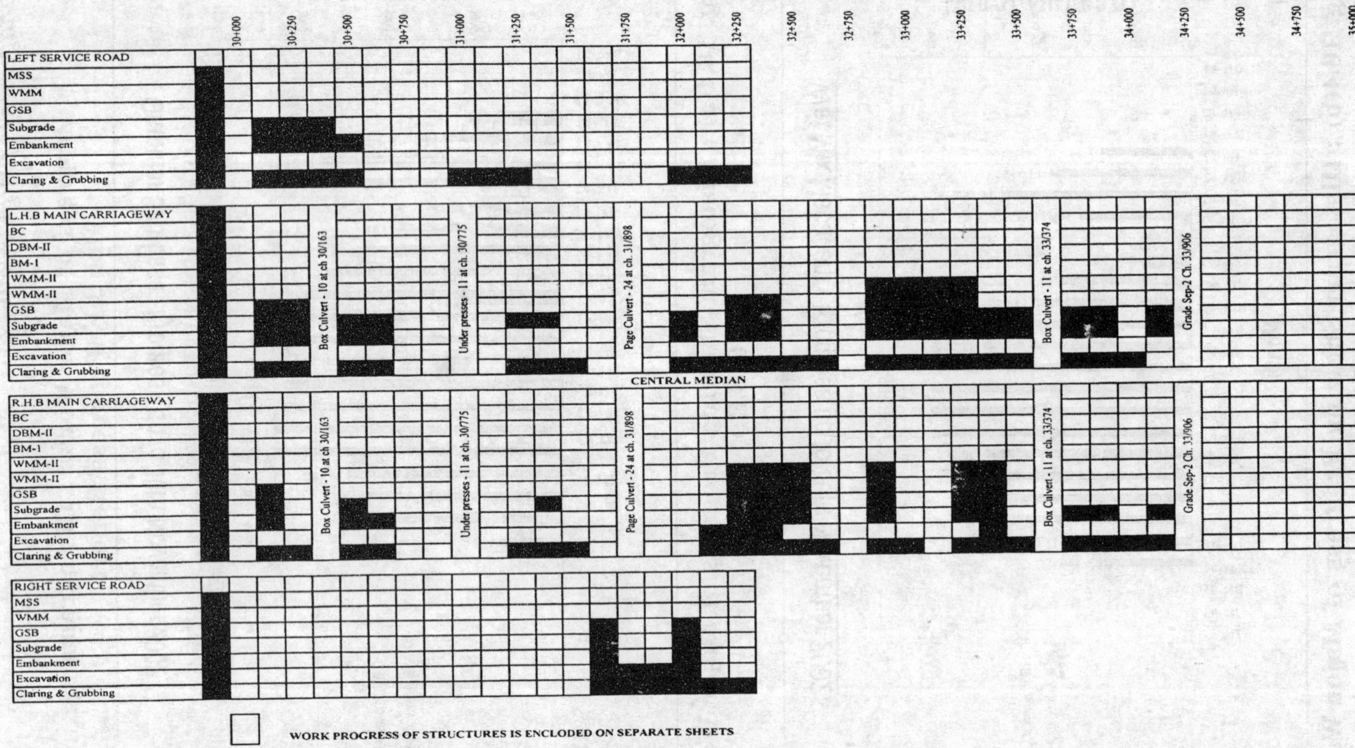

Fig. 16.5 : Bar Chart showing Physical Progress of Road Crust Works

Annexure - II

Monthly Progress Report

Name of Highway Project ___________

Progress Report for the month of _________

Contents

16.25 Important Points/Policy Matters on Highway Project Management

(1) As per study carried out by an authority on highway project construction during 1990-93 it was found that only 8% projects were completed in time. Considerable delay occurred due to non-finalisation of field investigation, design and drawings, awarding works to contractor, delay in constructions etc. It was also found that not only the contractors were at fault but the Departments were also at fault. Many changes in procedures are required.

(2) The project should be framed on the basis of accurate and reliable site investigations and data. As already stated in this Chapter, above depends on engaging a competent reliable reputed Consultant for preparing DPR. The designs and drawings should be complete and final and truthful to site conditions. Utmost care must be taken to prepare pre-qualified list of Consultants.

(3) The site should be handed over to the contractor free from all encumbrances. This implies that entire land acquisition, removal of services and encroachments and environmental clearances are got completed before awarding the work by the Department. Lot of problems are faced for these works and it takes considerable period. Hence special task force / team should be constituted.

(4) Posting policy of engineers should be streamlined so that the key staff is not disturbed or transferred half-way through. Senior engineers of Consultants should not be changed frequently without prior permission of department.

(5) The analysis of rates should be up-to-date and relevant to the technique of construction and use of equipments.

(6) Escalation clause should reflect the weightage assigned to labour, materials, POL on realistic basis.

(7) Mobilisation advance may be increased from 10% to 15%. To improve the "cash flow" of the contractor his running account and final payment should be made quickly.

(8) Decision making process should be improved and there should be sufficient delegation of authority to the site management.

(9) Contractor should engage competent experienced engineers at project site so that site management is handled efficiently. He should deploy necessary equipments and skilled operators to handle machines alongwith repair/ maintenance back-ups.

(10) Contractor should submit a "Method Statement" at the time of start of project explaining his scheme of execution and the type of equipment he wants to use for various activity.

(11) The "lowest acceptable" offer (tender) should be considered and not just be "lowest".

(12) The CPM chart, its frequent updation should be mandatory for important work like, bridges etc. A strong MIS (Management Information System) to anticipate bottlenecks, evolve solutions, monitoring of progress, monthly/ fortnightly site co-ordination/monitoring meetings should be adopted.

(13) Upgradation of road construction industry should take care of existing socio-economic conditions of the country.

(14) For large sized projects, efforts should be in the direction of optimum mechanization for construction works. Suitable status of industry to road construction should be given by the Government authority. Seed capital assistance to technical entrepreneurs could be effective in solving working capital problems.

(15) To meet the requirements of contractors, particularly small and medium, "equipment hire concept" needs to be encouraged. Independent equipment hire companies or subsidiaries of manufacturers should come up fast to fill up this gap.

(16) An integrated quality assurance system involving quality control, quality management should be evolved in the highway organization. This system should cover the activities from planning and design stage upto the time of maintenance facilities.

(17) There should be a high-level steering committee representing Central/State road construction Departments, contractors, consultants, equipment manufacturers that could consider broad issues from time to time.

(18) For reducing the environmental degradation due to road transport, necessary measures are to be identified at the planning stage to avoid future controversy during implementation of the project.

(19) Due to paucity of funds, possibility may be seen for giving road side land which is beyond the requirement on lease basis to the private participant for constructing highway on BOT basis.

(20) Traditional system of supervision by the Departmental organizations does not instill the confidence in contractor;

this is vital for successful execution of the projects. Supervision by independent consultants as adopted in developed countries should be encouraged. The attitudes of both the engineer and contractor require a lot of changes. FIDIC contract document which is acceptable to many foreign/western authorities should be adopted.

16.26 Application of CPM to Highway Project

(a) CPM network planning and programming of project constructions have been dealt under Chapter 6 and 7. In India this management technique is used in many large building projects. However, CPM network method is not generally applied in India in respect of Highway/Road project. Bar Chart is used for such works; because the same is simple and can be drawn quickly "kilometer-wise". Length of one highway project (package wise) may vary from 30 km to 300 km; project cost may vary from Rs.25 crores to Rs.600 crores or more. Charts (Bar Chart/Graphical Form/ CPM Network) are to be drawn kilometer-wise; hence, the same are to be prepared on many sheets to cover entire length of road (package/contract wise).

(b) But above stated practice i.e. not using CPM network techniques in highway project, is not desirable. There are many components of highway project construction, such as:-

Pre-construction stage –

Preliminary planning, Feasibility Study, preparation of Detailed Project Report (as dealt under this Chapter)

Construction stage –

Detailed working programme of various activities to be prepared by the contractor/ consultants in respect of various components of Highway viz., Formation, Road Crust, Cross-Drainage/culverts, Bridges, Flyover, Road Over Bridge/Under Bridge, Land Acquisition, Re-settlement, shifting service lines, etc.

There are a lot of inter-dependencies between many activities and components of project. Hence, proper planning/ programming can be done through CPM network. However, Bar Charts for certain matters can be prepared on the basis of CPM network diagram. Many components of highway projects are to be completed in time-bound manner. A network for many numbers of activities related to DPR should be prepared and monitoring can be done on the basis of same. Transportation Planning is quite intricate type and involves large number of activities. It is required to examine existing facilities, connectivity of different places, study of accident data, parking needs, formulation of future transportation facilities, etc.

(c) A lot of investigation/studies is required for bridge construction. This involves activities such as site selection, foundation exploration, analysis of hydraulic data, waterway fixation, selection of alignment etc. It is possible for proper control and monitoring over these inter-related activities through CPM network. CPM network is also useful for designing highway/carriageway, culverts and bridges.

Acquisition of Right of Way and land involves a number of stages; some of them are related to Land Acquisition Act. Critical activities on this matter should be given special attention.

(d) Application of CPM to highway projects results in many advantages over the conventional planning/reporting system. Some of these advantages are stated below :-

(1) It enables the engineers to make a logical programme with inter-dependence of various activities and restraints.

(2) It is easy to prepare various schedules viz., activity schedule, manpower schedule, materials, machinery, etc. alongwith calendar date/period.

(3) It is possible to know likely activities which may cause the trouble and delay in the project.

(4) Reporting system can be done correctly in time on the basis of CPM network.

(5) If there is delay or other problem during execution, the network helps for taking remedial measures in time.

(6) If required, it helps for changing original programme.

(7) It provides for easy and clear method of communication to others.

(8) It is possible to achieve time and money saving by applying CPM.

(9) Data available in CPM network can be useful for preparing suitable graph of total project cost, actual expenditure at different stages of project. Thus, it will be possible to know whether actual expenditure is likely to exceed the budgetary provision; and precautionary steps can be taken in advance.

17

Total Quality Management

17.0 Introduction

17.1 Conventional System of QC

17.1.1 In India, generally conventional "quality control" system is followed in Civil Engineering works. As per this practice, quality of works is controlled by supervising construction works, various services/utility installation works, works of other disciplines (viz. electrical, mechanical, electronics, lifts, A.C., fire prevention/ suppression works etc.) in accordance with design & drawings of architecture, structure and specifications as stipulated in the contract agreement. Samples of raw materials, mixed materials (viz. cement concrete, cement mortar at green stage, bituminous mix from laid courses and hot mix plant site etc.) are collected and tested in the laboratory as per procedure laid in the contract documents. Number of samples and tests depends on quantum of works of specific items. Quality of finished works is also checked like, undulation of finished road surface by certain tools/instruments, field density of road pavement for measuring compaction standard, proper finishing, levels/slopes in drainage works, floors of buildings, doors, windows etc. Generally, requirement of quality as laid down in the specifications in respect of certain items (like grade of G.I. pipes as "B") is given as minimum. But such minimum specification/grade may not be satisfactory in case of plumbing works in building constructions. It is advisable to use "C" grade GI pipes for building works and also for underground GI pipe work because life of underground G. I. pipe is less and it is difficult to repair G. I. pipes concealed in brickworks, RCC works.

17.1.2 Quality assurance/quality control of certain items of works depends on stipulation of proper specifications in the detailed estimate/NIT (tender documents). Finishing quality of certain items of works (viz. exposed concrete finishing on external walls, plaster/ finishing on surface of internal walls of buildings etc.) depends on selection of appropriate specifications by the architect. However, many manufactured materials (like wooden flush door shutters, flushing tanks for toilets, Cast Iron pipes, readymade tiles for flooring, wall cladding, RCC pipes etc.) are supplied by the vendors/ suppliers/manufacturers at project site. It may not, therefore, be possible to carry out certain tests in normal laboratory and to ascertain correctly about quality of these finished materials. Apparently, there is a gap/flaw in the conventional system/practice of quality control. More checks/safeguards are required for ensuring quality.

17.1.3 There is another example. If stone flooring (marble/granite) is to be constructed, then it is required to finalise/approve samples of slab/stone tiles. The colour, quality of stone will be as per this approved sample. Pattern/design of stone flooring will be as per detailed drawings. Thickness and size of stone slab/tile are to be stipulated in the nomenclature/specifications of items of the contract documents/NIT. Drawing showing the floor pattern and approved samples of stone slab/tile shall be part of item/specifications of NIT (tender document). Quality of stone cannot be controlled if approved samples are not made as part of tender document. Regarding ceramic/ vitreous tile flooring, it is required to state size of tile besides quality etc. in the nomenclature/specifications of the item. It is better to state names of a few reputed manufacturers in the item/ specifications.

17.1.4 Reliability of testing, laboratory, skilled operation of laboratory etc. will be dealt at the end of this chapter.

17.1.5 Quality has always been demanded by the customer/client. Likewise, every organization desires to churn out quality work. Modern methods of the advertisements have made the client aware of the availability of various new and aesthetic products. When he does not find the final product according to his conceptualization, he does not get satisfaction. Now a days, organizations are adopting some quality control measures like setting up separate quality control unit, testing laboratories, inspection system, management reports/reviews.

17.1.6 However, quality is not an absolute word but inter-related with "stated" and "implied needs". The client, therefore, must be very clear of stated and implied needs. Client/ users should be associated at the early stages of finalisation of specifications, detailed drawings and tender document prepared by the architect and project engineers of the organization.

17.2 Inspection Unit

17.2.1 Quality Control inspections should be done on regular basis by an independent Q.C. unit of the owner/government department. It means the Q.C. unit should not be under Project Manager. External/third party inspection should also be carried out occasionally by independent organization (government/private engineers of Consultant). For central government work, the Chief Technical Examiner's wing of Central Vigilance Commission carries out inspections of certain projects as per their policy/system.

17.2.2 These inspections should be carried out in planned way. Necessary tools, checking instruments etc. are to be carried by the inspecting engineers. Relevant records of the project constructions are also to be examined during inspections. Quality is checked on the basis of specifications, drawings, terms & conditions stipulated in the contract documents. It is important to collect sample of material correctly as per codes/ methods and to get it tested from approved and reliable laboratory. In some cases, like, RCC hume pipes, the material is required to be tested at the manufacturing factory as certain testing equipments are not available in other laboratories. A few sample test checking of materials which bear "ISI mark"/"DGS&D Inspection Stamp" should also be done by the inspection wing. Detailed lists indicating testing of materials and finished items of works, their permissible values and number of tests to be carried out for various materials/works are available in Q.C. Hand Book/Codes/BSI standards. Inspections by quality control units should make some positive contribution to achieve quality standards rather than just being endeavours for appraisal. After issuing inspection report containing observations, it is essential to review all the observations systematically with Project Manager and to obtain compliance report from him within a reasonable time. If there is delay in getting satisfactory compliance reports, Q.C. unit or external inspection organization should bring to the notice of senior management authority (viz. Chief Engineer/S.E.). Serious/ important observations should also be brought to the knowledge of top management/Chief Engineer by the inspection unit/wing for taking appropriate action.

17.3 Time Control

17.3.1 There is a certain relation between time and quality. Realistic completion period of project should be stipulated in the tender document so that proper quality can be achieved. If completion period is very less, then it may not be possible to achieve proper quality. On the other hand, if the work is prolonged too much, then contractor's tendency will remain to earn profit by using sub-standard materials and by deploying unskilled/semi-skilled workers for skilled jobs as the Escalation clause of the agreement generally do not provide full compensation.

17.4 Necessity of TQM

As per conventional/existing system, the quality checking/ inspection/supervision of construction works are carried out during/ after construction on the basis of design, drawings and specifications prepared by architects and structural designer. Planning engineers also prepare detailed estimate/tender documents on the basis of specifications and design prepared by other various consultants/ experts, including for water supply, sewerage, electrical services/ utility lines. In case quality of aforesaid design & drawings is not satisfactory and there are some errors/faults in the specifications/ tender documents, then construction works will also be defective. If services during pre-construction stage are sub-standard, then quality of production (viz. construction works) will also be below standard. Hence it is required to check the quality (i.e. errors, deficiencies, appropriate design standards, faults in the design/ drawing works, adopting appropriate specifications etc.) of various services rendered by the architect, structural designer, specialist consultants for various utilities, planning engineers of the owner, particularly during pre-construction stage (right from conceptual/ preliminary stage). However, conventional quality control system does not check the aforesaid design/drawing works. Many other related matters like, reliability of material testing laboratory, economy, completion period etc. as well as auditing of above stated system are also included in this new concept which is known as "Total Quality Management" (TQM). Conventional Q.C. system cannot detect these deficiencies.

The TQM management is being dealt as under:

17.5 Total Quality Management (TQM) (Principles)

17.5.1 After the Second World War, massive rise of the Japanese economy has been the biggest surprise. How could Japan make it? A detailed answer to this question is still being attempted all over the world. However, the most important one relates to Japan's emphasis on "T.Q.M.". There are many definitions of TQM. World famous management expert Dr. Juran states that – "TQM is really a collection of all the things that we must do to have quality leadership."

17.5.2 (a) What is quality? Earlier it was thought that quality is richer specifications. It was also meant to be product's ability to conform to specifications. New quality is strategic; it is led by management and is for everyone. TQM is a philosophy, a new way of life. TQM has relationship with organisation's culture and behaviour of people. It is about improvement for "Customer Satisfaction" For TQM to be successful, total work culture changes are required.

(b) As per another definition, "TQM is an approach for continuously improving the quality of goods and services delivered through the participation of all levels and functions of the organization".

(c) When a structure is to be constructed, the quality would mean ability to satisfy the "stated" and "implied" needs, conformance to requirements and fitness to use. Final product (like a building) depends on each member of the project team including architects, structural designers, planning engineers of the owner/department, various consultants, experts, contractor, sub-contractor, material suppliers, vendors etc.

17.5.3 (a) There are different models/concepts/approach to TQM. The ISO 9000 (Series) is one model/concept of TQM. It is achieved through development of documentations and their implementations according to the various guidelines developed by ISO as supporting standard. It is required to select particular ISO 9000 standard according to need and applicability. There are three ISO 9000 standards/ models on quality system:-

(1) ISO 9001 : 1994

(2) ISO 9002 : 1994

(3) ISO 9003 : 1994

(b) ***ISO 9001 : 1994***

The ISO 9001 relates to Service Sector like, architectural services, structural design services, other consultancy/ expert services for water supply, sewerage, electrical installation, A.C. works, fire safety installations etc. ISO 9001 : 1994 relates to Quality systems/model for Quality Assurance (Q.A. in design development, production, installation and servicing).

(c) A separate standard ISO 9004–2 :- "Guidelines for Services" has been published for facilitating suitable adoption of ISO 9000 series. Initially it was thought that ISO 9000 standards are for manufacturing industry due to its language. But later on, it has been found that these are also adoptable for construction industry.

17.6 Model of TQM

A leading consultant/adviser of Quality Management at Associated Chambers of Commerce & Industry of India has suggested an integrated TQM model which is reproduced below (vide figure No.17.1):-

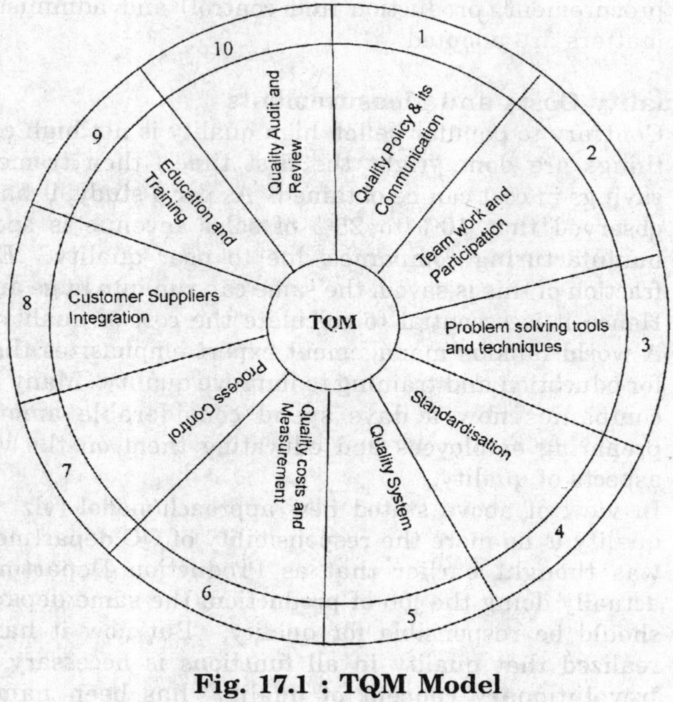

Fig. 17.1 : TQM Model

17.7 Quality Policy

(a) For TQM, a sound quality policy is a fundamental requirement. The TQM foundation consists of strategic quality management, executive leadership, a continuous focus on the consumer, linking with quality assurance in respect of various activities right from planning/conceptual stage of the product. These must be clearly stated, documented and communicated as a quality policy to every employee of the organization. The next step is to clearly define the key objectives and quality goals.

(b) Both Team Work and Participative Management are important for TQM model. To institutionalize both these concepts, Cross Functional Management teams, Quality Circles and Suggestion Systems are formed. Quality Circles are popular in Japan. Standardisation is a management tool for encouraging and securing optimum utilization of resources and maximum efficiency of operations through pre-determined solutions and answers to recurring of problems and needs. In each organization, certain standards in respect of technical matters (viz. specifications in design, procurement, production and control) and administrative matters are adopted.

17.8 Quality Costs and Measurements

(a) Contrary to popular belief, high quality is not high cost. If things are done "right the first time", then tremendous savings in cost can be obtained. As per a study, it has been observed that 10% to 25% of sales revenue is spent by manufacturing companies due to poor quality. Even a fraction of this is saved, the same can run into huge amount. Hence it is essential to calculate the cost of quality.

(b) A world famous management expert emphasizes the need for education and training to improve quality. Many Indian companies now a days spend considerable amount in preparing employees and educating them on the various aspects of quality.

(c) In view of above stated new approach/model (viz. TQM), quality is no more the responsibility of QC department. It was thought earlier that as Production Department is actually doing the job of production, the same department should be responsible for quality. But now it has been realized that quality in all functions is necessary. This "revolutionary concept of quality" has been named by

Japanese as TQM. As per their new model, the TQM is a customer-oriented philosophy and the profit is a by-product. It is a complete "bottom-up" management philosophy wherein "everyone in every function of the company is responsible for the quality of the project." It is the management approach of an organization centered on quality, based on participation of all its members and aiming at long term success through customer satisfaction and benefits to all employees of the organization and to the society.

(d) House-keeping is to be kept at high standard for high quality. Employees should be given responsibility and authority to make changes at their work place and should be treated with respect. The most commonly used techniques are Brain Storming, Arrow Diagram/Bar Charts, Control Charts, Histograms, Pie Chart, Cause & Effect Diagram etc. Giving green signals to initiate TQM by top management is not enough. Their support should be visible and demonstrative. TQM is to be well planned in advance. Problems should be used as sources of improvement.

(e) Total Quality Control (TQC) is one model/concept and the same is helpful for achieving TQM. There are five Ws and one H and the same are as under:-

(1) Why - Objective
(2) What - Object
(3) Who - Subject
(4) Where - Space
(5) When - Time
(6) How - Method

17.9 Errors and Preventive Action

(a) Many organizations (particularly government departments) are organized to inspect errors/defects of the system/ production (viz. construction works). This is followed by activities like, issue of detailed inspection notes, collecting samples, checking/testing and obtaining replies from concerned persons/incharges. All those mean re-working, re-checking and repetition which is expensive, frustrating and demotivating. It is a fire fighting exercise and involves reworking of mistakes already committed. In similar way, marketing/sales managers meet many customers to know about their complaints related to quality and problems faced by them. This entails travel, waste of time and extra expenditure.

(b) The wastage can be checked and considerable economy achieved by using the "technique of prevention". This means identifying and fixing problems before they can turn into actual mistakes. Prevention techniques lead to more productivity and lower cost prices. Prevention is an attitude of mind and requires detailed planning. Prevention techniques require the measurement (i.e. quantifying) of activities to generate data; this helps to identify the problem areas. Above techniques are parts of TQM. Lot of economy and customer satisfaction are achieved by saving time and man hours.

(c) Following are a few acts in TQM system:-
- Being polite on the telephone.
- Getting things "right first time" every time
- Aiming for zero defects
- Being flexible and responsible to change.
- Tackling the source, not the symptoms of a problem.
- Admit that the most valuable resource is people of the organization.
- Having cost effective and workable quality system.
- Measuring the progress of quality improvement.
- Co-operating and not confronting.

17.10 Cost of Quality

(a) This is an important aspect of TQM. Competitive and customer's acceptance cost depends on total cost of the finished product/services (in service oriented organization). Cost of quality is described as how much it costs the company to provide quality in everything the company does. They include the cost of "non-conformance" of the product. Majority of managers are unaware of how much it costs them "to do things wrong". Cost of quality is not shown in the balance sheet of the company.

(b) In western countries, the cost of quality is about 15 to 25% of total cost of manufacture and 40 to 50% of turnover for the service-oriented company. Companies without proper approach to quality may be incurring more than this amount. Cost of quality services is a yardstick to assess the progress and reduction in wasteful activities. Moto should be :- getting things "right first time" every time.

17.11 Assessing Cost of Quality

(a) Cost of quality is composed of certain key elements. The major part of cost of quality is the "Cost of errors". This involves rework with high costs on account of indirect labour. Sales/marketing managers spend lot of time in meeting customers due to defects/errors in quality of the product. Extra cost is added due to poor knowledge of some employees and the resultant mistakes/errors. As per a study, 85% of quality problems are created by people who never touch the product.

(b) Some companies boast that they have the largest customer service departments compared to their competitors. This means that the company is committed to getting it right "second time" and not the "first time" itself. This cost can be very high. This situation is not good.

(c) In a study done in USA, it has been found that the cost of quality as cost of doing things wrong is 12.5% of total project cost. If TQM is followed effectively and trained manpower is deployed, then aforesaid cost of quality (12.5%) can be reduced to near zero. It means that quality is achieved "free" as there is no non-conformance, there is no expenditure for rework or rectification.

17.12 Cost of Inspection

(a) Inspection costs include the cost of enforcing quality control, the cost of calibration of equipment and measuring instruments/gauges, the cost of tools etc. The cost of inspection is high, particularly in administrative areas. It is required to plan properly and to take preventive measures so that necessity of inspection is reduced considerably.

(b) Inspection activities are pursued by organizations which have major problems with product and service quality. If things are not right, then inspect and again inspect for review of previous inspections and action taken reports on inspections etc. As per TQM, the inspection does not bring quality into product. We must always remember that - "Prevention is the key".

17.13 Preventive Measures for Unsatisfactory Quality

(a) Concept/model of TQM is being dealt in this chapter. The concept/model (like ISO 9001 / other model) of TQM in respect of production and service sector has been stated on the foregoing pages. As per this concept, necessary policy, goals, detailed procedures/Quality Manual, Quality Plan, training & education at all levels of employees etc. are pre-requisites. Audit is also to be carried out in time. In respect of construction of a multi-storey building, some main activities related to quality are:-

To prepare reports of inception, feasibility, outline/conceptual design, detailed design (architectural & structural), proper engineering documents, proper specifications, tender documents, selection of competent/reputed contractors; to check design control, document control, purchase control, process control, control of non-conforming products (viz. design, drawings, software in respect of architectural, structural design and other services).

(b) There are many other activities as per model of ISO 9001 and compliance action is to be taken as per various documents of this model. Main aim is "to do right thing first time" through application of various techniques/tools for achieving quality as stated on foregoing paragraphs. A suitable system should be evolved for knowing in advance about possibility of errors, faults, defects in design, planning and execution of construction works. If output of the first stage (viz, conceptual scheme/ drawing of building) is wrong, then the same will become as input of the second stage and the detailed design/drawings will be prepared on non-conforming output of the first stage. As per TQM system, the error should be detected at this stage before proceeding further. Quality of third stage (i.e. detailed specifications, B.O.Q., tender documents) will depend on the output quality of second stage. Tenders should not be invited unless there is re-checking/auditing and confident clearance about quality of aforesaid first, second and third stages. Main responsibility for executing construction works depends on the contractors; but their selection depends on earlier stages, particularly pre-qualification of suitable contractors. Another important requirement is integrating all levels, categories, branches of the organization including top management / CEO/Chairman.

17.14 Cost Saving by Prevention

(a) Another component of cost of quality is "Prevention". Some preventive steps have been outlined as above. The best way to remember the role of prevention is "One hour of planning saves ten hours of problems and confusion."

(b) Unfortunately, little time, effort, energy and manpower are spent on prevention in today's organizations. The usual way of life in many organizations is oriented towards "problem solving" and "crisis management". Fire fighting becomes a way of life. Managers and other staff get used to such culture and some of them actually enjoy it. Their time could be spent more effectively in creating better work culture/ environment and opportunities for future as well as in anticipating problems, rather than on avoidable corrective measures. Every departmental manager should work with his team to identify key cost of quality measures responsible for rework. Production areas facilitate easy measurement of cost of quality. But measurement of cost in service area is a difficult task. It is generally found that problems are faced due to non-receipt of finally approved working drawings in time. If effective data review/monitoring is done, then it is easy to solve such problems in time. Sometimes, quality of work in service sector suffers due to non-availability of sufficient number of computers and relevant software/ programme. Cost of these items may be minor, but the cost of rework may be ten times more.

17.15 Quality Control in Construction Works

(a) It is required to explore how the TQM system could be made to work on an actual project. For this, quality manual, procedures and the work instructions are prepared. Client's brief and quality assurance (QA) systems of the organization are also required to be linked effectively. Some essential features of the project quality are:-

- Serviceability
- Safety
- Timely completion
- Maintenance aspects
- Ease of maintenance
- Modifications/Upgradation
- Aesthetic

(b) These aspects are to be integrated into Quality Plan. Quality policy should deal with key aspects of project including Contract review and the Contract/Sub-Contract Control System. Process Control (i.e. construction activities) is the most important aspect for developing documents. The clause on inspection and testing have many applications in construction works, including control of non-conformities arising out of incorrect execution (various defects etc.). Learning from past mistakes is key concept of ISO 9000 standards. Making a mistake is no crime, but repeating the same is definitely so. The internal/external audit plan should also be included. Training to various categories of manpower (from top to lowest level) is important. Statistical techniques need to be considered including studies on durability.

The basic concept of TQM is "Right first time" and "continuous improvement".

17.16 Customer Focus

(a) Customer focus is a corner stone for the TQM. Customer can either be internal or external. For example, in case of a housing society's building, project team of society is internal customer for the contractor. The flat owners are the external customers. Hence flat owners' satisfaction is most important and as per TQM, it is possible to achieve the same (at least major part of satisfaction).

(b) Certain problem is faced for achieving the best quality in construction industry particularly in government departments. General procedure of awarding the work to the lowest tenderer. This creates problem for proper implementation of TQM system. When main emphasis is given on completing the project in very short period then quality of construction works gets second priority treatment. Time is the essence of contract; penalty is imposed for delayed completion of project. Of course, this is justified, but no incentive is given for achieving high standard of quality in India and early completion of project. This affects adversely on the quality.

17.17 TQM in Construction Industry

Following are the important elements of TQM in construction industry:-

(i) ***Involvement of Top Management***

Top management should show interest in arranging the resources in time. Structure of project staff should be constituted with competent, experienced engineers. In manufacturing industry, the workforce is generally steady;

but in construction industry, lots of changes occur. Due to this, continued training to manpower of all categories is particularly important in construction industry. There is a Japanese axiom that :- "Quality begins with training and ends with training". Training should be arranged for management, engineers, technicians, skilled workers and other field staff. Useful tools for TQM are appropriate pre-planning right from conceptual stage, proper quality in designing and preparation of architectural and structural drawings, correct specifications and tender documents. Proper equipment/machinery planning is also essential. Check lists for various works during pre-construction stages (i.e. design etc.) and construction stage are useful to achieve good quality and avoiding errors. It saves extra time and cost involved in rectification of errors. Statistical tracking and data are useful for TQM concepts. Award of contract to competent contractor, team building and incentives are important for achieving good quality. Poor performance among the vendors and contractors is to be identified and eliminated. In private sector, partnering agreement may be possible.

(ii) ***Quality in Structural Design***

(a) It has been observed in many civil engineering projects that structural design contains some errors/faults. In one project, grade of cement concrete (viz. M-25/M-30/ M) – was too less or too high/reach concrete. It is to be decided whether the method of design adopted by the designer is preferable or not as per latest code/ practice. Special mix design/aggregate gradation, cement content, chemical etc. are required for water tight structures (like overhead/underground water tank), but in some projects, the same had not been designed correctly. If RCC work of such structure is designed suitably, then it should be leak proof even without further water proof treatment on RCC structure. However, it has been observed that leakage is found even after carrying out water proof treatment on RCC structure; this is due to faulty structural design, specifications, mix design/grade of RCC work, bad workmanship etc.

(b) Many structural failures are there due to wrong design. Another example is given here. RCC structure of three storeys below ground level were designed without taking into account the correct amount of upthrust / buoyancy effect due to subsoil water on the lowest level of RCC slab/raft. Highest level of underground water during the last some years should have been considered to work out the upthrust force on the bottom of lowest floor. Another serious mistake was also done by the designer. The design of structure of a high rise building (24 storey) was done without "Wind Tunnel" test. Subsequently, it was found that the proposed building failed structurally due to adverse effect of wind (as per Wind Tunnel test carried out later on).

(c) Type of steel reinforcements is also to be considered in the design correctly for sound and economical design. Data considered correctly for seismic force, particularly for high rise building is crucial. Correct/sound design of foundation also depends on correct soil investigation; guidance/ instructions are to be given by the designer to the soil investigation agency. Certain fault was found with foundation design as adequate numbers of bores were not done at correct locations by the soil investigation agency. Scour depth is very important in case of design of Piers/Abutments of bridges, culverts. In this context, design fault was noticed in many Highway projects.

(d) In view of above, good quality in structural design is a must. ISO 9000 series (particularly ISO 9001) is applicable in the discipline/service of structural design. Quality policy, planned documents for this service are to be prepared on the basis of aforesaid model. It may be stressed that structural design is an intellectual activity and falls under service sector. Huge numbers of humans are involved in service sector related to civil engineering works. Hence quality management and implementation of quality system requires special attention. "Consistency" in design of various disciplines of the project is important. The quality system/management also should be repeatable. Continuous training in structural design for all levels of designers is essential. Planning for training, as already emphasized, should also be done.

(iii) ***Review***

Field unit is the customer of Design unit. By reviewing with Field Unit, it is to be confirmed that their requisitions have been understood clearly.

(iv) ***Design Control***

After initial review, further verification is to be made about structural design & drawings to ensure that the field unit's requirements are fully met. Checking is to be done during design process at all stages i.e. Design input, Design output, Design Review, Design Verification, Design Approval and Design Changes.

(v) ***Document Control***

To ensure that correct/latest revised drawing is issued to the field unit and other concerned persons / offices / consultants. Obsolete drawings / documents are to be withdrawn in time.

(vi) ***Purchase***

Proper procedure is to be laid down for quality goods/ services.

Product (i.e. Design/Drawings) identification system should be set out properly so that the structural design, drawings, computer software etc. can be traced out easily and delivered to customer.

(vii) ***Process Control***

This is very important. Procedures in respect of various items, stages are to be prepared as documents of system.

(viii) ***Inspection/Scrutiny***

All incoming and outgoing data, drawings, documents are to be scrutinized properly.

(ix) ***Control of Non-Conforming Products*** (i.e. design, drawings, software)

- To make sure that no errors are committed in the use of drawings and software.

(x) ***Corrective and Preventive Action***

If any error/non-conformity is noticed, then corrective and preventive action is taken immediately.

(xi) ***Procedural Manual*** is also to be prepared.

(xii) ***Audits***

There may be internal audit (which is constituted within the department/ organization) or external audit (from other organization). Quality Manual, documents required as per ISO 9000 series, procedure/policy records are to be checked during audit.

17.18 Quality Systems in Architectural Design

(a) Architectural design is a complex process which requires right inputs at the right time in the right phase from a vast cross section of participants viz. Associate Consultants for water supply, sewerage/drainage systems, fire prevention system, electrical installation works, air conditioning system, structural designer, clients and clients' representatives (their project manager), contractors, suppliers, financial institutions, landscape expert and so on. This complex process is further constrained by the requirement of strict time schedules, cost limits and quality bench marks. Problem is also faced due to different opinions from above stated different participants but aims of all participants should be same as per principle of TQM. Sometimes, the participants may have conflicting interests. But, actually none of the participants can achieve their goals without inputs from the others. Main goal of all participants should be to satisfy customer's requirements of which quality is of paramount importance.

(b) However, there are no clear answers to ensure adequacy of the design development process. The ISO 9001 quality system is considered here for achieving TQM of building projects. Architect designer is required to face socio-economic aspects, pressure from final users/customers/ owner, government authorities, overall building quality (i.e. performance), to comply with cost limit and time limit. Designing has become a complicated and collaborative process. Hence it needs efficient communication and "collective control".

(c) Architectural design is a process involving basically the following three stages:-

(1) Pre-construction stage
(2) Construction stage
(3) Post-construction stage

The design development in the pre-construction stage essentially consists of following phases:-

(d) ***Briefing***

(1) *Inception*: Architect discusses with owner and his project engineer about preliminary matters, proposed work/ project and appointment of architect.

(2) *Feasibility*: Architect carries out site conditions, environmental aspects, preliminary planning, design, rough budgeting cost and submits feasibility/appraisal study report to the owner of project.

(e) ***Sketch Plans***

(3) *Outline Proposals*

Architect develops the brief for owner further to determine general approach of layout, deign to obtain approval of the owner/client. Conceptual design is finalized.

(4) *Scheme Design*

Architect prepares final/full design of the project for approval of the various statutory government authorities (like Municipal authority, City Development Authority, Fire Service Authority etc.). Preliminary cost estimate of project and full explanatory report, model of project, drawings etc. are submitted by the Architect.

(f) ***Working Drawings***

(5) *Final detailed working design / drawings*

Architect prepares and finalises full design of every part of building project in collaboration of all concerned associate consultants, experts.

(6) *Production Information*

Architect prepares final production information, i.e. drawings, schedules, specifications and to obtain final decisions of the owner to carry out the construction work.

(7) *BOQ*

To prepare BOQ and tender document.

17.19 Application of ISO 9000 Series in Quality System of Service Sector (Architectural Design)

17.19.1 To achieve proper quality in architectural design, it is required to adopt a good standard. The ISO 9000 series standards are one of these standards. Certification is done for the organization (including service sector) by Bureau of Indian Standards, ISO. This provides third party "assurance" to the customer in regard to quality of their products, such as:-

- Site analysis reports
- Work schedules
- Organisation structures and responsibilities
- Drawings (architectural, structural, plumbing, electrical, interior design)
- Design review reports
- Design change reports

- Specification data base
- Detailed data base
- Contracts
- Cost estimate, BOQ and Tenders
- Vendor lists and pre-qualification issues.

17.19.2 Basic quality management system for TQM of Service Sector as per ISO 9000 series consists of the following:-

(1) *Organisation*
Requires the clear definition of the responsibility and relationships.

(2) *Auditing*
To demonstrate that tasks defined in the responsibilities are continually carried out according to the stated methods.

(3) *Reviewing*
Continuous checks on process/methods and action/procedures if the stated requirements are not being met.

(4) *Feedback*
Requires tracking of causes of errors at every stage/sub-stage that generate defect, in measurable terms, so that process can be improved and non-conformance to requirements can be eliminated.

17.19.3 Quality System Approach in Building Design and Construction

(a) The plans of work, by definition, only suggest a pattern and sequence of events to be carried out during the process, whereas the requirements of quality management system are intended to control the "actual process" of the design. Flexibility of various parameters for such type of services and building construction project finally affects the end product. "Quality Policy" which is to be prepared in the organization would ensure commitment from everybody. However, for actual implementation of quality policy and various procedures regarding production process, continual checking is essential.

(b) Quality System elements under ISO 9001
These are listed as under:-
(1) Management responsibility
(2) Quality Systems
(3) Control review
(4) Design control

(5) Document and data control
(6) Purchasing
(7) Product identification traceability:- proper system, records with dates, revision of records etc. are to be maintained; name of specific person who is incharge for this should be circulated.
(8) Process Control:- detailed system, continual checking are to be stated
(9) Inspection and testing:- results are to be recorded in systematic way (e.g. documentation).
(10) Control of non-conforming product
(11) Corrective and preventive action at every stage
(12) Quality records
(13) Handling, storage, purchasing and delivery
(14) Internal audit at various stages of services (design, drawings etc.)
15) Training
(16) Statistical techniques.

Above stated elements have relevance and are inputs for the pre-construction stage. However, the focus will be on "Design Control" and Document and Data Control.

17.19.4 The design firm must maintain proper approach and procedures to design in order to ensure that the requirements are met. All stages of design as stated in the relevant documents shall be considered carefully. Entire activities of all the stages shall be identified and documented. As per ISO 9001, various requirements, plans, process, procedures, activities are required to be put on record. However, the actual implementation is to be ensured as per time schedules. Documentation does not mean that final product is automatically produced as per planned cost and quality; management is required to implement effectively and in time.

17.19.5 Activities

(1) Establishing the needs and converting them to requirement of space, materials, finishes, time and cost.
(2) Sorting out various options/alternatives and deciding on the most suitable.
(3) Organising people and materials and formulating the schedules.
(4) Developing plans, specifications, estimates and details.
(5) Verifying the plans and details against the requirements at each stage.

(6) Identifying different groups who must interact at each stage.
(7) Receiving the design against various inputs at each stage.
(8) Approval of design by the competent person.

17.19.6 Once the activities have been identified, when and how they have to be done is to be established. Persons required for different items of work have to be posted. Which groups interact with each other at what time in the design process must be established; also pinpointing as to who from each group would interact. Participants at each phase must be identified. Various participants for a multi-storey building could be from following groups:-

- Owner's representatives/Project Engineer
- Architectural group
- Plumbing group
- Electrical group
- Structural design group
- Civil Engineering group
- Project Management group, tender cell
- Site Survey/soil Investigation group
- Specialist Consultants

17.19.7 Design Input

Design input requirements need to be identified and documented for design process. Some of these are as under:-

- Purpose of project - Client may not be final customers.
- Aesthetic requirements - Owner may require maximum output at minimum cost or may have priority in particular style/appearance of the building.
- Load requirements - Extra on account of snow, wind, machinery, moving parts as per actual and special position.
- Information available - Feasibility study report, site data, soil investigation report, hydrology, existing services etc.
- Funds available - To enact a balance between excellence and economy.
- Required life time of the building - Reflects on structural design, materials, specifications.
- Codal necessities.
- Local bye-law requirements.
- Safety requirement.
- Interior works requirements.
- Energy saving - design/intelligent building.
- To achieve space efficiency of the building.
- Environmental requirements - Detailed study is to be carried out for this important aspect.

17.20 Quality Audit

17.20.1 Introduction

Position of conventional quality control and TQM concept/model on the basis of ISO 9000 series has been stated on foregoing pages of this chapter. TQM concept/model of quality in respect of service sector and construction/production is based on:-

ISO 9001-1994

Quality systems - Model for Quality Assurance in Design, Development, Production, Installation and Servicing.

Above ISO is a model/concept and detailed systems, planning, manual etc. are prepared for achieving TQM. However, these are just documents. Required standard of quality (TQM) will not be achieved unless various procedures, planning, steps are implemented correctly in correct time. "Quality Audit" is one important management "tool" for helping the owner/client to ensure that services (design, drawings, planning, selecting correct agencies, supplies etc.) and construction/production are carried out correctly in time as per standards and specifications. Quality audit should be carried out at different stages i.e. during pre-construction stage (for services), construction stage and post-construction stage. Quality auditor is different from financial auditor and consultant who may be engaged for checking certain technical matter like, Proof Consultant is for checking structural design carried out by another design consultant.

17.20.2 Quality audit is not limited to audit of system but includes processes, products and services. It is not only for the purpose of supervision over activities of service/ processes/construction/ manufacturing but it is also essential to check whether principles of TQM have been adopted or not. Success of quality depends on proper design, drawing (architectural, structural and various service lines/ utilities), preparations of appropriate specifications of various items and works and tender documents (NIT), awarding the work to competent contractor etc. Quality of these important service activities are also to be audited in time.

17.20.3 There are many practical complexities in the way of achieving good quality in totality. Construction organizations tend to manage these problems in an isolated manner without considering their relatedness to other activities. There is no formal approach to identify and resolve the problems occurring in various activities.

This results in a typical crisis management rather than proactive initiatives. Periodic appraisal through "quality audit is one managerial tool" for sorting out of above stated problems. A major tool which ensures continued maintenance and review of quality system is "Quality Audit".

17.20.4 Objectives

The objectives of quality audit may be outlined as under;-

(a) To determine conformance or non-conformance of quality system element.

(b) To determine and evaluate effectiveness of the implemented quality system as envisaged in the quality policy.

(c) To verify implementation of corrective and preventive action.

(d) To provide an opportunity to improve the quality system.

The quality audit is not a "fault finding" mission; it provides objective feedback and assessment about performance of the quality system.

17.20.5 Types of Auditing

There are two types of auditing

(a) Internal Audit

(b) External audit

In big organization, internal audit may be done through two auditors separately (who are not connected to the project). The external audit is carried out by an outside agency, not connected to the department/owner of the project. Auditor/lead auditor/external audit agency is engaged by the highest authority of the department/ owner. There are three different parties i.e. the client (department/ owner), the auditor and the auditee (audit objects).

17.20.6 Audit Procedure/Stage of Audit

(a) *Audit Plan Preparation*

This is prepared in consultation with auditee/client. It includes audit objectives & scope, audit schedule date, place, expected time & duration, check lists, formats etc.

(b) Opening Meeting

This is mainly to review the scope, methods of auditing, availability of facilities required for the audit team and to establish line of communication.

(c) *Audit Execution*

Relevant particulars are collected and documented through discussion, examining records, observing site activities, working conditions.

(d) *Evolution of Results*

The documents of audit finding so prepared should include the outstanding/ non-compliance reports, opportunity for improvements and remedial action report.

(e) *Closing Meeting*

This meeting is conducted between the audit team, senior management, auditee and personnel responsible for functions.

(f) *Audit Report*

This should include the following:-

(i) Scope, objective and details of audit plan
(ii) Observations on non-conformances
(iii) Audit team's judgment of the extent of the auditee's compliance.
(iv) Capability of the system to achieve quality objectives.

17.20.7 Role of Owner, Auditor and Auditee

(a) Their roles are important for successful/effective quality audit. The owner selects and appoints the auditor for auditing the quality of services carried out by the architect, structural designer and other consultants/experts. He is the main person for using the audit result/report; he takes necessary follow up action in a time bound manner. It is required to audit the aforesaid service related documents/output in time and to comply with the observations of audit party before inviting tenders.

(b) Auditor is required to plan and conduct the audit. The auditor possesses necessary qualification, practical experience for relevant types of construction works/products and also has experience in auditing. Pre-announced audit is better than surprise audit. Auditor is required to discuss with key persons of the project and also to observe operation/ process of products, organizational structure, staff actually deployed by the architect, structural designer, consultants etc. Co-operation of project teams (of owner and contractor), architect, and consultants with auditor is important for success of audit. Auditor should be considered by the auditee as guide, philosopher and friend. Hence auditee should not be afraid of the auditor. Necessary facilities for observing operation/process of products are to be arranged by the auditee. All relevant records as per list given by the auditor are to be produced by the auditee.

17.21 Laboratory's Role in Quality Control

(a) In the drive towards continuous improvement of quality, the role of laboratory is very critical. The laboratory has to provide accurate, reliable and reproducible signal. It is, therefore, essential that the laboratory is technically equipped and has a well established "Quality System Management".

(b) In civil engineering projects, it is required to carry out at site itself certain routine tests like, cube test of cement concrete used at works, testing of sand, stone aggregates, bricks etc. very frequently. Number of tests depends on the quantum of cement concrete work done and other materials used. The cement concrete mix (at green stage) is filled in the steel cube moulds; three cubes are required for testing (one cube for 7 days' curing, second cube for 21 days' curing and the third cube for 28 days' curing). There will be huge number of such tests almost daily at site. Hence it is essential to set up a field laboratory at project site. Relevant test instruments (viz. compressive strength equipment, sieve sets, balance etc.) are required at such field laboratory. Correctness and calibration of the instruments are to be checked by skilled laboratory incharge. If necessary, this may be got done from an approved/ accredited central laboratory.

(c) Various other tests of materials and completed works are done from the central laboratory of the department or approved public laboratory. Reputation of the laboratory should also be ascertained. In fact, reliability is very important. In addition to the tests carried out by J.E/Lab. Assistant, independent test checking on sample of same source on random basis by an experienced, competent senior person is essential; but generally this is not done in most of the laboratories. Reliability/accuracy of the results also depends on other factors like:-

   (i) Whether instruments/apparatus/equipments have been calibrated by reputed approved agency.

   (ii) Whether various testing equipments/instruments of good standard are available in the laboratory in working condition.

   (iii) Whether appropriate sampling procedure has been followed.

(iv) Whether testing personnel (junior and senior level) are skilled in this field. Their practical experience in testing of relevant materials / water is important.

(v) Whether the laboratory has been accredited by the approved authority.

17.22 Laboratory Accreditation

Laboratory accreditation is formal recognition of the technical competence of a laboratory, inclusive of quality system management, which is based upon the third party assessment.Throughout the world, accreditation is based on the international guide known as "ISO IEC guide 25". The accreditation body itself maintains the quality system in accordance with the ISO IEC guide 58. This provides a high degree of assurance to the validity of test results for benefit of the users. There are two types laboratories:-

(1) Testing Laboratory

(2) Calibration Laboratory

Accreditation improves the quality of laboratory services to users. It develops and maintains criteria for laboratory. It provides an independent, impartial and objective system.

All the laboratories for testing building materials should be accredited with National Accreditation Board for Testing and Calibration Laboratory (NABL), Ministry of Science & Technology, Government of India.

Test methods are required to meet the client's needs. Laboratory itself should select methods of testing/calibration, if the client does not specify. The criterion for laboratory accreditation spells out all the requirements in terms of quality management system as well as technical matters. Internal audit of laboratory should be conducted to ensure that the laboratory is complying with all the accreditation parameters.

18

Safety in Construction

18.1 Introduction

With large scale construction activity, it has become necessary to adopt measures for minimizing the likelihood of accidents in the building industry. Safety in construction is an important part of management. But this is neglected in the construction industry. New problems on safety are being faced in view of wide range of construction with new and complex techniques and use of new machinery. Number of accidents (major, minor and fatal) in construction industry is much more than in manufacturing industries/factories. Second largest manpower (after agriculture sector) is involved in construction industry in India. But unfortunately this is an unorganised sector. There is lot of exploitation of labour in this industry. In certain western advanced countries there are in-built elaborate safety unit which ensure the requirement of occupational, safety and health care. Such arrangement does not exist in construction industry of India. There is lot of women and child labour though engaging child labour is illegal. There is no effective penalty clause for imposing against contractor for fatal/major injury cases; compensation is to be obtained from insurance company. Generally contractor goes scotfree.

Safety should be not only for accident during construction activities but it should include the loss of life, property due to fire accident, illness due to bad hygienic condition at work site/labour camp. Accident also occur due to structural failure. Structural failure may be due to wrong structural design, wrong centring & shuttering arrangement for casting RCC beams, slabs etc. Management authority is responsible for such type of accidents resulting death/ injury of workers.

18.2 Safety and Health Acts/Codes/Rules

Unfortunately there is no specific legislation/Act for safety during construction works at site. There are some laws/Acts on labour regulation like, Workman Compensation Act, Minimum Wage Act,1948, Bonded Labour System (Abolition) Act 1976, Child Labour (Probation & Regulation) Act, 1986 etc. There are some codes published by B.I.S. (Bureau of Indian Standards) on safety of a few construction activities like, Prevention measures against hazards at work place (I.S. No - 13416), working with construction machinery, excavation work etc. Some conditions on safety are stipulated in the contract agreement. One chapter of National Building Codes is related to safety aspect. But unfortunately most of the provisions made in above stated documents are not implemented. There is no effective clause in the agreement for imposing good amount of penalty against contactor. Action is also not taken by the department/ owner's organisation against their staff for negligence. Negligence is found in designing of structures, designing centring & shuttering for large RCC structures and its errection/to make these temporary structures including scaffolding for multistorey/high rise building etc. Caving/side slides of deep excavation occur due to negligence and not following laid down technical safety methods; many fatal accidents occur due to caving/side slides during earthwork in trench, basement area etc. Accidents occur during operation of heavy machinery/construction equipments on account of negligence/ unskilled operation.

18.3 Safety Measures Peculiar to Different types of Construction

There are different types of project constructions like Multistorey/ High Rise building, Sewer construction, Road/Highway construction, Dam construction across river etc. Safety measures depend on type of project construction. It is required to plan/design in suitable way keeping in view of safety of public, workers/labour etc. Dam construction project may be for hydro-electricity, irrigation purpose, drinking water supply to city etc. Huge areas on upstream and down stream side are inundated during storing water as reservoir on upstream side and releasing surplus water on down stream side. Areas are identified for inundation of water; villagers, public are to be informed properly with advance notice. If there is wrong planning and design in releasing/storing water then there will be loss of properties and human beings. In case of sewerage works (connected with existing sewerage system in city) frequent accidents occur and labours die due to poisonous gas (viz. Methane). Special type of safety measure are to be taken for such type of works. Practical problems are faced in respect of certain underground works (viz. deep trenches for sewer work etc.) for taking laid down safety measures. However it is required to find out solution for practical problems at site of work. Electrical safety is another type. Safety measures for road/highway works are of another type; special safety measures are essential for bridge works across river.

18.4 Safety of Labour

National Commission on labour recommended in one report for a comprehensive legislation covering health, welfare, safety and other aspects of working conditions in the construction industry. There is another type of safety problem of life; during working long time under certain bad environment the workers may suffer disease of tuberculosis or some other cronic disease (i.e. occupational disease). Safety measures are to be taken in such type of works which are different than normal safety measures. It has been observed that although many common type accidents are known to construction engineers, constructor yet necessary preventive measure are not taken. Continuous educative and participative effort is required for essential safety and health measures. Management authorities should know the detailed scope so that appropriate measures can be taken as preventive action. Project managers/supervisors are required to arrange safety measures and to train workers, machinery/equipment operator, labours about way of working. Many accidents are occurred due to lack of knowledge and lack of necessary skills for the work to be done.

18.5 Measures for Avoiding Accidents

There are financial and physical adverse effects on the project on account of major accident (fatal or major injury). There is a cost of accident. Besides medical treatment it is required to pay compensation to injured person or next of kin of deceased person. Progress of project will also be hampered due to major accident. Training on safety measures should be given to project team, supervisors, workers. The central Labour Institute at Mumbai arrange such training courses. Comprehensive construction safety manual should be prepared by certain organisation for various types of works. One supervisory person/manager of project team should also perform duty as "Safety Officer" on additional duty. The Indian National Council may help for making publicity in creating awareness on industrial safety.

18.6 Safety Measures

18.6.1 As stated on foregoing paras certain safety measures are common for various works. It is also required to take certain safety measures as per special nature/type of works. Some of these safety/ pre cautionary measures are stated as under.

18.6.2 Preventive measures should start at the planning and designing stage itself. The planners, architect, structural designers, engineers should bear the safety of the works in mind at all stages of project. The contractor's role is most important. Safety must be a basis component of the management philosophy just as profit making because cost of accident itself may drain away the profits. The construction manager is required to mark the layout plan of site materials, stores, field laboratory, machinery space for repair etc. on the plot for safe working arrangement as shown on figure 14.1 (Chapter - 14).

18.6.3 There is another aspect of safety from functional point of view (i.e. during use of building). It has been found that due to certain wrong architectural design the building become unsafe. As example : - If architect does not provide grills on window and fixed glazing areas of the building then it is not safe for residing. Low height and large gaps are unsafe in balcony railings as children may fall down. Due to defective architectural design/planning the proper security is not ensured. Similarly highway consulting engineer should take special care for designing safe curves with correct super elevation, safe sight distance, proper traffic segregation arrangement etc. This will ensure traffic safety.

18.7 Safety Measures to be taken during Building Constructions

(1) Suitable scaffolds should be provided for workmen. If a ladder is used for short height then the ladder should be held by a mazdoor or tied up with the wall. Ladder should be strong enough to withstand the weight of the workman.

(2) Scaffolding or staging more than 3.6 meter above ground are erected with stationary support, shall have a guard rail outside, properly attached, braced and extending along the entire length. Such scaffolding or staging shall be so fastened so as to prevent it from swaying from the building or structure. Working platforms, gangways and stairways should be so constructed that they should not sag unduly or unequally. This should have adequate width and should be suitably fastened.

(3) Openings in every floor of the building under construction or in a working platform shall be provided with suitable measures to prevent the fall of a person or material by providing fencing or railing.

(4) Safe means of access shall be provided to all working platforms and other working places. If this is not done properly it may not be possible for the site engineers and

senior inspecting engineers to check the work, and to collect the sample materials for testing. Every ladder shall be securely fixed. Uniform spacing for steps not more than 30 cm shall be kept. Adequate precautions shall be taken to prevent danger from the electrical equipments. No materials particularly heavyweight in heaps shall be stacked or placed so as to endanger any person. The contractor shall provide necessary fencing and lights to protect the public from accident; contractor shall be bound to bear the expenses of legal proceedings of law and to pay any damages and cost which may be awarded in any such case.

(5) Earthwork in excavation is done in foundation or basement of building. Accidents causing death of workers sometimes occur due to slides from vertical sides of pit/trench. In case such excavated basement/foundation trench is very near to the road then there will be side thrust due to the traffic movement. Any leakage of water from underground water/sewer pipe lines is dangerous as it causes slides of excavated trench/pit. For draining out the water of heavy rains, temporary drains of suitable capacity should be made along the trench/around the pit excavated for the basement. Heavy rain may also damage round RCC retaining piles constructed for deep pit of basements and slides may occur. It is required to excavate foundation trench/pit for basement by giving suitable slope and steps (magnitude of slope depends on type of soil). Barricading is also to be done around the site with cautionary signboard. It is better to put barricading with movable stand type sheet fixed on steel frame and supported by steel joists/channel sections as per Fig. 18.1 (a) , (b). The height of the same may be 1.5 metre so that screening purpose is also achieved. It may be stated here that leakage of water from underground pipelines and improper drainage to divert heavy rain water are the main causes of undercutting and slides. This creates lot of problems viz., accidents/injury, filling pit/trench with water, extra cost for restoring site in workable condition by pumping out of water/slush and hampering of the progress of construction. Fatal accidents may occur at site and police case is registered against the contractor and site engineers of the department.

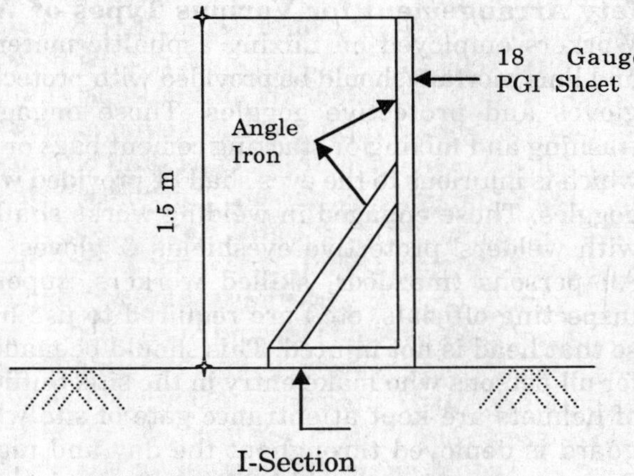

MS Joist (20 CM × 12CM)
SECTION - A-B

Fig. 18.1 (a) : Barricade Unit

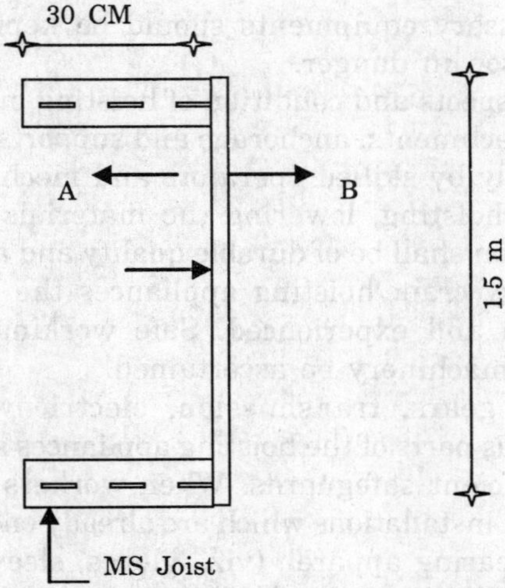

PLAN

Fig. 18.1 (b) : Barricade Unit

18.8 Safety Arrangement for Various Types of Work

(a) Workers employed on mixing asphaltic materials, cement and lime mortars should be provided with protective footwear, gloves and protective goggles. Those engaged in white washing and mixing or stacking cement bags or any material which is injurious to the eyes shall be provided with protective goggles. Those engaged in welding works shall be provided with welders' protective eyeshields & gloves.

(b) All persons (mazdoor, skilled workers, supervisory staff, inspecting officials, etc.) are required to use helmet at site so that head is not injured. This should be made compulsory for all persons who make entry in the site. Sufficient number of helmets are kept at entrance gate of site where security guard is deployed throughout the day and night.

(c) Proper precaution should be taken for lead painting. White lead, sulphate of lead or product containing these pigments, shall not be used in painting operation except in the form of pastes or paint ready for use. The "overall" cloth shall be worn by painters. Care is to be taken for lead poisoning.

(d) When worksite is near a pond, river, stream and lake then all necessary equipments should be kept ready to rescue any person in danger.

(e) Safety aspects and condition of hoisting machines including their attachments, anchorage and supports are to be checked frequently by skilled operators and mechanics. Every rope used in hoisting, lowering the materials or as a mean of suspension shall be of durable quality and adequate strength. For every crane/hoisting appliances the operator shall be qualified and experienced. Safe working load/capacity of various machinery be ascertained.

(f) Motors, gears, transmission, electric wiring and other dangerous parts of the hoisting appliances should be provided with efficient safeguards. When workers are employed on electrical installations which are already energized, insulating mats, wearing apparel, (viz., gloves, sleeves, boots) should be used. Workers should not wear any rings, watches and carry keys etc. on energised electrical installations.

(g) Above stated safety provisions should be displayed on a notice board at a prominent place at work site. The person responsible for compliance of the safety Code shall be named therein by the Contractor.

(h) Smoking should not be allowed in godowns where the inflammable materials are stored.

(i) Acids, spirits & other toxic liquids should be stored separately in proper leak proof containers.

18.9 Safety Measures for Sewer Rehabilitation Works

(i) Manhole covers of required length of sewer line are opened and ventilated at least one hour before the workers are allowed to go into the manholes. The work site shall be barricaded suitably along with warning signals and caution signboards to prevent accidents to the public.

(ii) Entry for workers into the manhole/sewer line shall not be allowed except under supervision/presence of the Junior Engineer or higher official.

(iii) 4 to 6 manholes on upstream and downstream sides should be kept open at least for one hour before worker is allowed to enter into the manhole for working inside. Problem may be faced due to broken/missing iron ring steps. In such cases the ladder and safety belt with rope are to be used. New steel steps (coated with plastic material) are to be fixed in place of damage/missing steps.

(iv) Before entering into the manhole, the presence of toxic gases should be tasted by inserting wet lead acetate paper which change colour in the presence of such gases.

(v) Presence of oxygen should also be verified by lowering a detector lamp into the manhole. In case, no oxygen is found inside the sewer line, workers may be sent only with the oxygen kit.

(vi) Safety belt with rope should be provided to the workers. While working inside the manholes/sewer line, such rope should be handled by two men standing outside to enable him to be pulled out in case of emergency.

(vii) No smoking or open flames shall be allowed near the blocked manhole being cleaned.

(viii) Gas masks with oxygen cylinder should be kept ready at site for use in case of emergency.

(ix) Air blowers should be used for flow of fresh air through the manholes. The motors for these shall be vapour proof and of totally enclosed type.

(x) The workers shall be provided with gumboots/non-sparking shoes, helmets, gloves, non-sparking tools, safety lights, gas masks, portable air blowers. Barrier cream should be given to the workers for anointing the limbs before working inside the sewer lines.

(xi) The malba/silt obtained on account of cleaning of blocked manholes and sewer lines should be immediately removed to avoid accidents on account of its slippery nature.

18.10 Safety Measure in Construction of Sewer/Drain Lines

In India most of the sewer/drain lines are laid through conventional method i.e. to excavate trenches along road/lane. Normally sufficient space is not available for making the steps/ benching or giving required slope on both sides of the open trench. It is very dangerous to work in deep sewer line. Problem will be faced if there are earth slides on the vertical sides of trench. This is aggravated when trench is very near to the carriageway. There will be horizontal force on the trench side due to plying of the vehicle. This will accelerate the earth slides. Fatal accidents do occur at many such sites. It is essential to provide timbering on both sides of trench (with planks supported by horizontal runners and struts and also by inclined shoring). But unfortunately labourers remove some part of this timbering protection during laying pipes manually. RCC pipes are rolled on the road by tying with rope and dropped in the trench by rolling on the edge of the trench. This crude method should not be allowed. The RCC pipe should be placed on the bed by chain and pulley system or by crane. Proper barricading as stated in para 18.7 of this chapter is to be provided around/along the work site. Red light blinkers should also be provided for safety during night along with the watch & ward to check the same. If there is fatal accident, then police registers case against Contractor and Department's officials. As per Labour Law/ Act the Department who award the work to the contractor is "Principal Employer"; hence Department is also to face the police case and to pay compensation as per order of the Court. Department must issue letters/notices at correct time to the Contractor for lapses/deficiencies in safety arrangement.

18.11 Accident due to Collapse of Centring and Shuttering

(a) As per codal requirement it is required to design centring & shuttering particularly for heavy RCC structures. One major accident occurred when centring & shuttering of heavy RCC structure consisting of girder, deck slab of large span of a Flyover collapsed after a few hours of casting. A few persons died. There was considerable amount of loss. The work was held up for some period. Technical and criminal investigation were carried out. It was essential to design the centring & shuttering from structural stability point of view but the same was not done. Centring & shuttering was not structurally strong enough to take the load. This technical aspect was not checked by any engineer of department and contractor (who was a Government undertaking) before and after completion of centering & shutter.

(b) Another major accident of heavy RCC roof (girder and slab) of a building was occurred because the centring & shuttering were removed before attaining the required strength. Contractor removes the centring & shuttering quickly for using on another part of roof. There was loss on account of this accident. The work was held up for some time. Aforesaid collapse of one part of roof also damaged the adjacent part of roof. Permission was not taken from site engineer of department for removing centring & shuttering before required period for gaining strength of C.C. It depends on weather i.e. if it is too cold then it takes more time; it also depends on type, quality of cement. Hence shuttering is to be removed after considering these factors also.

18.12 Protective Barriers in and around Buildings

18.12.1 Certain protective barriers are to provided in and round the buildings and other construction site. Special care is to be taken for effective barriers around foundation/ basement, trench excavation, demolition of old buildings. Red light points are also to be installed and to be kept on during night. Cautionary sign boards with reflective tape words are also to be fixed at appropriate locations. Above arrangement will minimize the risk of accident. Risk is more if the work site, building construction sites are in urban area or adjacent to road which is used by public and vehicles.

18.12.2 It is required to take precautionary measures for falling materials from construction site. There is another hazardous working place. This is scaffolding/stage which is erected on the external walls of building. This is very tall for multistorey building. The scaffolding (single or double vertical pipes/bally/bamboo system) is to fastened with wall properly. Cross bracings are also to be done for more strengthening the fixing arrangement to prevent it from swaying from the building or structure. Accidents occur due to failure of scaffolding. In case multistorey building is constructed or renovation work is carried out and the building is located on main road or in congested area of city then it may be required to cover the building with protective barriers like, safety nets, protective coverings, covering the passage (with wooden planks/board) along footpath.

18.12.3 Old building/structure are demolished so that clear site is made available for new work. Demolition is to be done systematically after technical planning so that untoward incident does not occur. Electric and water connections are to be disconnected before starting demolition. Entire building is to be vacated. Required numbers of labours, skilled workers with welders, gas welding machines for cutting steel bars/frames, ropes safety belts, steel helmets, chisels, hammer of different weight/sizes etc. are to be arranged. Site engineer must be present for ensuring safety measures. Special care is to be taken for unstable portion of structures where demolition work is in progress. Heaps of debris should not be kept on floor. Demolition during night should be avoided. JCB machine with suitable attachments/accessories should be used for certain part of demolition, removing debris etc.

18.13 Safety Measure for Working with Construction Machinery

18.13.1 Experience has shown that many of the accidents occurred in the construction industry are attributable to faulty selection of construction machinery or, more often, to incorrect due to unsafe practices during using and operating the construction machinery. Hence management of construction machinery has become important part of project construction. There are many types of construction machinery, equipments, plants (stationary and mobile types). Certain equipments/machinery are technically intricate/ sophisticated/modern types; special care for operation and safety aspects for such machinery are to be taken.

18.13.2 Certain safety care is common for various machinery and some of these are stated below : –

(1) Drivers/operators must be skilled, experienced and trained for relevant machinery/equipments. Initial training should

be given by the manufacturer not only for operation but also for day to day maintenance, oiling, lubricating, and safety aspects.

(2) If machinery is parked along road during night then it is required to put red light lantern, red flags. Machinery should not be parked on slope ground. If it is on mountainous road then the same should be parked on hill side and extra precaution should be taken for gradient/slope of road. Manual parking brake shall be used.

(3) Parts of machinery shall be cleaned when the machinery is stopped. No part of machinery while in motion or in operation shall be examined/lubricated/ adjusted or repaired except by a duly authorised mechanic.

(4) *Drivers Cabin : –*

The driver shall be adequately protected from the weather by a cabin; if enclosed, shall be provided with windows that give unrestricted maximum possible view. Windows shall be provided with toughened safety glass.

(5) In placing construction machinery for operation, attention shall be given to the strength of foundation or structure on which the machinery may stand or move or be supported and to the proximity of hazard and in case of cranes, overhead electric wires/cables are to be checked. If ground is slushy/ soft it may be required to cheek bearing capacity for heavy machinery/plant.

(6) Fencing/cover of dangerous parts of machinery shall not be removed while the machinery is in use or in motion.

(7) Driver shall not leave the machine while the engine or motor is running or the load is suspended.

18.14 First Aid and Fire Fighting arrangement

(a) First aid box with certain medicines, bandage etc. are to be kept at work site.

(b) When operation of construction machinery is confined to one place or is within a restricted area of about 30 meters radius, "one fire point" shall be established at a suitable location. The fire point shall be provided with the following equipments (which are to be red painted) and the same are to be checked every week : –

(i) 5 kg Dry Powder Extinguisher conforming to IS : 2171 - 1972 – 2 Nos.

(ii) Sand Buckets – 2 Nos.

Where mobile type construction machinery is used then one 5 kg. Dry Powder Extinguisher shall be carried on the machine itself.

Operator, security guards should be trained about method of using aforesaid fire Extinguisher.

18.15 Safety Measures for certain Construction Equipments/ Machinery

18.15.1 Cranes/Hoist/Lift

(a) Cranes, hoisting/lifting machinery are used at site for hoisting, carrying materials at different heights and direction (horizontally). Vertical lift (temporary type) is installed on external wall of the building so that persons including supervisory engineers can reach at required location during construction. Special safety measures must be taken for above stated machinery (viz., cranes, hoist/lift).

(b) All practical steps shall be taken to prevent earth moving and lifting and hoisting machinery being operated in dangerous proximity to live overhead power line. If the power line is 400 kv then the machinery shall be at least 6.50 m away from the power line. Location of underground power cables are also to be located and safety measures are to be taken.

(c) Steel wire ropes and other ropes including terminal fittings shall be examined at least once a week to check that these are not frayed or damaged. Record for such checking is to be maintained with signature of checking/inspecting person. Above checking is most important as accidents occur due to failure of ropes or terminal fittings/fixing arrangement.

(d) All components, parts, accessories, fittings of machinery (crane etc.) and safety devices shall be inspected by authorised person periodically. This is to be recorded on inspection register.

(e) Driver shall not wear loose clothing and shall cover loose hair. Steel helmet as conformed by BSI code shall be used by all workers, drivers at work site.

(f) Erection and dismantling operations at site for lifting and hoisting machinery are to be done under supervision of competent person. The erection crew shall wear safety helmets and where necessary, safety harness.

(g) Stability of cranes :

Cranes shall be so operated as not to be liable to over turning by : –

(i) Taking load in excess of safe working load.

(ii) Rapid deceleration and by sever application of brake.

(iii) Sudden release of heavy load.

(iv) Pulling loads from the sides.

(v) Operating the crane without sufficient counter-weight.

(vi) Positioning the crane at a steep slope.

18.15.2 Batch type Mixer

Hoppers or batch type mixers shall be guarded by grating.

18.15.3 Vibrating needles of poker type vibrator shall be completely sealed against concrete. Electrically operated compactum vibrators shall be totally enclosed units.

18.15.4 Hot Mix Bituminous Plant

(a) Piping for hot oil and bitumen shall be adequately inserted to protect workers from injury by burns.

(b) Workers handling hot bitumen shall wear gloves, rubber boots, goggles, suitable protective clothing.

(c) A chimney or other suitable exhaust arrangement shall be provided to remove the combustion gases and dust.

(d) No open light shall be used for ascertaining the level of bitumen.

18.15.5 Electric Power operated Machines

Bodies of all electric power operated machines shall be well "earthed". This should be checked from time to time.

PART III

★ Environment Management related to Civil Engineering Works (Basic Aspects, Water Supply, Sewerage, Drainage and Solid Waste Management)

19

Environment Management related to Civil Engineering Projects

19.1 Basic Aspects

(a) It is required to know the issues of environment management related to Civil engineering projects. It is also required to know the environmental problems. It is no longer possible to ignore the needs of the society. The problems facing the environment are vast and diverse. The biggest impetus for developing a worldwide effort to monitor and restrict global pollution is the fact that most forms of pollution do not respect national or state or other boundaries. Issues concerning projection, conservation and management of the physical environment are to be addressed. Every activity generates unavoidable environment impact of some kind or the other; but all are not adverse. Ability of people and societies to adopt themselves and cope with the change is varied. Milking earth is akin to milking a cow. We have to feed and care for the cow before we can milk it. Sustainable development is important. There is tremendous pressure of population increase on the environment resources. As example, natural resource of wood has been badly affected for civil engineering construction works; due to this reason substitute for the same has been made in respect of many items/activities. The development initiatives be limited in such a way that the future generations can enjoy the benefits of nature without any compromise.

(b) 27 principles of environment management were adopted in an international conference arranged by the United Nations Organization held at RIO in 1992. These large numbers of principles are not dealt here. A few principles are mentioned here.

(c) As per principle – 4 : –
In order to achieve sustainable development, environmental protection shall constitute an integral part of the

development process and can not be considered in isolation from it. As per Principle – 8, – the state should reduce and eliminate unsustainable patterns of production and consumption and promote appropriate demo-graphic policies. As per Principle – 9, sustainable development should be done through exchange of technology, knowledge, transfer of technologies. As per Principle – 11, the states shall enact effective environmental legislation. Environmental standards, management objectives and priorities should reflect in these legislations. As Principle – 17 : – Environmental impact assessment, as a national instrument, shall be undertaken for proposed activities that are likely to have a significant adverse impact on the environment and are subject to a decision of a competent national authority.

(c) Following are some factors which can promote sustainable development : –

(i) Population stabilisation and health care.

(ii) Integrated land use planning and watershed management ensuring availability of adequate area for use as cropland, woodland and grass land for food, fuel, timber and fodder.

(iii) Water pollution control in rivers.

(iv) Air pollution control.

(v) Waste recycling and reuse.

(vi) Slum improvement.

(vii) Environmental education and awareness.

19.2 Major Environmental Problems

(i) Over use of chemicals, fertilizers in agriculture has degraded ground water quality.

(ii) High rate of increase in population has created lot of problems in environment. Various factors has created problems in reducing forest cover, increased soil erosion, silting of reservoirs and lakes etc.

(iii) The lopsided development policies have resulted in degraded soils, depletion of underground water tables, increased in water logging and salinity, pollution of water and air.

(iv) Adoption of less environmental friendly technologies has resulted in air and water pollution, which has made most of our rivers dirty.

19.3 Natural Resources

Physical environment provides resources for life like, – raw materials, energy, water, air, land, plants, fish, animals etc. Most of these resources are in the biosphere of the earth i.e. a 6 km thick layer on the surface of the earth which includes the atmosphere. Resources that the environment provides may be : –

(a) *Tangible* – Like, air, water, minerals, food, material etc.

(b) *Intangible* – Climate, regulation, storage, degradation of wastes etc.

(c) *Aesthetic* – Scenic, pleasant recreational and other pleasing features.

Resources may also be renewable and non-renewable.

19.4 Basic Aspects of Management

Environmental management is not "management of the environment". It is the management of activities within some constraints imposed by the government with full consideration of ecological factors. Environmental management includes the preparation of plans and legal evaluation of administrative and technical solution to various environmental problem in terms of both preventive and remedial measures, taking into account the multidisciplinary approaches to development. Environment management is understood to mean prudent or optimal use, maintenance and enhancement of both the quantity and quality of national and local resources, – primarily natural and mostly renewable such as land (which includes stone, earth etc.) water, air and inclusion of human.

Besides above environment management refers to all the systematic planned efforts by the policy makers, directed towards regulating and managing the utilisation of natural resources. So the term, environment, is used to refer to the immediate physical environment which serves as the base of drawing natural resources for human activities into which flows the wastages of residential, industrial, developmental activities.

19.5 Law Related to Environment

For effective Government on environment it is essential to enact laws through legislations. Many Acts and Rules (more than 15) had been framed and passed in the Parliament. On the basis of these Acts/Laws the Ministry of Environment and Forest Government of India had frame many rules, regulations, manuals. Detailed requirements related to environments, enforcement, penalty for non-compliance, monitoring, inspection, auditing have been stipulated in these Acts, Rules, Regulations. Notifications on above matters are also issued by the aforesaid Ministry. Certain legislations have also been passed by the State Assembly for certain local requirements which are not specifically covered under Acts passed by the Parliament. A few important Acts passed by the Parliament are stated as under :

(1) ***The Environment Protections Act of 1986 –***
This Act covers many aspects. Definitions of many terms related to environment have been stated in this Act. Section 3(I) of the Act empowers the centre to take all such measures as it deems necessary for the purpose of protecting and improving the quality of the environment and preventing, controlling and abating environmental pollution. The Ministry has adopted rules for effluent discharge and emissions for 24 designated industries.

(2) ***Environment Protection Rules, 1989 –***
As per this amendment the air quality standards in respect of "noise" had been brought under this rule. These standards lay down the day and night time limits of noise.

(3) ***Violations and penalties under the Act –***
As per the Environment (Protection) Act authority has been given to the Central Govt. to issue direct written orders, including orders to close, prohibit or regulate any industry, operation or process or to stop/regulate the supply of electricity, water or any other services.

(4) ***The Water (Prevention and Control of Pollution) Act, 1974 –***
The Act vests regulatory authority in the State Boards and empowers these boards to establish and enforce "effluent standards" for factories discharging pollutants into bodies of water. The boards control sewage and industrial effluent discharges.

(5) ***The Air (Prevention and Control of Pollution) Act 1981–***
This Act concern for the detrimental effect of air pollution on the health of people, animal life, vegetation or property.

19.6 Water Pollution

The quality of water is described under following four categories:

(A) Drinking after disinfections.
(B) Bathing, Swimming and recreation
(C) Drinking water after conventional treatment
(D) Any irrigation, industrial cooling and controlled waste disposal.

It may be stated here that in many reaches of rivers "coliform" count and high BOD (Biological Oxygen Demand) values are a common problem. Considerable quantity of sewage discharge in the rivers. River water is also polluted heavily due to effluent discharge from industries in the rivers. Several rivers and lakes are little better than open drains. Ground water pollution data is alarming. Another important sector is pollution due to solid waste. There is tremendous pressure on the environment due rapid increase of quantity of solid waste; the same is due to rapid increase of pollution particularly in urban areas. The Hazardous Wastes (Management and Handling Rules), 1989 is important.

19.7 Environment Impact Assessment

(a) Legislature and institutional regimes are to be developed and applied for the promotion of environmental management. In the absence of familiarity with environmental Legislation and the environmental impact of human activities, there is likely to be no spontaneous observance on Environmental protection. Environmental management on development of projects is achieved not only through laws, rules, regulations but also through administrative mechanisms, technical standards etc. Environmental impact assessment in respect of development projects and procedures are to be followed at initial stage of planning. Hence Environmental legislation, rules regulations, technical standards are the main tools for project planning and implementation.

(b) Environmental Impact Assessment (EIA) may be defined as: It is a formal process used to predict the environmental consequences of any development project. EIA thus ensures that the potential problems are foreseen and addressed at an early stage of the project planning and design. The purpose of EIA is:-

(1) To support the goals of environmental protection and sustainable development.

(2) To integrate environmental protection and economic decisions at the earliest stage of planning.

(3) To predict environmental, socio-economic and cultural consequences of the proposed project and to assess plans to "mitigate" any adverse impacts.

(c) As a planning tool, EIA has both an information gathering and a decision-making component. Sustainable development by protecting the natural resources, which is the foundation for further development. If problem is faced to approve a proposed developmental project on account of environmental adverse effects then it is required to work out alternative proposals which are environmentally sound after taking mitigation action. It is essential to prepare EIA particularly for certain, large public investment, in areas of infrastructure and this should be linked to the cost-benefit analysis. The objective of EIA is to ensure that environmental aspects are addressed and potential problems are foreseen at planning design stage.

19.8 Initial Environmental Evaluation Report and Categorisation of Project

(a) All proposed projects should initially undergo an Initial Environmental Evaluation Report (IEE). Thereafter categorisation of project is made from point of environment. The categorisation of each project is made according to the physical and ecological characteristics of the project site and its surroundings, the current land use and landscape character, cultural resources, socio-economic characteristics of the affected public, infrastructure services, existing social and community services and facilities, health and safety risk, current and predicted services and facilities, current and predicted pollution level and availability of cost effective mitigation measures. Following are three categories of projects : –

Category "A" –

Projects with significant adverse environmental impacts; these require full I.E.A.

Category "B" –

Projects with adverse environmental impacts but of a lesser degree than Category "A" impact. In this case, the IEE will determine what are the issues the EIA should address.

Project "C" –

Projects unlikely to have adverse environmental impact or those that will improve the environment. An E.I.A is not normally required for such category of project.

The inclusions of the IEE and project classification will be included in the Project Proposal Report.

(b) Environmental Assessment (EA) covers project – specific and other environmental impacts in the area of influence of a project. Environmental Considerations are to be given adequate weight in all stages of the project planning & design. The IEE should be completed during project identification and prior to submission Project Proposal Report.

(c) The capability of the implementing agency, the adequacy of their environmental procedures and their ability to comply with environmental requirements should be examined at the time of appraisal. During implementation of the project the "environmental mitigation" and "management plan" as developed during process should come into operation. Ministry of Environment and Forest, Govt. of India have developed procedures, rules, regulations, Manual, Guidance, Check List, Questionnaires for preparation E.I.A. and various parameters, standards/ norms. EIA clearance is required for 29 categories of industries (latest list is available in the website / notification of the ministry). The regulation governing EIA should indicate which projects are subjected to EIA procedure and which are not so that project process is not delayed. On the contents of EIA, the laws/rules may provide for submission of written documents to designated body. Some of these documents are as under :–

(1) Description of proposed project/activity.

(2) Description of the potentially affected environment.

(3) Description of practical alternatives.

(4) Assessment of the likely or potential environmental impacts of the proposed activities and alternatives including the direct, indirect, cumulative, short-term and long-term effects.

(5) Identification and description of measures available to mitigate environmental impacts of the proposed activity and alternatives.

(6) Indication of whether the environment of any other state are likely to be affective by the proposed activity/ project.

(7) *Regional master planning :* –

In this case, an entire river basin (a watershed), an air quality district, a managed ecosystem, a costal zone, or an island is analysed for the probable environmental impacts of all types of economic development. Current and future pollution controls are estimated.

(8) Selection of technologies and development strategies are important factors for environmental assessments.

EIA is important to play an increasing role for ensuring the projects meet the sustainability criteria. EIA of the future should take a leading role in playing various projects for waste minimization, pollution prevention etc.

19.9 Procedure for E.I.A.

One notification dated 27.01.1984 was issued by the Ministry of Environment and Forests, Govt. of India about procedure to be followed for obtaining their environmental clearance of the proposed project. There is a schedule of projects for environmental clearance which are to be obtained. The application on prescribed proforma is to be submitted to the Ministry of Environment or to their department /agency. Project report alongwith E.I.A, Environment Management plan, details of public hearing (if it is required) are to be prepared as per guidelines issued by the Ministry. Environmental clearance may not be required for widening and improvement of existing highways. The Impact Assessment Agency which has been set up by the Ministry prepared recommendations based on technical assessment of documents and data furnished by the project authorities. The aforesaid agency is also to monitor the implementation of their recommendations. The EIA need a significant amount of primary and secondary environmental data. Primary data are collected by carrying out field survey on the environment (viz., air and water quality data etc.). The secondary data are those data which have already been collected by various organisations over the years and can be used to understand the existing environmental scenario of the project study area. In many EIA studies, the secondary data needs could be as high as 80% of the total requirement. The Ministry has also set up Environmental Information Centre for such purpose. It is required to ascertain potential impacts on account of proposed projects during construction and operation phases such as : –

- Vegetation destroyed or disturbed.
- Wildlife habitat reduced and broken up
- Wildlife destroyed or displaced.
- Bird and wild life migration disrupted.
- Animal behaviour disrupted by noise and activities (like, due to trains, aeroplanes, heavy vehicular traffic etc.)
- Soil erosion
- Effects of air and water pollution on plants and animals.
- Fisheries destroyed.

19.10 Questionnaire for Environmental Appraisal

The formations of questionnaire for various types of projects are prepared by the Ministry/ their department/ agency. These formats are lengthy. As an example following is stated in respect of Road/Highway projects.

Many information, data, particulars are to be stated on the prescribed formats alongwith annexure (Statements etc.). A few of these are mentioned below : –

Length of road/highway, names of villages, tehsil, district, state alongwith kilometer, geographical information, total area of project, nature of terrain, soil, current land use, alternate routes/alignments, reason for selecting proposed alignment, land use plan, site preparation, gradient details, cutting of trees, tunneling (if any), effect on CRZ area, position of flood plain of river (if applicable), location of water bodies, sea, land reclamation (if any), effect of existing drainage pattern, effect on migratory path of animals, project details (with summary of project proposal), raw material and water requirement for construction, present and future effect on ground water, certain environmental effects within 7 km of proposed project site, solid waste, air quality, noise level, storage of construction materials, deployment of manpower during construction, socio-economic profile (human settlement, economic activity), rehabilitation and resettlement plan etc.

19.11 ISO 14000 and Environmental Management

The international environmental standards are intended to provide organisations with the elements of an effective environmental management system and the same can be integrated with other management requirements to assist organisations to achieve environmental and financial goals. These standards specify the core elements of an environmental management system to support environmental protection in balance with socio-economic needs. The current international standards on environmental system cover the following areas : –

- Environmental management system.
- Environmental auditing
- Environmental labeling
- Environmental performance evaluation

ISO 14000 builds a single global management system that allows effective management of environmental responsibilities, liabilities, costs, document commitment to Government and promote concern for the society. Lack of an effective Environmental Management System and proper guidelines cost the company /organisation. The task of management is to control internal costs and maximize market opportunities by enhancing the product and company's image.

ISO 14000 – Guide to environmental management principles systems and supporting techniques.

ISO14001 – Environmental management systems : –

Specification with guidance for use. This is the only standard intended for registration by third parties. All other ISO 14000 series are for guidance. ISO 14001 is a management standard; it is not a performance or product standard. It is comprehensive, proactive, system approach. As per this ISO requires an environmental policy to be in existence in the organisation.

ISO 14010, 14011, 14012 – Environmental auditing.

ISO 14031 – Evaluation of environmental performance.

19.12 Benefits of ISO 14001

(a) Intangible benefits : –
- Continuity of operations
- Clear objectives and targets
- Clear job procedures, responsibilities and accountability
- Increased communications
- Improved community goodwill

(b) Tangible benefits : –
- Potential cost savings.
- Reduced raw material usage.
- Reduced waste disposal costs
- Recoverable resources
- Reduced penalties.

19.13 Planning/ Environmental Management System (EMS) as per ISO 14000

There are two phases in establishing as ISO : 14000. It is required to create the EMS and getting registered. As first step it is required to constitute a Project Committee. This committee is to take action for creation of EMS. This committee prepares briefing for top management, EMS, cost, benefits. Then top management is required to commit for processing EMS and implementation. It is also required to prepare draft environmental policy. E.I.A is to be prepared and the same is very important. Procedure is to be prepared for identifying environmental aspects of the operation. It is also required to identify legal and other requirements. It is also important to prepare detailed programme for achieving targets and objectives. EMS is to be approved by the competent authority. Detailed action plan is prepared activity wise for implementation and actually implemented on ground. Necessary monitoring, review of progress is to be carried out during implementation. Audit function as per ISO is to be established. Organisation is to apply for certification.

20

Water Supply Management

20.1 Introduction

Water is an important part of environment. Broadly following are the two categories of water in the planet of earth : –

(i) Sea water

(ii) Non-sea water

(viz., rivers/streams, lakes, water falls/snow in the mountaineers regions, ponds, underground water/ aquifer, surface runoff of rain water etc. are non-sea water).

The sea water can not be used for most of our requirement. Hence we have to manage our various requirements from other aforesaid sources of water. As dealt under section of Rain Water Harvesting one basic source of good water is rain water. Rain water flows over the earth surface and also percolates in the underground water reservoirs (i.e. aquifers). Requirements of non-sea water is grouped as under : –

1. ***Rural Areas*** – Mainly for irrigation purpose and also for drinking and other purpose.
2. ***Small Towns*** – Mainly for drinking purpose and also for industrial and other purposes.
3. ***Large Towns, Cities, Mega City*** - Mainly for drinking purpose and industrial and other purpose.

Practical Aspects

Management of water supply is dealt broadly on ground in respect urban and rural areas in India as under : –

20.2 Water Supply in Rural Areas

Agriculture sector is largest in India. Huge quantity of irrigation water is required in the rural areas through out the year. Agriculture programme is made specially during rainy season so that rain water is used maximum. Rain water flows over the earth surface and enter in the ponds, lakes, rivers, streams, surface reservoirs, water also flows down from snow bound mountainous areas. Huge quantity of surface water flows through rivers and discharges into the sea; this part of water goes as waste. Canals/channels are made from source of water to crop fields as gravitational flow. At crop field if required water is pumped from the canal/channel. It is required to check the levels of ground from source of water to crop field.

In hilly area catch water drain is made at right angle to vertical streams at certain interval. Water flows from these drains to crop fields (or for drinking purpose). During summer season or certain period of year adequate quantity of surface water (from river, stream, ponds etc.) is not available then it is required to obtain under ground water by installing deep/shallow tubewell or dug well. Water is arranged by pumping. Availability of adequate quantity of underground water depends on aquifer size. It is also required to check the depth of underground water table. The water table is depleted during summer season and some problem is faced for pumping huge quantity of water for irrigation purpose. For raising underground water table the rainwater harvesting will be dealt under this chapter. Informations regarding quantum/size aquifer may be available from the department of Central Ground Water Board, Ministry of Water Resources, Govt. of India.

In India rural drinking water systems are primarily dependent on ground water, with break-through of "India Mark -II" hand-pump in late 1970, the programme of provision of safe drinking water to its rural populations was predominantly implemented through installation of hand-pumps all over the country. Taping the aquifer more than 100 meters below ground level is possible through India Mark-II hand pumps.

Safe source of drinking water in rural area is from underground source. Deep tubewell is safe for drinking water. Shallow open dug well may not be safe for drinking water. However precaution should be taken to cover the open dug well. Some simple type treatment (like, chlorination/boiled water) should be made to water collected from open dug well. Contaminated water is main reason for many diseases in rural areas. Shallow dug wells, ponds are main source of contaminated water and the same should not be used for drinking purpose without some treatment. It may not be difficult to install tubewells for drinking water of small population in villages. Shallow hand-pumps are also not safe for drinking water. However hand-pumps are useful for other purpose.

In many villages electricity is not available; in some villages low voltage electricity is available for very short period. Hence diesel pumps (with 100% standby arrangement) of suitable capacity is important. Its operation and maintenance are most important.

The operator should have practical experience in maintenance and repairing works also. Sufficient spare parts should be purchased in advance. Sufficient length of hose pipes should be available for irrigation purpose.

From above position it may be seen that underground water source is important in case of water supply in rural areas for purpose of irrigation, drinking etc. Problems faced in certain areas/regions during summer and some other period due to non availability of underground water at reasonable depth and from surface source. This aspect will be dealt separately.

20.3 Water Supply in Small Towns

Requirement of water is mainly for drinking , cooking, washing purposes and also for industrial purpose. Surface water (viz. rivers, stream, lake, ponds) and underground water are sources. It is not economically feasible for filtration of river/pond water for supplying drinking water in small towns. Hence water supply in small towns depends on underground water source. Water supply system/scheme of the town is prepared after making necessary surveying and level. Quantity of drinking water is worked out on the basis of projected population for suitable design period and standard norm of water consumption. Number and capacity of tower reservoirs and its locations are designed and decided. Chlorination treatment of underground water is done at inlet points of distribution network. Public hydrants are also installed at suitable locations with special type of taps. Special precaution is to be taken for reducing wastage of water from these public hydrants. It is essential to make proper drainage arrangement at these public hydrant points to maintain good sanitary condition. Problem may be faced during summer season due to less underground water in certain region/area. This aspect will be dealt separately. If population of town is more and adequate quantity of underground water is not available particularly during summer season but water is available in nearby river/lake/ large ponds then it may be required to install filtration plant. Quantity of drinking water shall be as per standard norms as approved by the Central Public Health & Environment Organisation, Govt. of India. Regarding requirement of water for industrial purpose (including construction industry) aforesaid quantity norms are not required. Certain lower standard of norms related to quantity as laid down for various industries is relivant. Effluent of sewage treatment plant (after certain treatment) can also be used for boiler, cooling tower system in industries.

Capacity/size of town reservoir is to be designed after working out its designed requirement of water supply for the command area of reservoir. It is important to find out supply pressure at the tail end of distribution network; it should adequate at least for two storey buildings. If the head loss is more at tail end then it may be required to install "booster pump". Number of tower reservoirs, its locations and height with suitable command areas are to be decided carefully so that water pressure at tail end of network is sufficient. Levels/contour map of the town are required for aforesaid purpose. Design and selection of pumps are important. There should be 100% standby pumps (electric and diesel pumps) of suitable capacity. The operation and maintenance of pumps are most important. Sufficient spare parts are to be purchased in advance. The pump operators and mechanic should have practical experience in such type of job.

Quality of drinking water is to be checked frequently after collecting samples from the taps of consumers and public hydrants. Chlorination of tubewell water is to be done before supplying to distribution network.

Water Supply in City

20.4 General

Practical experience and technical knowledge are essential for planning and designing of water supply scheme in respect of city. This is a specialised subject. Such scheme is prepared for new township, subcity of existing mega city, augmentation/ improvement of existing water supply system of city, large town, some colonies/ zones of existing city. Practical aspects of planning, management of conventional and non conventional projects have been dealt under Part –II of this book. Basic principles, certain management tools (like CPM network planning, Bar chart planning, progress charts/ network etc.), monitoring of various activities during pre-construction period, construction stage etc. are also applicable in water supply scheme/management. There are special features and special/ emergency activities in the operation and maintenance of water supply system of city. These activities of post construction period are different than other construction industries. Success of water supply scheme depends on efficiency of operation and maintenance. There is special role of mechanical and electrical engineers also for operation and maintenance of pumping system, operation of filtration plants and other machinery/equipments. Maintaining adequate water pressure in entire system/area (particularly at tail ends of distribution network) is very important and the same depends on correct designing of the scheme. Designing of scheme will not be dealt in this book. Certain practical problems and its solutions will be dealt very briefly from management points of view. Special stress should be given at planning and design stage. If there is design fault then it will be very difficult to rectify the same after construction/ installation. Practically serious problems are faced by consumers for non-supply of water or supplying water in very less pressure at consumer's premises (i.e. in over head tanks of two or three storey buildings); such serious problem is faced particularly during summer season. Such problem is due to faulty design of the water supply scheme. Another important matter is "quality" of drinking water with reference to standard norms as approved by Central Public Health and Environmental Organisation, Govt. of India. The quality and its management will be dealt at the end of this chapter; contamination of water after treatment (i.e. in distribution network and in the premises/service connection line of the consumer)is also important. Consumers may fall sick due to collera, tyfoid, jaundice, diarrhoea which are on account of contaminated water. Certain aspects of pre-construction stage from management points are stated briefly below.

20.5 Planning

After finding out the necessity of potable water for certain area/ zone (new proposed township/sub-city/ new colony/augmentation of existing city) it is required to obtain map of the city from city development authority. Detailed map of proposed area showing all roads, streets, lane, by-lane are to be arranged. The same is to be checked at site and necessary survey is to be carried out for updating the aforesaid maps, area plans. It is required to prepare feasibility study report. Broad particulars are stated in this report. After approval of the same it is required to prepare the "Project Report" which contain several informations. At this stage it is to be decided about standards/norms on various matters like, design period, per capita supply, population forecast method, method for measurement of yield/flow, type of water treatment, specifications of materials (particularly pipes). There are many components of water supply project, like : –

Source of raw water and its transmission to water treatment location (if necessary, by constructing dams), pump house and pumps/ motors, water treatment arrangement/plant, raw water and clear water conveying mains, clear water reservoirs, balancing tanks, service reservoirs, distribution system/network. Design period varies for different components from 15 to 50 years; water treatment units is 15 years (as per manual of CPHEEO). Per capita requirement varies for use of different purpose like domestic needs, institutional needs, industrial and commercial uses, for livestock, likely waste/ leakages in the system (which is very high / alarming in India, – may be 35% to 40%). Domestic need for communities with population 20,000 to 1,00,000 is 100 to 150 litre per capita per day (lpcd) and population of above 1,00,000 is 150 to 200 lpcd. Norms for various uses (like, Hospital, Hotels, Hostels, Restaurants, Schools, Offices, Factories etc.) have been laid down in the manual of CPHEEO.

20.6 Water Pressure

For design/planning the water pressure at supply point (i.e. taps of consumers) is very important. Continuous supply (i.e. 24 hours) is preferable from technical point and consumers' requirement. But due to shortage of water the intermittent supply (viz. morning and evening) system is followed in India. The minimum residual pressure at ferrule points should be 7 m for direct supply. For two storeyed buildings it may 12 m and for three storeyed buildings it may be 17m.

20.7 Quality Standards

(a) The objective of Water Works Management is to ensure that the water supplied free from pathogenic organisms, clear palatable and free from undesirable taste and odour, neither corrosive nor scale forming and free from undesirable minerals/chemicals. Standards/norms of various elements of potable/drinking water have been laid down in the manual of CPHEEO, Govt. of India under following four groups : –

(i) ***Physical and Chemical –***
Like turbidity, taste & odour, pH value, total hardness, chloride, fluorides, iron, magnesium etc.

(ii) ***Toxic Material –***
Arsenic, Cyanides, lead, chromium, mercury etc.

(iii) ***Radio Activity –***
Goss alpha and beta activities

(iv) ***Bacteriological –***
Faecal coliforms, coloform organism etc.

20.8 Designing

(a) It is important to work out estimated requirement of water for various purposes on the basis standard norms as state above for the project area. For this purpose it is required prepare forecast demand of water for the designed period (which has also been stated in foregoing page). There are different methods for projecting population. There are losses of raw water and treated water during transmission through pipe line networks. The loss of leakages of potable water in distribution system is very high in India (viz. 35% to 40 % of total actual treated water quantity). This aspect must be considered as super-critical in water supply management. This depends on proper design, specifications of materials (particularly pipes) and maintenance of distribution system. This important matter will be dealt under this chapter.

(b) It is essential to carry out detailed study on the resources of raw water; its locations, types of sources, quality, quantity of raw water available throughout the year etc. It is crucial to know the availability of quantity of raw water particularly during summer season. Practically it was found in certain projects that no raw water could be drawn from the source as stated in the project report after completion of construction of project (in a period of 4 to 8 years) or very little quantity of could be drawn after 3 years of completion of project. Due to this reason the project could not be commissioned/operated for long period. Hence it is essential to obtain written commitment from competent authority of State Govt./ District Authority about allowing for project authority for drawing specific quantity of raw water by specific date. This commitment letter should be made as part of project report. As this was not done serious crisis was faced due to non-availability of water and infructuous expenditure was incurred. Percentage of losses of raw water due to various factors/reasons is to be decided and effected on the aforesaid raw water quantity. Similarly percentage of realistic losses of treated water due to leakages and other reasons in the distribution system is to be assessed properly. Considering above stated position it is required to work out correct total quantity of raw water which is to be transmitted (i.e. conveyed from the resources). Long term development plans should be prepared in consultation of city development authority. It is essential to prepare detailed "Operation and

Maintenance Manual". Success of water supply also depends critically on the "O & M" of the entire water supply system. Type/specification of pipes and its fittings, valves and jointing etc. are most important for reducing losses and to increase life and to reduce contamination of water. Environmental impact (beneficial and adverse) is to be worked out and to be included in the project report. Mitigation measures are to be taken on adverse impact.

(c) Yield /availability of water from various sources (viz, rivers, streams, spring, lake, ponds, dug wells tubewells etc.) should be measured so that correct quantity of raw water is considered. There are various methods of measurement of flow. The capacity of flowing streams and natural lakes is decided by the area and nature of the catchment, the amount of rainfall and alied factors. The yield of subsurface sources is decided by the hydrological and hydrogeological features relevant to each case. There is huge quantity of ground water resources in India; but the same is not evenly distributed. In certain areas this source is meagre and some where it is not fit for drinking. Ground water is mainly available in the sedimentary rocky strata and topmost soil layer of earth. It is stated that transmission of water accounts for an appreciable part of the capital outlay/project cost. Hence careful consideration of the economics is called for, before deciding on the best mode of conveyance. While water is being conveyed, it is necessary to ensure that there is no possibility of pollution from surrounding areas.

(d) There are different pipe materials such as : –
Cast Iron, Steel, RCC, Prestressed Concrete, Asbesters Cement, Polythylene, Rigid PVC, Ductile Iron, Fibre Glass Pipe, Glass reinforced Plastic, Fibre reinforced Plastic.
Ductile Iron pipe is better than cast iron pipes. Steel pipes are used at certain places. Prestressed pipes are recommended for conveying raw water and main water supply pipe (of larger diametre). While RCC pipes can cater to the needs where pressures are upto 3.0 kg/sq. cm. and C.I. and steel pipes cater to the needs of higher pressure around 24 kg/sq. cm. The P.S.C. pipes cater to intermediate pressure range for which the metallic pipes are expensive while RCC pipes would not be suitable. Main advantages of PVC pipes are : –

(e) Resistance of corrosion, light weight, toughness, rigidity, economical in laying. During detailed designing special care should be taken to provide various appurtenances like, sluce valves, butterfly valves, globe valves, scour valves, air valves, pressure relief valves/ pressure reducing valves, check valves, ball float valves, automatic shut-off valves etc. Design should be made correctly on account of "Water Hammer", Achorages, expansion joints. Special care should be taken for "water hammer" during design process. It is important to select, design and locations of above stated appurtenances carefully so that operation and maintenance of entire water supply system can be done effectively and easily. Entire system is to be divided in some maintenance/operation zones/ pockets.

20.9 Water Treatment and disinfection

In case of ground waters and surface water of good quality (with protected storage) and where the water has very less turbidity (below 10 NTU scale) and they are free from odour and colour, then the plain disinfection by "Chlorination" is adopted before supplying to distribution system. In case of surface waters with turbidity at medium level (not exceeding 50 NTU scale), the plain sedimentation followed by slow sand filtration and disinfection are done. Conventional treatment including prechlorination, aeration, flocculation (rapid and slow mixing) and sedimentation and post chlorination are adopted for highly polluted surface waters laden with algae/other micro organisms.

After testing the raw water of source above stated treatment plan is to be designed by competent experienced water engineering designer.

Modern disinfection process include use of :

(i) Physical methods like, thermal treatment (i.e. boiling water)

(ii) Chemicals including oxidising chemicals such as chlorine and its compounds, bromine, iodine, potassium permanganate ozone.

(iii) Radiation

Disinfection and sterilisation are different processes. Disinfection aims at selective destruction of disease producing organisms. The sterilisation is employed to completely destroy or inactive all micro-organisms including bacteria, algae, viruses.

20.10 Loss of Water and Remedial Measures

20.10.1 General

Limitation in the growth of source of raw water and the importance of potable water due to line loss and leaks has increased considerable. The problem can worsen further if the leaks are not plugged in time. In certain position it is required to replace very old/outlived corroded pipe lines on priority basis. Leakage of water can take place at :

(i) Intake and conveyance system from main water source(as raw water)
(ii) Treatment plant
(iii) Service Reservoir
(iv) Distribution Network.

Out of all above the leakage/loss of water in the distribution system is the biggest culprit (may be 60 to 70% of total leakage/loss of water). Total loss/leakage of water in the systems in India is very high (about 45 to 50%).

Following are the effects of leaks in the water supply system :

(i) Lower pressure of water supply
(ii) Less quantity of supply to the consumers
(iii) Contamination due to reverse flow in the treatment low pressure lines.
(iv) Deterioration of roads causing traffic hazards and other nuisances.
(v) Soil movement/erosion and adverse effects on adjacent structures and buildings.
(vi) Loss of revenue

The defective installation / construction coupled with inefficient operations and insufficient maintenance has resulted in very high leakage levels, as high as 50% in some cities in India (based on certain records). Leakages and loss of water is considered here as total loss/leakage in city water supply system. High leakage rate coupled with low tariffs and high subsidies affect the financial performance badly. Total requirement of potable water in large cities may vary from 600 MGD for population of 10 million to 1295 MGD for population of 18 million. Hence, if there is total loss/ leakage of water is 50% then loss of quantity of water is about 300 MGD to 647.50 MGD. Such huge loss of potable water is a serious situation when there is acute shortage of water in cities / towns particularly during summer season. Besides loss of revenue, there is another important aspect. Serious problem is being faced for production of potable water by treating raw water due to scarcity/ non-availability of raw water. Problem is also faced to convey raw water from a far away place from city (may be more than 600 km). Thus it is difficult to get raw water even after incurring huge expenditure. There is also shortage of underground water. There may be disastrous situation (viz., water famine) in the city during summer season due to acute shortage of potable water. It may be stated here that loss / leakage of water in many developed countries may vary from 8 to 12% only. Hence, it is essential to take necessary action to "detect" leakage / loss of water in potable water supply system "scientifically". There should be a separate Leak Detection Unit in the Water Supply Department. Proper management and continuous monitoring of Leak Detection Unit and its remedial measures are required. Suitable instruments are to be used for detecting leakages.

Types of Leakages/Losses

(a) *Physical Leakage*

These are identified by visual inspection at regular intervals. Underground leakages which are not visual are to be detected with suitable leak detection instruments.

(b) *Commercial Leakage*

This is due to inaccurate metered connections.

(c) *Unauthorized Connections / Theft of water*

Physical survey should be carried out.

(d) *Wastage of water*

There is wastage of considerable quantity of potable water from public toilets / conveniences and other public taps. There is heavy overflow from overhead tanks of buildings due to non-functioning of float valves, and also due to excess pumping of water from underground / surface tank. There is misuse of potable water, like – using potable water for horticulture/gardening purpose. As bills are prepared on flat rate basis, consumers do not bother about the above stated wastage and misuse of water. Wastage / loss of water is reduced if water meters are installed in the premises of consumers.

(e) Public hydrants are installed at streets of small towns, urban villages, JJ clusters / slums. Taps / valves of these hydrants are not closed by public and there is huge wastage / loss of water. It is required to reduce the number of such hydrants and to provide automatic closing type taps.

Leakage may also be classified under following two types :

(i) *Ground leakages*

Such leakages are seen on the ground and can be detected visually by normal water supply / road maintenance gang / staff who are in-charge of various areas / sub-zones. Leakages are found on ground / surface from sluice valves, air valves, flowing public hydrants, tampered / damaged pipes etc. and the same can be detected easily by maintenance labour gang and staff during their daily inspection in the areas.

(ii) *Underground leaks*

For detection of underground leaks it is required to set up special "Leak Detection Unit" in Water Supply Department. This task is to be carried out scientifically with suitable instruments.

It is essential to prepare / update the detailed maps of water supply distribution networks showing all branch and main pipelines, service reservoirs, boosting stations, various valves, etc. Survey is also to be carried out for service water connections (area / zone / sub-zone wise).

20.10.2 Records

(a) To achieve the objective of planned work of leakage detection it is necessary to sub-divide the entire zone in convenient numbers of zones. It may be preferable to have sub-zone of area covering about 250 water connections. Close network of distribution pipes with necessary valves should be there so that leakages of a particular sub-zone / area can be checked by operating valves during supply/non-supply hours.

(b) Reports and Records related to field works on leak detection and remedial works

Proper permanent records are to be maintained in respect of field checking of various zones / sub-zones, instrument readings. Reporting at the time of carrying out the test either by sounding or stop tap method is to be done; this will enable the engineering authority to know the following:

(i) Zones / sub-zones tackled

(ii) Improvement in pressure after the work of leakage arresting is over

(iii) Types of leaks arrested and statistical data

(iv) Number of buried sluice valves and its repair / replacement

(v) Particulars of unauthorized water connections (including size of connection)

(vi) Cost of materials, labour, transport etc. for carrying out field checking and rectification works.

20.10.3 Measures / Methods to Reduce Wastage / Leakage of Water

After carrying field checking, using leak detection instruments (zone / sub-zone wise) the same are recorded properly. Sources of leakages and wastage of water are identified. Certain measures / rectification of leakages are taken immediately. If required further planning in respect of certain leakage, damages, replacement of pipeline, valves as permanent / preventive measures is to be made for future implementation. The possible methods for remedial measures to stop leakages, wastage of water (as immediate and preventive) are stated as under :

Location	*Possible methods to reduce the wastage*
A. *Head Works*	
1. Evaporation in lakes / reservoirs	Spreading chemicals such as Alcohol, spreading floating plastic balls etc.
2. Loss of Run off due to precipitation	Judicious solution of catchment areas and afforestation
3. Overflow from one of the two neighbouring catchments.	Appropriate drainage channel across the ridge position
4. Wastage through dams / weirs (i) Construction joints (ii) Sewage through imposition galleries (iii) Expansion joints (iv) Inefficient under-screwing arrangements	Proper designing using good material and adequate supervision In certain large project raw water is conveyed through canal/channel (particularly when source of water is far away). If proper lining is not constructed then there will be lot of loss water. Repairing of lining is also essential.
B. Conveyance System	
Pipe joints e.g. cement joints lead joints	Good workmanship and use of required quantity of lead
Expansion joints of steel Trunk Mains	Provide good packing materials
Pipe fittings and apparatus	Properly maintaining of these apparatus
Corrosion due to corrosive nature of soil	Coating pipe with anticorrosion paints, fibre glass wrapping

Location	*Possible methods to reduce the wastage*
C. **Theft of water and tampering**	Keeping good and strict vigil
D. (i) Isolation of Transmission Main (ii) Pumping Station	Before programme of work can certainly save repeated isolation Excessive and uncontrolled pumping should be stopped or else sumps will overflow.
E. **Treatment Plants** Sludge bleed in purifiers and bank wash of filters.	Proper selection and jurisdiction operation use, compressed air and water for Back wash, intermittent destroying may be helpful. Discharging wastes in the lakes and recovered after proper treatment.
F. **Distribution system, Water mains and distributing pipe lines**	Replace, repair or condition these mains.
G. **Plumbing works**	Corroded surface and overhead tanks should be replaced. Replace tap washers. Corroded permits boards to be replaced. Close taps when not in use.

20.10.4 *Some Important Measures for Leakage Prevention*

There are following three categories of measures for leakage / loss of water :

(A) Fundamental measure
(B) Symptomatic measures
(C) Preventive measures

Various items under above stated three categories are stated as below :

Department	*Item*	*Measures*
A. Funda-mental measures	Preparation for leakage prevention	Security of sources and organisation consolidation of documents (piping drawing, section drawings, etc.), establishment or area, consolidation of metering equipment
	Investigation of real situations	Analysis of distributed water volume and of leaked water volume, measurement of water pressure, determination of cause of leakage, analysis of leakage

Department	*Item*	*Measures*
	Study improvement and development of materials	Material of distribution pipes and service pipes, joints and ancillary equipment
	Development of technology	Method of measurement of leaked water volume, method of detection, method of repair
B. Symptomatic measures	Mobilized work Planned work	Immediate repair of ground leakage.Early detection and repair of underground leakage
C. Preventive measure	Planning of water works	Planning with leakage detection taken into account
	Design and construction of water works facilities	Earthquake resistance, durability, corrosion resistance, water tightness
	Replacement of old pipes (from which leakage frequently occurs)	Replacement of distributing pipes and service pipes (including change of pipe)
	Improvement of structure of water service installations	Strengthening of pipes which cross the roads
	Protection of pipelines	Location of water meters in positions which are as close as possible to government-private boundaries, provision of corrosion protection and leakage protection fittings, and reinforcement of bends
	Disposal of remaining pipes	Complete disposal at branched points, through management of water service installations
	Patrol of pipelines	Guidance and supervision for preventing damage caused by construction works executed by other enterprises
	Adjustment of water pressure	Division of distribution system, provision of reducer valves

20.11 Management of Water Quality

(a) Various sources of water have already been dealt. Quality of raw water of these sources are to be checked by testing in the laboratory. Impurities are under different types viz., Physical, Chemical, Toxic materials, Radio activity and Bacteriological. Maximum impurities are found in the river particularly in the plain areas and when it passes through populated areas. Minimum impurity is in the rain water collected in the roof top. Ground water is less contaminated. Physical impurity is less in the water of lake, surface reservoir as it works as sedimentation tank. Before taking up projects of water supply chemical & process industrial project etc. It is required to find out data on water quality of relevant/available water sources. Then type of treatment of water and main project planning, technology, type of project, industrial type of process etc. are decided at project planning and designing stage. Quality of raw water may be deteriorated during its conveyance/transmission from source to project location. If it is conveyed through canal then contamination will be much more than the pipe conveyance system. If the source of water is from deep aquifer of ground water then good quality is found; only disinfection (viz. Chlorination) treatment will be required. If the aquifer is at shallow depth then there is possibility of some contamination. In coastal area there is possibility of intrusion of sea water in the underground water source and river water; due to this reason salinity of water is to be checked. It is costly to treat for removing salinity.

Contamination of potable water is found at receiving ends i.e. in the taps of consumers' premises. These contaminations are occurred in the water supply distribution system and also in the service connections after ferrule point and also in the underground/surface and overhead tanks of the consumers. If such contamination is due to entry of sewage in the potable water supply pipe line then it can create serious problem to consumers on account of water-borne diseases. Samples of waters are collected from taps of consumers and also from the municipal distribution system. It is essential to find out residual "free chlorine". Overhead and underground tanks of premises are to be cleaned 2 or 3 times in a year by the consumers. On many occasions G.I. pipes are found porous/damaged and the same are laid adjacent to sewer line and it passes through sewer manholes. This position is serious and water pipe lines should be shifted away from sewer line/manhole. Potable water is also contaminated in the municipal water pipe lines from leakage points. Leakage locations are to be traced out by checking with leak detection instruments. Then leakage are to be rectified. If condition of old pipe is very bad then the same is to be replaced.

20.12 Water Testing Laboratory

The laboratory tests of water are carried out in respect of various elements/items under categories of physical, chemical, bacteriological and radio activity as laid down in the manual of CPHEEO, Govt. of India. These tests are carried out in respect of raw waters of various sources and at the end of final treatment/ disinfection (i.e. at the point of supplying treated/potable water in the distribution system containing minimum residual chlorine. The role of laboratory is very important. It is required to set up the laboratory properly with necessary instruments, equipments (as per lists given in the manual). It is most important to deploy experienced competent technical persons of different levels; they must be qualified on this specialised subjects (viz. in the fields of bacteriology, microbiology, toxicology) and they should have sufficient practical experience to carry out tests independently. As per ground reality it has been found in many laboratories that junior Laboratory Assistant carry out the tests but test checking of same sample water is not carried out by a senior laboratory person independently for comparing result with testing carried out by the junior Laboratory Assistant. In such cases it is difficult rely the test results. Occasionally various activities particularly collecting water samples from premises and its testing should be checked by an independent specialised person. Administrative control is generally placed under Civil Engg./ Public Health Engg. Department who are required to interact with health department closely. Health department may also collect water samples independently and get it tested from an independent approved laboratory.

Necessary guidance should be taken from following organisations:–

(i) National Accreditation Board for testing and Calibration Laboratories (NABL) Dept. of Science & Technology. Govt. of India, New Delhi.

(ii) Institute of Public Health & Hygiene, Kolkata.

(iii) CPHEEO, Ministry of Urban Development, Govt. of India, New Delhi.

Laboratory should be got accredited from NABL. The instruments/equipments should be got calibrated occasionally from NABL. Quality control audit of the laboratory should be done by an independent organisation (like, NABL). There are many tests for potable water. These tests are in two groups. Tests for first group are essential; tests for second group may be carried out as per actual situation; special care should be taken for arsenic prone ground water, fluoride affected water; repeated tests should be carried out from same and extra samples. In case of new tubewells, it is required to collect samples of water at the stage of initial boring and to carry out the tests. If test results are favourable then tubewell work is carried out; if hardness factor is beyond permissible limit or it contains harmful elements/polluted then the bore is abundant.

20.13 Certain Aspects of Potable Water Quality Management

This involves three major issues viz. : –

(1) Drinking water quality testing and monitoring.
(2) Quality surveillance
(3) Sustainability of water quality and community participation.

20.13.1 Ground Realities

To be tested for Biological, Physical and Chemical characteristics particularly Fluoride & Arsenic in certain areas of states. Arsenic has been found in 7 districts of West Bengal. Due to present poor position it is required to provide necessary infrastructure of testing laboratories. Attitudinal changes are required among the laboratory personnel and Managers. As the laboratories are presently governed by the Public Health Engg. Department there is very little interaction with Health professionals. Regular interactions between two departments are required. Fluoride contamination may create various diseases like, loss of appetite, pain in the stomach, diarrhoea, anaemia, pain in joints, cholera etc.

The community will derive the attention from Medical Officer as he/she is able to diagnosis of the disease and prevent the same through provision of safe drinking water. Public Health Engg. Dept. will be able to prevent water borne diseases by supplying safe drinking water. There should be intellectual interaction between the professionals of PHED and Health Dept. Above stated approach would ensure that Health Dept. personnel get involved in the results generated by the water quality testing laboratories of the PHED.

20.14 Unfilter/Raw Water Distribution System in City

Water is required for horticulture and street washing purposes. It is required to develop and maintain greenery in the city and towns. There are many public gardens, parks. Horticulture / greenery development is also done along roads in the city. Gardens are also maintained in the houses, institutions, office complexes. Many water bodies (fountains, water falls, channels etc.) are also maintained in the city. Considerable quantity of water is required for aforesaid purposes. In case underground water table is very low (particularly during summer season) then water may not be available from tubewells. People use filter water for aforesaid purposes; the same is to be considered as misuse of filter water. To solve this problem a separate distribution system is installed in the city for supplying raw water. The source of raw water is nearby river/stream/lake. High water pressure is not required in this raw water distribution system. Generally water is pumped in this system directly from the river "intake".

20.15 Environmental Aspects of Ground Water Quality

Causes affecting the water quality –

The excessive use of chemical fertilisers in agriculture caused considerable damage to soils in India. This has reduced the productivity in the agriculture. The industrial activity has caused considerable damage to the water both underground and at surface due to release of effluents. Mining for lead and zinc affects the quality of water the human activity in the urban areas generates waste materials which gets mixed up in the municipal water supply. The large livestock and poultry operations can pollute fresh water. The over exploitation of ground water in the costal areas causes intrusion of sea water in the fresh water aquifer.

Technology development in drilling and pumping methods have resulted in massive exploitation of ground water mainly for irrigation. Considerable advancement has been achieved in agriculture sector in India. Exploration of ground waters is causing imbalance between over withdrawal of ground water and different recharge which has resulted in rapid lowering of water table. The indiscriminate over-drawl have changed Geo-Hydro-Chemical environments of the aquifers and in general enhanced toxic and undesirable chemical constituent of water beyond the permissible limit viz. fluoride, arsenic, nitrate etc. This has caused diseases, suffering and miseries to the community. The surface and sub-surface sources are prone to biological contaminations mainly because of the open defecation habits.

Periodic assessment of both surface and ground water potential and continuous monitoring of quality is essential to adopt remedial measures. District wise resources evaluation is more useful for planning and implementation. In drought prone areas, first priority should be given to domestic water supply. In planning the aspect of sustainability is to be kept in view.

Renovation of existing huge numbers of village ponds in India would generate additional ground water and surface water potential. Awareness amongst people regarding scares nature of water resources is required and needs for its quality protection. People's involvement and part ownership is essential for smooth functioning of any water supply scheme.

Pollution from industrial and domestic sewage disposal and from the chemical inputs used in agriculture is slowly on the increase; if this is not controlled at this stage, it is going to convert these essential life saving fresh water resources into a rare commodity.

To improve the management of water resources, greater knowledge about their quantity and quality is needed. There should be only one state Ground Water Organisation dealing with evaluation, exploitation, augmentation and control on use of ground water. Major policies should be laid down by the Central Government agencies like, CGWB, Central Water Commission. It is essential to control fearful population growth in India.

20.16 Rain Water Harvesting (RWH)

20.16.1 Introduction

Existence of our green earth, human being, animals, birds, production of many materials, power for use of human beings etc. depends mainly on water. Due to rapid increase of population the requirement of water for various purposes (like, for vast agriculture sector, drinking etc.) has increased considerably. Serious problem is being faced due to shortage of water particularly during non-rainy seasons/summer season. Period of rainy season is short (may be one or two months) in many places in India. Sea water can not be used for aforesaid purposes. Due to above stated position there is excessive stress on "ecosystem". In the face of growing problems, the society has begun to realise that it can no loger subscribe to a "use and discard" philosophy with any natural resources. Hence action is to be taken for conservation and recycling of natural resources including surface and underground water. Detailed studies have been carried out on International Hydrological Cycle under UNESCO. Effort is being taken by various authorities for finding practical solutions to real water resources problems.

Central Ground Water Board, Ministry of Water Resources, Govt. of India deal this matter including Rain Water Harvesting (RWH)/Artificial recharge.

There is considerable shortage to irrigation water for vast agricultural areas and due to this reason huge quantity of underground water is obtained by pumping from tubewells. Position in urban areas is bad and very bad in large cities for acute shortage of drinking water. Level of underground water has depleted very much; quantity of available underground water has been reducing considerably. It affects on quality of water i.e saline/brakish water is available at very low level. Technology development in drilling and pumping methods have resulted in massive exploitation of ground water. Exploration of ground waters to meet in increasing demand is causing imbalance between overwithdrawl of ground water and deficient recharge.

Drinking water supply in large cities mainly depends on raw water obtained from the surface sources viz. rivers, streams, lake, man made large reservoir on the ground by constructing dam, weir across river, stream. It may be required to obtain raw water for treatment (drinking purpose) from far away places (may be at a distance of 100 to 500 km) through canal/large size conveyance main pipes/ducts. Interstate problems are faced for such arrangement. Certain country obtain raw water from neighbour country. Treatment of raw water is done at city through water Treatment plants for supplying drinking water as per standard norms. Availability of required minimum quantity of raw water from aforesaid surface sources during summer season has become very difficult in many cities and water crisis/famine situation is faced. Rivers, streams become dirty nalla during summer season; surface water is contaminated due to mixing of sewage flow and industrial wastes. Such heavily contaminated raw water may not be suitable for proper treatment.

20.16.2 What is Rain Water Harvesting (RWH)

This is a principle of collecting and using precipitation from a catchment area. This is an old technology gaining popularity in a new way. In ancient time impounding of rain water on ground surface and the same was used for irrigation and other purpose.

RWH is two types : –

(i) Natural way of percolation from ground surface and

(ii) Artificial Recharge to ground water.

Artificial recharge to ground water is a process by which the ground water reservoir is augmented at a rate exceeding that obtaining under natural conditions of replenishment. Any manmade scheme or facility that adds water to an "aquifer" may be considered to be an artificial recharge system. Water percolates below ground level and it remain in the interstices of underground soil/sand particles. This underground storage of water is known as "aquifer". Top of this underground water is known as underground water table (level). This level is important for many engineering/ geological/ hydrology purpose.

Rain is the first form of water we know in the hydrological cycle; hence it is a primary source of water. Rivers, lakes, streams, ponds, ground water are all secondary sources of water. In the present times, we depend entirely on such secondary sources of water. It is forgotten that rain is the ultimate source. Below a certain depth in the ground, the earth is saturated and this is known as ground water level. The ground water extracted is replant shed by the annual rain.

20.16.3 Why RWH

RWH is essential because : –

(i) Surface water is inadequate to meet our demand and we have to depend on ground water.

(ii) Due to rapid urbanisation, infiltration of rain water into the sub-soil has decreased drastically and recharging of ground water has diminished.

(iii) Over-exploitation of ground water resource has resulted in decline in water levels in most part of the country.

(iv) To arrest sea water ingress in coastal area.

(v) To improve the water quality in aquifers by dilution with freshrain water.

(vi) To improve the vegetation cover.

(vii) To raise the water levels in dug-wells and bore wells that are drying up; and to reduce power consumption.

20.16.4 RWH Techniques

There are two main techniques of RWH : –

(i) Storage of rain water on surface for future use.

(ii) Recharge to ground water.

The storage of rain water on surface is a traditional technique and structures used are underground tanks, ponds, check-dams, weirs etc. Large size reservoir is created on the upstream side of river/stream by constructing dam with regulatory gates arrangement. This water of large reservoir is used for drinking water (after treatment, irrigation, production of hydro-electricity etc.

20.16.5 Artificial Recharges in the Ground Water

Following are different types of RWH/Recharging ground water:

(i) Recharge pits

Recharge pits are constructed for recharging shallow aquifer. Pits are back filled with boulders, gravels and coarse sand.

(ii) Recharge Trenches

These are constructed when the permeable strata is available at shallow depths. Length of trench may be 10 to 20 meters. These are back filled with filter materials. Trench is constructed across the ground slope for intercepting surface flow. This is shown in figure 20.1.

(iii) Dug Wells

Existing dug wells may be utilised as recharge structure. Rain water is channelised and pass through filter media before putting into the well.

(iv) Hand pumps

The existing hand pumps may be used for recharging the shallow/deep aquifer. Water should pass through filter media before diverting it into the hand pumps.

(v) Recharge Trench with Bore wells

Recharge wells of 100 to 300 mm diameter are generally constructed for recharging the deeper aquifers and rain water is passed through filter media to avoid choking of recharge wells. Recharge Trench with Bore wells are shown in figure – 20.1

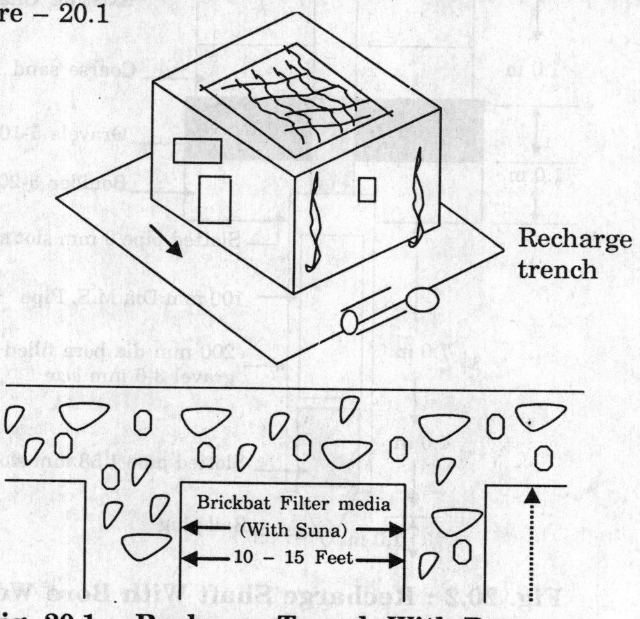

Fig. 20.1 : Recharge Trench With Bores

(vi) Recharge Shafts

For recharging the shallow quivers which are located below clayey layer shaft of 0.5 to 3 m diameter and 10 to 15 m deep are constructed and back filled with boulders, gravels and coarse sand. Structure of shaft with bore well (depth 10 to 30 m) is shown vide figure – 20.2.

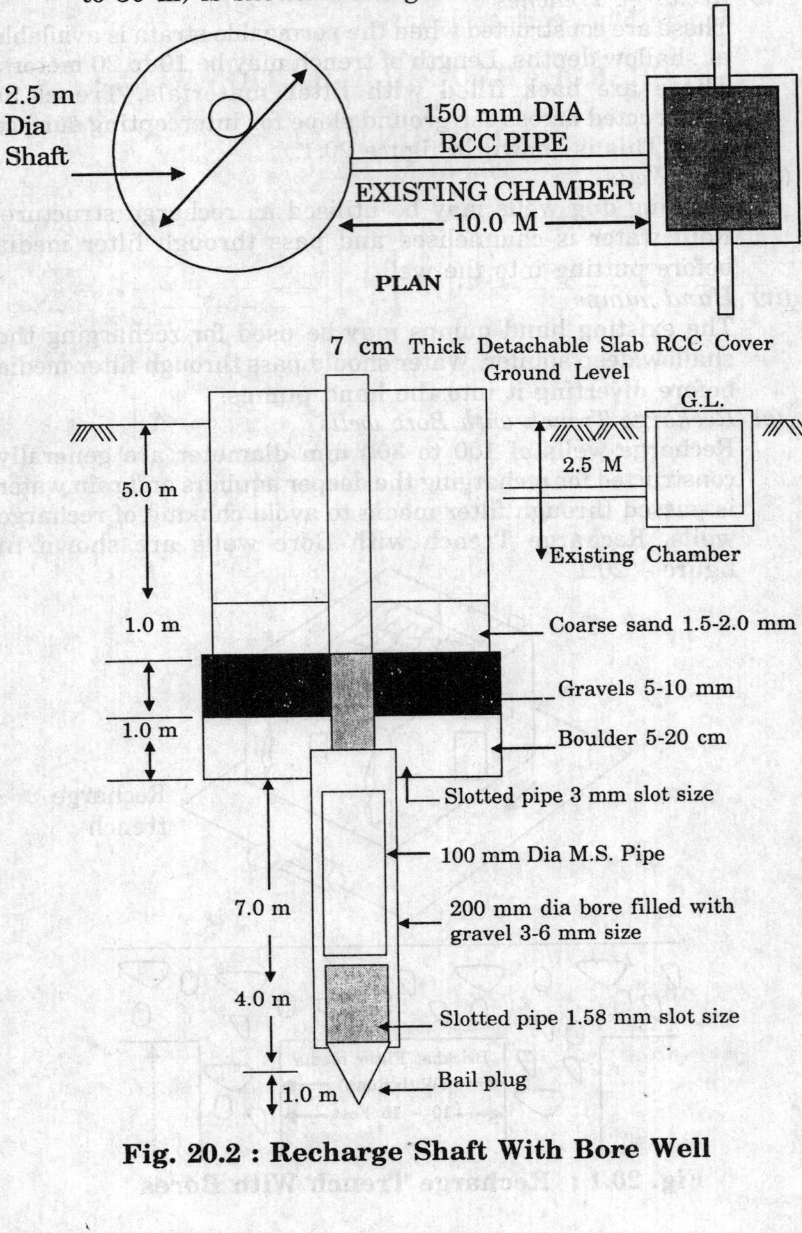

Fig. 20.2 : Recharge Shaft With Bore Well

(vii) Recharge through Injection Well

Structure is shown vide figure 20.3 Diameter of well is 203 mm and depth may be 25 m. Rain water is collected from trench, pits, roof top (through filter media) and put in the injection well through pipe. Ground water level is raised in the well. Lower portion of pipe is perforated for charging underground permeable layer.

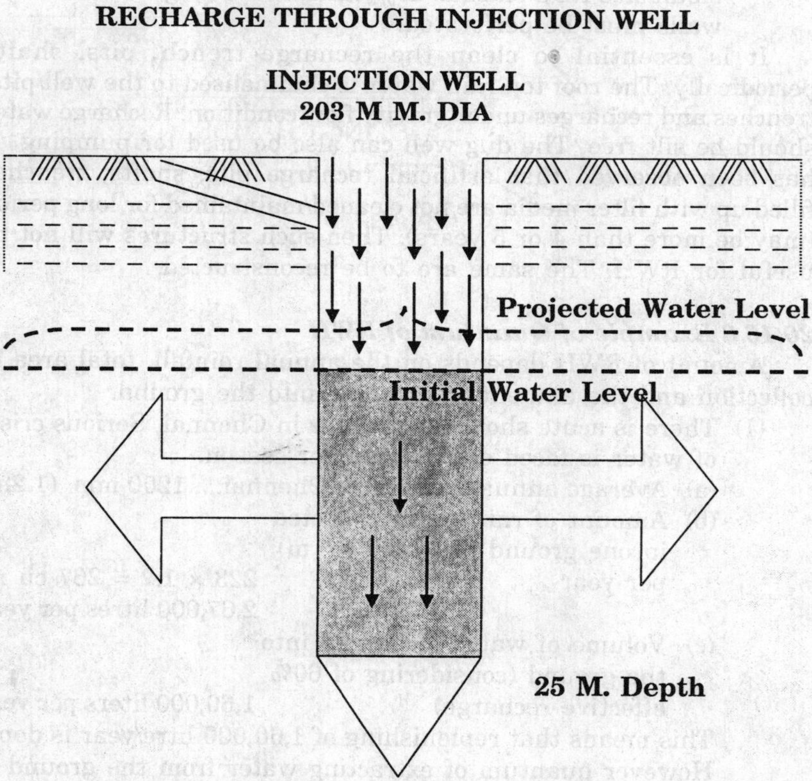

Fig. 20.3 : Recharge through Injection Well

(vii) Spreading Technique

Storing water on ground surface by different techniques/ structures is important as huge quantity of rain water can be stored. Spreading the water in streams, nallas, rivers by constructing small/large dam, nalla bunds, gaboin structures or a percolation pond may be constructed.

(viii) *Roof Top RWH*

Wherever open wells, bore wells are available, the roof top water can be used for direct recharging of those wells. Rain water is collected from roof top through pipe drain system and the same is connected with any above stated RWH structures like, pits, trenches bores filled up with filter media. Top of pits/trenches should be covered with precast small size RCC slabs in dry joints. If it is dug well then side walls must be perforated.

It is essential to clean the recharge trench, pits, shafts periodically. The roof top rain water is channelised to the well/pits/ trenches and recharges under gravity flow condition. Recharge water should be silt free. The dug well can also be used for pumping. It has been observed that artificial recharge pits, shafts, trenches filled up with filter media are not cleaned/maintained for long period (may be more than 4 or 5 years). Then such structures will not be useful for RWH. The same are to be reconstructed.

20.16.6 Example of Quantum of RWH

Amount of RWH depends on the annual rainfall, total area of collection and the amount of recharge into the ground.

(1) There is acute shortage of water in Chennai, Serious crisis of water is faced during summer season.

(a) Average annual rain fall of Chennai.....1200 mm (1.2m)

(b) Amount of rain water collected in one ground plot (223 sq. m) per year — 223 × 1.2 = 267 cu m
2,67,000 litres per year

(c) Volume of water recharged into the ground (considering of 60% effective recharge) — 1,60,000 liters per year

This means that replenishing of 1,60,000 litre/year is done. However quantum of extracting water from the ground is less than above.

(2) As per report RWH was done in JNU and IIT area of Delhi earlier (1997 - 98) through "spreading technique" (by constructing check-dams) RWH of 76,000 cu. m was done with a rise in water level from 1 to 13.7 m.

(3) In Jalgaon district (Maharashtra), the recharge through six percolation tanks was about 681 thousand cubic metre. Rise of water level was from 1 to 5 m. The benefited area was 545 acres.

(4) Recharge through two injection wells was done (before 1999) in IIT, Delhi and quantum of 830 cu. m was achieved. Level of water raised from 0.29 to 0.87 m; benefited area was one hectare.

(5) RWH was done in the President's Estate, Delhi (before 1999) through recharge shafts. Rise of water level was upto 2.58 m.

20.16.7 Attributes of Ground Water

- Ground water is less expensive and economic resource.
- Ground water is sustainable and reliable source of water.
- Ground water is relatively less vulnerable to pollution.
- Ground water is usually of high bacteriological purity and free from pathogenic organisms (not from shallow depth).
- Ground water needs little treatment before use.
- Ground water resources can be instantly developed and used.
- There is no conveyance losses in ground water based supplies.

(A) Artificial recharge techniques are adopted where : –

(1) Adequate space for surface storage is not available especially in urban areas.

(2) Water level is deep enough (more than 8 m.) and adequate sub-surface storage is available.

(3) Permeable strata is available at shallow/moderate depth.

(4) Where adequate quantity of surface water is available for recharge to ground water.

(5) In case ground water is brakish then same can be improved by recharging.

(6) Where evaporation rate is very high from surface water bodies.

(B) In other areas spread techniques RWH may be adopted.

21

Sewerage and Drainage Management

21.1 General

Sewerage and storm water Drainage in City/towns are vast specialized subjects. Some important points will only be highlighted here. These are part of Public Health Engineering. Sewerage and S.W. drainage systems are interlinked. Environment and sanitation of city/ towns depend on proper and adequate arrangement of these systems. There will be adverse effects on environment and sanitation if proper systems of these utilities do not exist or not operated properly. There are following three types of systems in different cities/towns : –

(1) ***Separate System –***
There are two separate systems i.e. Sewerage system and storm water drainage system.

(2) ***Combined System –***
There is one system for Sewage and storm water.

(3) ***Mixed System –***
Separate system may exist on some part of city and combined system may exist on some other part of the city.

If rainy season is for short period or rainfall is very less then separate system is better (like Delhi). If rainy season is longer and rainfall is high then combined system is better (like Kolkata). Type of systems in small/medium size towns is different than the type of systems in city/large town. Brief position on the same is stated as under.

21.2 Planning

21.2.1 The objective of a public waste water collection and disposal system is to ensure that sewage/excreta and sullage discharged from communities is properly collected, conveyed through pipe networks, treated to the required degree and finally disposed off without causing any health or environmental problems.

There are following stages of planning and designing : –

(a) Pre-Investment Planning
   (i) Identification of a Project.
   (ii) Preparation of Preliminary Project Report.
(b) Appraisal and sanction of project.
(c) Preparation of Project Report including Engineering Design.
(d) Construction Stage.
(e) Operation and Maintenance
(f) Monitoring and Feed Back.

Following three reports are prepared as first stage for Project Report : –

(i) Identification Report.
(ii) Pre-Feasibility Report.
(iii) Feasibility Report.

Basic principles of various stages have been dealt briefly under chapter of Water Supply Management and same are applicable for Sewage and drainage management. But engineering details are different than water supply project. Since project preparation is quite expensive and time consuming all project should proceed through above stated three stages and at the end of each stage a decision should be taken whether to proceed to the next planning stage. Report at the end of each stage should include a time table and cost estimate for the next stage activity. The basic design of a project is influenced by the authorities/organizations who are involved in approving, implementing, operating and maintaining the project. Therefore the institutional arrangements through which a project will be brought into operation, must be considered at the project preparation stage.

21.2.2 Design Periods and Components of Sewerage System

Sl. No.	*Component*	*Design Period (in years)*
1.	Collection System i.e. Sewer Network	30
2.	Pumping Stations (Civil Work)	30
3.	Pumping Machinery	15
4.	Sewage Treatment Plant	30
5.	Effluent disposal and utilization	30

It is essential to carry out field survey for primary data and also to collect data of existing system from secondary sources. Various problems being faced on the ground will be known from above stated sources. Some of these existing problems of city are stated as under : –

Sewerage System in City / Town

21.2.3 Some basic problems

(a) Rapid population growth coupled with unplanned habitation has resulted in the Sewerage facilities falling far short of requirements. This has serious adverse effect on sanitation, environment, pollution in the city / town. Some problems are stated as under.

Sewerage system in the congested areas of old part of city / town has no capacity to receive further sewage. It is also very difficult to augment the sewerage system along very narrow lanes / bye-lanes / streets. It is difficult to approach the existing sewer network of such congested areas of old part of city / town. Due to unsafe structures and space constraints it is difficult for cleaning, repairing, rehabilitation and augmentation of sewer lines.

Several trunk sewers and subsidiary sewerage systems are heavily silted (50% to 75% of diameter/depth of sewer line). The conventional methods of de-silting / rehabilitation are not successful.

(b) Reliable information of gradients, invert levels, specific length, alignment and condition of the largely "unmapped" network is not readily available. It is also difficult to carry out repair works of existing sewerage system including branch lines and sewer service connections without maps. Hence it is essential to take up separate task on priority basis for carrying out detailed survey of levels of existing sewerage systems and to prepare maps/drawings.

(c) Dependence on other organizations for inputs such as land and power coupled with infrastructural and resource deficiencies affect the provision of services, levels of efficiency thereof, response time to basic as well as emergency requirements.

Generally total capacity of sewage treatment plants in the city/ town is far less than the total sewage quantity. Huge fund is required for constructing sewage treatment plants. Due to this deficiency considerable quantity of untreated or partly treated sewage is discharged in the river / streams which is objectionable.

(d) Sewerage services, after the initial payment for infrastructural facilities, are provided free of cost with consequent fund problems for operation and maintenance activities. Hence sewerage tax should be imposed.

(e) Schemes for rehabilitation/ replacement/ de-silting of badly affected sewer lines (particularly in chronic problem areas) should be taken up on priority basis. Chronic problems can be identified by maintaining records of day to day complaints/ over flowing of sewer and action taken reports. Thus the overhauling of the existing sewerage system/ construction of new sewer lines should be identified separately as short term/ long term actions. ·

21.2.4 *Augmentation of Existing Sewerage System*

(a) Planning and designing are done for laying of new sewer line under scheme of augmentation of existing system. As per such scheme it is proposed to connect new sewer lines with main sewer lines (new or existing). Due to faulty planning it is not possible to connect branch sewer lines to main sewer lines as the main sewer lines are not completed or capable to carry additional discharge. Due to this faulty execution plan (or delay in progress of main sewer work) the entire augmented new sewer line cannot be commissioned and remains as unused for very long period (may be more than 5 to 10 years).

Under the scheme of augmentation, it is required to transfer the existing sewer service connections from huge numbers of old sewer lines. This part of job is very important and difficult as public/occupants are directly concerned. This transfer (i.e. by laying new services-sewer lines) is to be done simultaneously alongwith commissioning of new main sewer line (i.e. on the same day) otherwise serious problem will be faced by the occupants of building (as toilets will not function at that time). Users should be well informed in advance; if need be, temporary pumping from the last Man-hole of the premises should be carried out to avoid inconvenience.

(b) It has been found in India that most of the sewer lines are silted heavily and the same were laid many years back. In many areas silt remains for very long period. There is no proper preventive system of maintenance (i.e. annual desilting operation) in respect of entire sewerage system. Major length of sewer lines are silted by "30% to 80% of depth of sewer". This is alarming. This is the main cause of many problems. Preventive maintenance i.e. annual programme of de-silting and its correct compliance report for entire sewerage system is essential; this should be cross checked by superiors and recorded. This is done effectively (in video CD/Cassettes) in many foreign countries (eastern and western). Silt and sewer condition survey are carried out every year through CCTV film method. This is done before starting the de-silting work and again after completion of de-silting. It has been found that silt is about 3% to 5% depth of sewer pipe in many foreign countries. In these countries, the above is done every year through contract (payment is made to the agency at the unit of metre length of sewer line size wise). In India, it will be difficult to do complete de-silting of entire sewerage system. Thus, it is seen that de-silting of entire sewerage system is a huge task and a major project involving considerable amount. This may be done phase-wise and priority-wise during a period of 5 years in a big city. After doing this the future maintenance cost will be reduced. Sanitary condition of city will be improved considerably and public will be benefited. For desilting sewer lines, the Gully Emptiers (Suction type machine mounted on a vehicle) are effective.

21.3 Aspects of Management

There are five important aspects of management which are as under : –

(i) General administration.
(ii) Personnel administration.
(iii) Inventory Control.
(iv) Financial Control.
(v) Public Relation.

The management should aim the following : –

(i) Proper collection of waste water / sewage / sullage discharged by the community.
(ii) Adequate treatment to waste water / sewage to achieve the prescribed effluent standard.
(iii) Safe and efficient operation.
(iv) Sound Financial Management.

Estimate of flow in the sanitary sewers should include certain flows due to infiltration of ground water through joints. The quantity will depend on workmanship in laying of sewers and level of underground water table during rainy season / immediately after rainy season.

21.4 Sewer Works through Trenchless Technology

Broadly there are two categories of major works which are as under : –

(1) Rehabilitation of existing bad conditioned sewer line through trenchless technology.

(2) Laying new sewer line through Trenchless technology.

21.4.1 Social Cost

Justification for higher cost on account of trenchless technology is as under :

During laying sewer / drain pipelines along/across roads at deeper level, many problems are faced. Main problem is disruption of traffic due to partial closing of road or diversion of traffic through longer route. There is possibility of accidents of vehicles; accidents due to slides of vertical sides of trench may occur and workers/ pedestrians/ commuters may be injured or fatal accidents may occur. Police Case, claims etc. may crop up due to such accidents. As per research, it has been found that there is heavy social cost on account of following elements : –

(i) Traffic
(ii) Environment
(iii) Commerce and Industry
(iv) Citizens and Society

Social cost under category of "traffic" is due to : –

(a) traffic hindrance
(b) traffic obstruction
(c) probability of damage- as a result of less safe situations
(d) damage to diversion roads

Regarding commerce and industry, the same is due to : –

(a) Costs as a result of loss of sales for the shops along the trenches.
(b) Cost of production

Citizens and society cost is due to : –

(a) Increase of maintenance cost for road surface.
(b) Compensation of damage
(c) Public hindrance

21.4.2 Rehabilitation

In India, presently, main problems are being faced for repair and replacing very old sewer/drain lines in cities.

Population of cities has increased manyfold after Independence. Old parts of cities/towns are very congested and roads are very narrow. Old sewers (of pre-independence period) cannot carry discharge of present population. It is also not possible to lay new sewer line of required size by excavating open trench along narrow streets passing through congested business/ commercial areas. It is essential to carry rehabilitation of sewer line through trenchless technology. It may be stated that augmentation/ rehabilitation of sewerage facility in such old parts of the cities is essential, otherwise sanitary condition may go out of control and there may be disaster in such areas.

A specific problem is stated as an example. A large size (man-entry) very old brick sewer exists in congested area of a metropolitan city. The structural condition of sewer is bad. There is heavy deposit of silt (60% to 75% of depth of sewer). Vacant space is not available in this area for laying a new sewer line. As there is no proper flow of sewage through this trunk sewer, the back flow problem of sewage is faced by the occupants of buildings of command areas of the trunk sewer. Hence, there are unhygienic and environmental problems in the area. It is, therefore, essential to rehabilitate the old trunk sewer through trenchless technology i.e. by lining on inside surface of old sewer with suitable material which will also take care the structural stability. In case narrow space is available along the road for constructing a new sewer in place of old sewer, then also there will be problems of public transport, environmental problems, very high social cost, safety hazards, etc. Road cut permission is also not generally given by the authority. In certain cases, it has been found that cost of rehabilitation work is less than the construction of new sewer line (without considering social cost of various elements), besides many other advantages as stated above.

21.4.3 Various Types/Methods of Rehabilitation of Sewer

As per situation in India, following are some types/methods of rehabilitation works :

- (i) Cured in Place Pipe (CIPP)
- (ii) Slip lining.
- (iii) In-Line Replacement
- (iv) Close Fit Pipe
- (v) Point Source Repair.
- (vi) Sewer Manhole Rehabilitation.

21.4.4 Laying of New Sewer Pipe Line through Micro-Tunneling and Pipe Jacking Technology

This method of trenchless technology for installing underground services pipes (including gravitational flow viz., sewer pipeline) is very useful and the same is also applied in combination with some other method. This is an important method.

(a) *Pipe Jacking*

It is a process of lining an excavated underground passage (formed by tunneling machine or other means) by pushing especially designed RCC pipe or other pipes from jacking shaft to receiving shaft. The pipe itself functions as product pipe (like, sewer line or other service pipe).

(b) *Micro-tunneling*

Micro-tunneling and Pipe jacking technology is an art of installing underground pipelines without digging roads, busy streets and to lay pipe across railway tracks, streams, etc. This is laser controlled and computer aided with tunneling machine (known as shield). Tunneling is done by mechanical shields or remote controlled tunneling machines to a designed level and alignment between two shafts (known as working shaft). These shafts are converted as manholes after laying underground service pipelines.

There are many advantages of this method; some advantages have already been stated. This method is fully remote controlled technology which keeps workers out of harms. It is possible to achieve accuracy of (±) 25 mm in line and grade (level). Certain problem like, to lay deep sewer line along narrow/ congested streets, to lay pipes across railway tracks, etc. can be solved by adopting this method. It has also been found economical in certain special/ critical cases. It performs under a wide range of soil conditions, like treacherous soil, high water table, rocky strata etc. Transportation of city area is not disrupted. There is no hazard for workers, traffic, pedestrians; there are many other advantages.

21.4.5 Planning and Site Investigations

Successful and trouble free micro-tunneling operation mainly depends on good engineering planning, geo-technical soil investigation, soil characterization, equipment selection, highly skilled and experienced operation of remote controlled equipments. Precision in tunneling work, gradient of pipe, vertical level control is to be achieved by operating remote controlled equipments correctly through monitoring computer system.

21.4.6 Survey of Existing Service Lines

(a) It is required to ascertain the existing various underground service lines along the proposed new service line (sewer, drain, etc.) and to mark the same on survey plan of relevant area strips. As far as possible, these may be ascertained from maps/area plans and other records. Pits/shallow trenches may also have to be made across the proposed alignment of new line for locating existing service lines and the same are to be marked on survey plan of the area, indicating depth of these existing service lines on the cross sections. Most of these service lines (except sewer line and some other lines) are generally laid within a depth of 1.5 metres.

(b) *GPR Method*

Besides above stated survey/site checking, the underground survey upto depth of the proposed new service line may be carried out through GPR (Ground Penetration Radar) method. This is based on passing a multi-frequency radiometric detective device. It is difficult to locate non-ferrous pipes. Hard stone boulders can be located by GPR method.

21.5 One Example of Ground Reality about Major Pollution of Storm Water Drainage Discharge with Sewage, Excreta in Delhi

Separate system (i.e. sewers and storm water drains) exists in Delhi. S.W. drainage system consists of covered and open drains. Many open & covered nallas are part of S.W. drainage system. Size of these nallas may vary, top width – from 2 m. to 75 m. and depth 1 m. to 6 m. Many jhuggy/slum clusters exist along bank of these nallas. There are many unauthorized colonies JJ/slum clusters and urban old villages having about 50% population of Delhi. There is no proper sewerage and S.W. drainage systems in these unauthorized colonies and urban old villages. Streets, lanes of these colonies, function as channels of drainage. As per judgements of High Court and Supreme Court it is essential to construct sewerage and S.W. drainage systems before regularization of unauthorised colonies.

Due to above stated ground position the storm water drains (including all large nallas) discharge liquid waste containing lot of sewage, excreta and other biodegradable/toxic solid wastes in the river Yamuna. Serious pollution problem has been created due to above situation. There is another problem. Total capacity of sewage treatment plants in the city is much less than the total quantity of sewage of the city (based on total quantity of potable water supply). As per judgements of High Court and Supreme Court it is required to install adequate numbers of sewage treatment plants for total quantity of sewage including sewage flow through S.W. drains. As per records the river Yamuna remains a "Sewage Canal" mainly because of the 143 unauthorised colonies and 1,080 slums and rural villages which have created serious problem in collection of sewage that flows into the river untreated. Despite 14 years of Court monitoring of the steps taken by the authorities to clean up the river with nearly Rs. 1800 crores spent on the 22 km stretch of the river the condition remains more or less a "sewage canal". Above position has adverse impact on the environment of the city. Court added in the judgement that planning and designing should be done as per actual ground position and not remain confined to paper. In a land mark judgment of the Delhi High Court dated 24.11.2008, two weeks jail has been ordered for former Chief Executive Officer, Chief Engineer and two other persons of Delhi Jal Board for their failure to prevent sewage flowing into the storm water drains and Yamuna river. This order has been issued in a public interest litigation case initiated by one residential colony association.

There is no sewer lines in above stated unauthorized colonies, slums, urban old villages and hence sewage, excreta, sullage liquid etc. are discharged in the nearby S.W. drains and nallas unauthorisedly. The main route cause of above stated bad situation is due to ineffective control over unauthorized colonies, slums, jhuggies in the city by violating master plan of Delhi during last five decades. Central Ground Water Board, Govt. of India has also pointed out that underground water of this zone of Yamuna Valley has been polluted due to dirty water mixed with sewage flowing through the river.

22

Elements of Solid Waste Management

22.1 Introduction

Management of solid waste involves development of an insight into the impact of waste generation, collection, transportation and disposal methods adopted by a society on the environment and adoption of new methods to reduce the impact. The growing accumulation of garbage reduces the land values, increases truck traffic in the city, towns and ruins health, aesthetics and necessities of life (viz. the air we breath and the water we drink). There is lot of adverse effect of the solid waste on the environment. People suffer on account of diseases. The solid waste management is a vast subject. Hence elements of certain principles will be dealt here briefly.

22.2 There are different solid wastes which are as under : –

Domestic/residential waste, Municipal Waste, Commercial Waste (viz., offices, wholesale & Retail Stores, Restaurants, Hotels, Markets, Institutional Waste (viz., Schools, Universities, Research Centres, Temples etc.), Garbage (animal and vegetable wastes which attract rats, flies and other vermin), bulky wastes, garden wastes, street sweeping, dead animals, construction and demolition wastes, industrial waste, hazardous wastes (like, Bio-Medical wastes). Here management of municipal solid waste will be dealt briefly. It may be stated here that some industrial wastes, hazards wastes (medical wastes) are thrown in the municipal solid wastes. Waste may be toxic or non-toxic. Air and water are contaminated due to unsatisfactory management and handling of solid wastes in the city, towns. Methane gas which is dangerous is formed from old dumped wastes. It gives ugliness and bad odour. Preventive action is not taken in many countries including India. In USA the system of mandatory recycling has been introduced. In certain countries tax incentatives are given on equipments which uses second hand materials or for processing solid wastes.

Causes of growing solid waste : –

- Industrialisation
- Urbanisation
- Higher standards of living.
- Change in life styles.

Quantum of solid wastes in a city /town depends on present and future population. Government has enacted certain laws/acts/rules which are related to solid wastes, like--

The Environment (Protection) Act, 1988, Hazardous Wastes (Management and Handling Rules), 1989. Authorities like, Pollution Control Board (State and Central Govt.) have been set up for control on Environmental Pollution.

22.3 Environmental Impact of Solid Waste Disposal on Land

- Ground water contamination by the leachate generated by the waste dump.
- Surface water contamination by the run-off from the waste dump.
- Bad odour, rodents and wind-blown litter in and around the waste dump.
- Generation of inflammable gas (viz. methane).

22.4 Principles of Municipal Solid Waste (MSW) Management

MSW management involves the application of principle of integrated solid waste management. Application of suitable techniques, technologies and management programme/planning for various types of solid wastes from all sources in the city/town. The aim should be for reduction of waste and effective management.

As per law the municipalities for smaller towns and municipal corporations for city/large towns are set up. Elections are held at the interval of 4 or 5 years for electing members of the municipality/ municipal corporation from different wards.

There is a large and important department in the municipal corporation/municipality for management of solid wastes. Health and engineering departments are responsible for MSW management. The management should be sustainable for : –

- Environmentally sustainable. Adverse impact of Environment is to be minimized.
- It should be economically sustainable. Considerable amount (major amount of total budget of the municipality) is required for solid wastes department. Lot of difficulties is faced by the municipalities on account of shortage of fund.

To achieve the above stated principles it is required to follow an integrated approach i.e. to deal with all types of solid wastes of various sources. A flow diagram on this matter is given of figure No. 22.1.

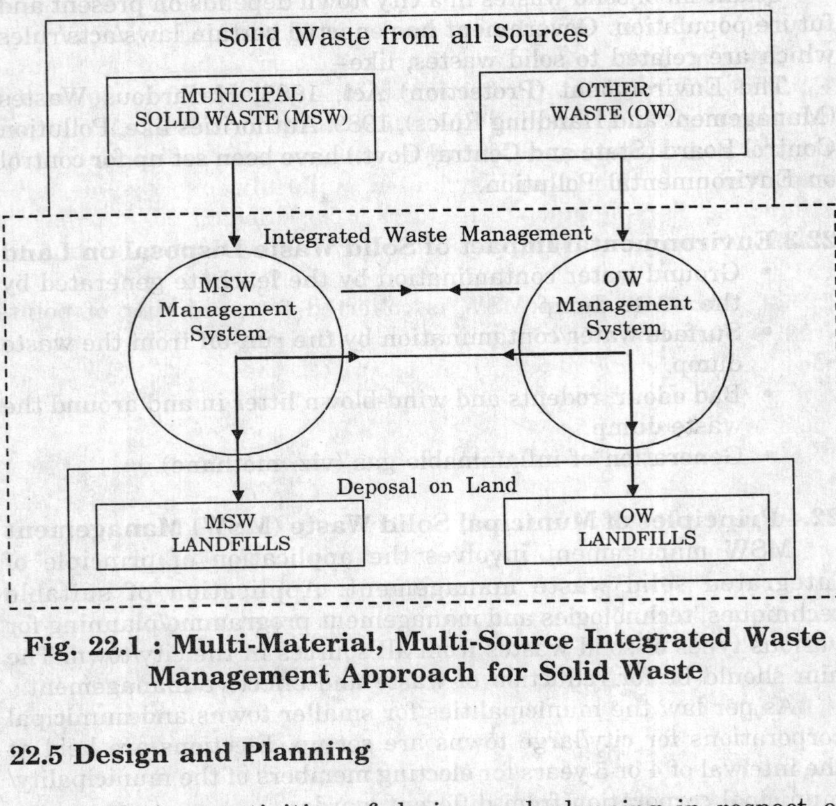

Fig. 22.1 : Multi-Material, Multi-Source Integrated Waste Management Approach for Solid Waste

22.5 Design and Planning

22.5.1 Various activities of design and planning in respect of construction projects have been dealt under various chapters. Similar way planning and design are to be carried out as first stage of SWM. Areas and population of small towns are not large. Considerable areas and population are involved in mega-city, city, large towns and lot of problems are faced for SWM of these cities & towns. Large cities and towns are divided in suitable / manageable zones. These zones/circles are marked on the map of city indicating different category / use i.e. residential, commercial, institutional areas etc. Existing populations of each zone / circle are to be ascertained from concerned Government department. Projection of population increase in future (i.e. design period) is to be worked out. Volume of wastes (present and design period) depend on the populations and many other factors. This part of the design and planning is important because subsequent detailed design, working out of manpower, vehicles, machinery/equipments depend on the correct data related to present and projected quantities of solid wastes.

Field investigations, surveys are to be carried out for field data. Collection of samples of solid waste(zone/circle wise) is important for planning and design SWM particularly for deciding types of recycling, obtaining compost/manure and energy (thermal, electricity, gas). Sampling points are distributed uniformly zone wise over the study area. Sampling points are further classified based on economic status of population like, high, middle and low income group.

22.5.2 About 10 kg of MSW is collected from number of points from outside and inside of the solid waste heap. Samples collected for physical and chemical analysis are put in plastic bags and sealed and sent to the laboratory for analysis. Reports are prepared on the basis of test results and their analysis. It has been found that many projects, design and planning programme failed during implementation at field due to wrong method of sampling and incorrect test reports. As examples it is stated that one large plant was got installed through a foreign company for generating electricity by using MSW. But as various characteristics (viz. calorific value, moisture content, organic matter, inert matter etc.) of solid wastes were wrong the scheme / plant for energy recovery was failed miserably. Serious problem were faced in this energy recovery scheme.

22.5.3 ***Composition and Characteristics***

(a) The composition and characteristics of solid wastes vary from place to place within a country as social customs, standard of living, geographical location, climate etc. are variable.

The major constituents are paper and putrescible organic matter. Broadly there are two types of solid wastes, viz. (i) Bio-degradable and (ii) Non-degradable. Variation is also observed during different seasons. As example patterns of composition, characteristics and quantities in respect of middle income country are stated below which is based on the book of Managing solid waste in Developing countries (by Mr. Holmes) :

Middle Income Countries

	Composition of solid waste	Percentage (by weight)
1.	Metal	1 – 5
2.	Glass, Ceramics	1 – 10
3.	Food, Garden Waste	20 – 60
4.	Paper	15 – 40
5.	Textiles	2 – 10
6.	Plastics, Rubber	2 – 6
7.	Inert materials	1 – 30
8.	Density	170 – 330 (KG/Cu.M.)
9.	Moisture content (% by weight)	40 – 60
10.	Waste generation	0.5 – 0.9 Kg/Capita/Day

Contents of above stated items will be more in respect of low income countries and less in respect of high income countries. These particulars are required for designing and planning of method and technology to be adopted for disposal, selection of recycling / energy recovery. The calorific value ranges between 800 — 1000 kilo calory per kg. Knowledge of the chemical characteristics is essential in selecting and designing the waste processing and disposal facilities.

Total compostable matter of solid wastes depends on population range. If population range is 2.0 to 5.0 million then total compostable matter is 56.67% and for population of greater than 5 million the compostable matter is 30.84% (which are based on report prepared by NEERI, Nagpur).

(b) Following refuse generation rates are suggested (as assessment for the purpose of planning / designing) : –

(i) Residential refuses – 0.3 to 0.6 kg/cap./day

(ii) Commercial refuse – 0.1 to 0.2 kg/cap./day

(iii) Institutional refuse – 0.05 to 0.2 kg/cap./day

(iv) Street Sweepings – 0.05 to 0.2 Kg/Cap./day

22.5.4 Handling of Solid Wastes

(a) Reliable data on composition and quantity of solid waste including projection for design period are important. The method and capacity of storage, the correct type of collection vehicle, the optimum size of crew and the frequency of collection depend mainly on volume and density of wastes. Climate is important. Disposal method depends on many factors.

Regarding recycling it may be stated that European steel industry reuses the scrap metal, copper and aluminium, resulting in energy saving upto95%. Netherlands runs the world's largest composting operation. Japan recycles 95% of its newspaper, 55% of its steel and 66% of its bottles. In fact "recyclables part of solid wastes are not perceived as garbage in some countries". Burning the tyres causes noxious fumes. Infectious medical wastes is a growing problems. The increasing flow of medical wastes pose the problem for municipal authority. Medical wastes are to be handled in specialized way. Decomposing garbage emits volatile methane gas.

(b) The most dangerous factor in the growth of landfill is its contamination of nearby water supply. Toxic chemicals are mixed in the garbage and the same may flow to nearby aquifers, streams, rivers, lakes and other water bodies. Another popular measure of handling the garbage is through incineration. But there will be air emissions of chemicals, sulphur and nitrogen oxide which are dangerous. The toxic ash is to be disposed in special sites carefully.

(c) An important option for managing the solid waste is the option of "source reduction". This will reduce the amount and toxicity of waste generated. This option should be made popular and effort should be made to identify the items. Industrial solid waste dumped at public places / with municipal wastes poses peculiar problems, such as : –

- Problem of contamination
- Identification of sites
- Monitoring

As per report of the Committee on Urban Wastes, Govt. of India (1997) it is required to adopt different method of disposal as per position, data, characteristics of wastes, climate, local culture. Excepting incineration, all other methods are being widely practised in India. During last some years incinerators are being used in hospitals for disposal of hazardous medical wastes. For improving environment and sanitary condition in the city, towns it is required to improve solid waste management and method of disposal. Burning of garbage on open ground is very harmful; air is polluted by the same.

Bio-degradable garbage should be used in the compost plant for producing organic fertilizer and manure.

(d) Following are some disposal methods of garbage : –

(1) *Source reduction* –
Awareness and co-operation from public are required. Environmental education is required.

(2) *Recycling* –
There are many methods of recycling and recovery of materials manually (with or without conveyor belt).

(3) *Resource Recovery* –
Rag pickers recover certain materials and sell to the traders.

(4) *Land Filling* –
Though there are some problems and adverse effects yet this method of disposal is popular in India and other countries. Ground water and nearby surface water (river, streams, lake, ponds etc.) are contaminated by the leachate formed in the landfill site. Remedial measures for adverse effect is to design proper "Sanitary Land Fill". But this is not done in practice in India mainly due to extra cost.

(e) By adopting western methods of disposable articles, we have increased the plastic & paper wastes in our cities. If we adopt our traditional methods and reduce the use of non-bio-degradable plastics, it will go a long way in reducing the wastes. People should be taught about segregating wastes into bio-degradable and non-degradable wastes. It has been observed that interdepartmental co-ordination is very poor viz, co-ordination amongst various departments like, civil engineering departments, mechanical engineering branches, health & medical department, State & Central Govt. Pollution Control Board etc.

(f) Field investigation, data collection, analysis of data, test reports are important for preparing proper scheme; success of solid waste management depends on these factors. It provides the basic data on which the management system is planned, designed and operated. The changes/trend in composition and quantity of waste over a period of time are known which help in future planning. It provides the information for the selection of equipment and appropriate technology. It indicates the amount and type of material suitable for processing, recovery and recycling viz., composting/electricity generation/gas production, manufacturing steel iron, copper ducts from scraps available as wastes.

22.6 Construction, Purchase of Vehicles, Equipments, Installation of Plants and Operation

22.6.1 Different Steps

(1) Segregation of solid wastes at sources (i.e. premise, transfer stations, collection bins.) is done Broadly in two categories viz. (a) Bio-degradable and (b) Non-degradable. Newspapers, metallic scraps (steel, iron, copper etc.) are picked up at source and the same are sold to kabariwala/traders who used for recycling purpose.

(2) Collection of wastes from houses and other buildings, lanes, bye-lanes and placing in municipal garbage bins/dhallaos/ containers/stationary compactor bins/garbage stations.

(3) Transportation of garbage with tipper trucks/compactor/ dumper placer (Skip Lift type) garbage dump ground / Land Fill site / Sanitary Land Fill / Compost Plant / Transfer station.

(4) Final disposal of garbage, bye-product (inert material) of plant, building rubbish, ashes from incinerators etc. is done at land fill site / sanitary land fill.

22.6.2 Collection of garbage from areas at narrow street/lane/ bye-lane

Access of truck/mobile compactor is not possible to lift garbage from narrow street, lane, bye-lane. At such places, properly designed garbage bins may be constructed at appropriate places. Finishing may be done with glazed ceramic tiles on brick walls. Floor may be made with Kota stone or with similar type of locally available stone. There should be an easy access upto garbage dumps/garbage bins for the handcart. Garbage will be carted from bins to nearby garbage collection centre and then the same will be lifted by Tipper truck/ mobile compactor. It should be easy for public to put garbage in these bins. The bins should be covered suitably so that during rainy season, problem does not crop up. Garbage bins should be designed aesthetically. Specifications should be good and easy to maintain. There should be proper drainage around these bins. Approach to these bins should be constructed with cement concrete.

22.6.3 Selection of exact location of garbage bins/containers/ trolley in city/ towns

It is quite difficult to select correct location for placing bins/ containers and to construct masonry bins/dhalao. Size of bins/ containers/dhalaos is also to be decided correctly on the basis of quantity of garbage generated in the area. There will be huge quantity of garbage at markets/commercial place. Problem is faced when nearby occupants of buildings object for certain locations.

Planning should be made in phased manner for lifting garbage from collection centres / bins / containers mechanically i.e. with mobile garbage compactor / Dumper Placer (Skip Lift type), tipper truck. Steel containers should be designed suitably so that lifting and unloading are done mechanically (as per figure No. 22.2 and 22.3). Bins/containers are also lifted by mobile compactor and garbage is put in the hopper of compactor. Cost of Dumper Placer (small size - 4.5 cu.m. and larger size - 7 cu.m) is not much; the same are being used in some cities effectively.

Non-degradable (i.e. inert and stabilized waste and building rubble/malba) are to be disposed to land filling site. Bio-Degradable waste should be sent for composting or to sanitary land fill site.

Fig. 22.2 : Garbage Dumper Placer

22.6.4 Transportation of Garbage

(a) Transportation of garbage from public places, garbage collection points (viz., garbage bins/dhalaos/containers/

garbage heaps on ground etc.) is most important part of solid waste management. Success depends on positioning of fleet and role of mechanical engineering branch/workshop and close monitoring.

The waste is generally loaded manually into the trucks/ lorries for transporting to dumping sites. This poses a great degree of health risk to workers handling it. It is transported generally to dumping sites in open flat bed trucks which spread foul smell and drop garbage on the way. Optimisation of route for plying trucks is to be done so that there is no wastage of mileage and manpower. All trucks should be "Tipper" type so that lorry beldars are not required at disposal ground for unloading. Tipper truck should be covered by fabricating during body-building with steel sheet and frame in suitable designs (curved/parabolic shape) with facility for tipping unloading.

(b) *Dumper Placer (Skip Lift)*

Transportation of large garbage containers (7 cu. m or 4.5 cu. m.) from collection place to disposal ground is very efficient and economical. Photos of Dumper Placer and garbage containers are as per figure No. 22.2. The containers filled with garbage is lifted mechanically (simple arrangement with chain system) in the Dumper Placer and the garbage is unloaded at disposal ground by tilting the container mechanically. Only one lorry beldar is required for lifting the container and unloading the garbage. Mini Dumper Placer can pick up containers from narrow streets. This arrangement is strongly recommended. It is also economical and simple to operate and maintain.

(c) *Mobile Garbage Compactor*

This is another very efficient and hygienic system of loading garbage and compacting garbage and unloading at dumping ground without any lorry beldar. Photos of garbage compactor (in operation) is as per figure No. 22.3. It is a mechanized equipment mounted on a truck chassis. It is equipped with various hydro and electro-pneumatic controls for operation. It lifts the garbage bins /containers from the road side and empties the same into the hopper of compactor vehicle, where the garbage is automatically compressed. After loading garbage fully the compactor travels to disposal ground and unload the garbage automatically by ejector method. Since it is a totally mechanized operation, the workers are not required to touch the garbage at all. Large quantities of garbage can be accommodated by compressing method. Since the bin made with steel sheet is emptied in a hopper of compactor, there is no need of transportation of bins to disposal ground. This itself achieves 50% savings in fuel, wear and tear of vehicles apart from reducing pollution. It is possible to transport 11 to 12 tonnes of garbage in one single trip whereas normal type of truck transports only 2.5 to 3.5 tonnes of garbage. Smaller capacity (viz., 6.5 cu. m.) garbage compactor is useful for lifting garbage from narrow streets. Medium capacity (viz., 14 cu. m.) compactor should be deployed on main / wider road / heart of the city.

Fig. 22.3 : Mobile Garbage Compactor

It is inconvenient to ply large capacity (more than 14 cu. m.) in the heart of the city / heavy traffic area/ during peak hours of traffic flow; hence medium capacity garbage compactor should be used at such congested places.

(d) The transportation fleet should be mixed with different types i.e. some tipper trucks, some Dumper Placers (smaller capacity of 4.5 cu. m. container and larger capacity of 7 cu. m. container) and mobile garbage compactor (of 6.5 cu. m. and 14 cu. m. capacity). These are to be worked out on the basis of quantity of garbage, number of shifts, locations from where garbage is to be lifted.

In certain areas of city it is feasible to collect garbage from house to house. Group of persons with handcarts (wheel barrow type - one for each labourer/ mazdoor) is deployed for certain specific area. This group move along street/lane giving whistle call from 8 am to 10 am. Occupants will put garbage in the handcarts. If any person throws garbage on street / lane / open piece of land after 10 am, a fine will be imposed. The garbage collected in the handcarts are put in the nearby garbage bins / containers which will be lifted by tipper truck/ mobile compactor/ Dumper Placer (Skip Lift).

(e) There is modern mechanical type garbage collection centre; this is a "Stationary Garbage Compactor" mounted on wheels (trailer type). This is placed at appropriate location of crowded market/ shopping area. Public will put garbage in this trailer type compactor bin. Capacity in terms of weight is much more than normal type of steel containers. The compactor is carted away with a tractor as trailer to disposal ground. Such stationary compactors as garbage collection centre are being used at Mumbai.

At certain places (particularly at old markets) where huge garbage is generated daily; the above stated bins/ containers may not be suitable. At such places masonry "Dhalao" is constructed with three side walls and front side is with large door and roof on top. Capacity of such collection centre is much more. However, aesthetical look of "Dhalao" is not good; hence, location should be selected carefully.

Cash incentive should be given to field workers (particularly labourers / mazdoors and drivers) for removal of garbage after certain fixed quantity (in weight). Incentive for mechanics/staff of workshop should also be paid on certain basis. Computerised weighbridge of 20 or 30 tonnes capacity should be installed at the disposal ground.

22.6.5 Privatisation

Privatisation of waste management including deployment of equipments and transportation is another solution to cope with the problem. A pocket of the city may be entrusted to a private agency. Cost will be much less than the expenditure incurred by the municipal authority. It may be possible to enter into a contract with private agency for all activities of garbage collection and final disposal. Specific area is to be demarcated for this agency. Approximate quantity of garbage generated in this area is to be indicated in weight (tonnes). A plot of land may be allotted for treatment/re-cycling of garbage on lease basis. The agency will produce manure or electricity. The agency may also pay some royalty to the municipal authority. Such arrangement has been done at Kolkata.

22.6.6 *Sanitary Land Fill*

Degradable garbage is dumped at this site and spread with Bulldozers. Sanitary land fill ground is made properly. Suitable lining (brick pavement or other suitable materials) is made on the sanitary land fill. Proper drainage arrangement is to be made in and around the sanitary land fill ground. Proper approach road and circulation space for vehicles with proper pavement are to be constructed. Garbage is to be spread with Bulldozer in layers. After laying one layer of garbage, half metre earth is to be spread over the garbage layer. Then next layer of garbage is to be spread. Compaction is done with Bulldozer or other equipments (like, Trashmaster). Gas can be obtained from sanitary land fill site by installing gas production equipments. Weighbridge should be installed at this site for weighing garbage.

22.6.7 *Organisational Set-up for Solid Waste Management*

Lot of problems are faced due to faulty organizational set up. Functional problem is faced during daily management/ operation. This consists of multi-disciplines, mainly, Civil Engineering, Mechanical Engineering, Management Techniques. In many municipal corporations, the head of this department is Chief Engineer (Civil) with one Deputy Chief Engineer (Mechanical); however, in certain municipal corporation, the head of this department is Chief Engineer (Mechanical) with Deputy Chief Engineer (Civil). But it is a rare case where head of this department is Medical Officer. Various supporting staff are to be placed under Superintending Engineer (Civil)/Executive Engineer (Civil) and SE/EE (Mech.). The role of workshop is very important. In city there may be two workshops (zone-wise) exclusively for Solid Waste Department. EE(Mech.) is in-charge of a workshop and responsible for purchase of new vehicles, machinery, equipments, spare parts, tools for workshop, etc. Management techniques and efficient operation are very important. Chief Engineer is responsible for effective management, scientific monitoring and coordination etc. Simple system of reporting on progress at different levels (on daily / weekly / monthly basis) is to be maintained. Operation and Maintenance Manual must be prepared properly alongwith inventory records. Detailed checklists of O & M are part of the Manual. Operation and maintenance of solid wastes are most important. Close co-ordination, monitoring are to be carried out effectively at different levels daily / weekly / monthly.

22.7 Medical Waste and other Hazardous Wastes Management

(a) Hazardous wastes can cause death , illness or injury to people and adverse effect in the environment if improperly treated, stored, transported or discarded. Medical wastes are hazardous wastes. This can produce toxic containing poisonous element. If such medical wastes are mixed with ordinary wastes then the same should be treated under category of medical waste.

(b) Besides medical wastes there are other hazardous wastes like, – industrial wastes, agricultural wastes, household wastes containing medical wastes, toxic batteries, pesticides, drugs, mercury, asbestos particles etc. Such wastes are to be handled and disposed carefully as per certain special method.

(c) Hospital is required to take special care for disposing wastes contaminated with blood and tissue. Hazardous wastes should be separated from ordinary waste. People come to hospital may be infected due to hazardous wastes; this may cause serious illness. Medical waste has been a growing concern because of incidents of public exposure to discarded blood vials, needles, empty medicine bottles and syringes. This is mainly due to improper hospital waste management. Some unscrupulous operators may resell the syringes etc. Medical waste includes wastes generated by health care organizations such as hospitals, physicians' clinics, dental clinic, veterinary hospital, medical laboratories, medical research units etc.

Quantum of general refuse by weight from hospital may be 75 to 85% and remaining is infectious / toxic wastes. Medical waste is bio- hazardous waste, which are contaminated with infectious elements. Ministry of Environment, Govt. of India have notified the management and handling rules 1998 related to medical wastes. Management is to be done as per aforesaid rules. Such wastes are not to be dumped in uncontrolled / ordinary / general landfill site. It is essential to segregate hazardous medical waste from normal waste in the premises of hospital.

Bio-medical wastes are leading to alarming increase in the spread of infections, disease like, cholera, HIV, Hepatitis "B", Skin disease, body allergies. Special precautions are to be taken by the persons who handle such waste.

(d) Incineration method has been adopted in many hospitals for disposing medical wastes. But it is required to take mitigation measure so that air is not contaminated. Residual ash, inert materials of incinerator should be disposed in separate landfill site. The incineration is effective only when waste is completely combustible. There are other methods of treatment of certain medical wastes such as : –

Sterilization, microwaving, land filling (in sanitary land fill). Biological methods are more eco-friendly. Effort should be made to reduce the quantum of wastes in the hospitals.

22.8 Construction and Demolition Waste

(a) Construction and demolition waste is generated when any construction / demolition / maintenance / repair works of building, roads, bridges etc. are carried out. It consists mostly of inert and non-biodegradable materials (such as concrete, debris, metal, wood, plastics etc.) This waste is found on the road (as heaps and at garbage bins). Quantity can be reduced at source by recovering certain materials (like, steel, iron, plastic, wood, fittings) and to sell the same to kabariwala / agent for recycling purpose. Building debris / rubbish (containing concrete, bricks ballast etc.) can be used for road construction at lower course and marshy land. This can be used for land filling. Concrete and masonry waste can be recycled by sorting, crushing and sieving into recycled aggregate. It has been found that 70% of the construction industry is not aware of recycling techniques in India.

(b) Above stated solid waste creates lot of problem for municipality as the same are dumped on the footpath / berm of road in the city, towns. Drains are also chocked due to this waste. Such construction / demolition / repairing materials should be stored within the construction site / building premises. Proper screen should be provided so that the waste does not get scattered and does not become eyesore. Above type of wastes are to removed, transported and disposed in the land fill / sanitary land fill site. After spreading garbage in the land fill with Bulldozer for a layer of about one metre, the debris / building rubbish mixed with earth is spread with a Bulldozer. If the construction debris is stored in skip type container then skip lifters (figure 22.2) fitted with hydrolic hoist system should be used for prompt removal. Effort should be made to reuse of brick, stone slabs, timber, pipes, steel railings etc. to the extent possible. Some materials can be disposed through sale / auction.

(c) Municipal authority should laid down procedures for obtaining permission related to construction waste materials. Low lying areas should be located and marked on the map and direction can be given to the owner for disposing such waste. It is important to set up separate section under civil engineering department for operation / control of such wastes. Solid waste management department should be made responsible for entire matter of construction / demolition / repairing wastes. Private party can also be employed for collection and transportation of this waste in addition of department' labours, vehicles. As per building bye-laws the building owner is to submit Affidavit and deposit money at the time of approval of building plans for above stated wastes.

22.9 Industrial Solid Waste

Industries may be small scale industry or large industry. Type of production items is relevant about generation of wastes. In certain industry the wastes may contain harmful chemicals etc. The small scale and certain plants may be located in the city / town; the same may be in conforming or non-conforming areas of city / town. As per rules local bodies are not responsible for collection, transportation and disposal of industrial wastes. But it has been observed that some wastes of small scale industries are thrown in the municipal's dust bins / containers alongwith municipal garbage. This creates some problems. It is required to check whether such industrial wastes are hazardous or not. If hazardous then suitable treatment is to be done and disposed properly.

Environmental pollution is the major problem associated with rapid industrialization, urbanization and rise in living standards of people. "Thus wastes may be considered as by-product of industrialization". Efforts should be made for conversion of unwanted wastes into utilizable raw materials for industries. State Pollution Control board is required to check the matter of industrial wastes and proper mechanism is to be evolved. Industrial estates located in city limit do not have adequate facilities so that industries can organize their collection, treatment and disposal. There should be regular interaction between Pollution Control Board and local body for controlling treatment and disposal of industrial wastes. If thermal power plants exist within city limit then problem is faced due to huge quantity of "coal ash" / fly ash.

Waste management should be carried out by adopting two approaches i.e.

(i) Prevention

(ii) Control of environment pollution.

Prevention aims at minimization of industrial wastes at source; while the latter stresses on treatment and proper disposal of wastes.

22.10 Street Cleaning

(a) Solid wastes are generated from street sweeping. This is a part of municipal solid wastes. Separate sweeping/ scavenging branch exists in the municipality. Sanitation of city, town also depends on proper cleaning / sweeping of streets. In India this is done manually because adequate labours are available at less cost. But dust pollutes the air during sweeping. In developed countries machinery (mechanical sweeper) are used for street cleaning with water.

(b) Effort should be made for following : –

  (i) Reduction of street litter by public education and awareness.

  (ii) To design and use of effective tools and equipments.

  (iii) To design strong litter bins and to install on the footpath/ road berm, parks with sufficient number.

(c) Daily sweeping of public streets is essential. This should be done before 8 a.m. Generation of waste is a continuous process. Hence it is required to deploy sweepers during Sundays and other holidays also. It is essential to prevent burning of tree leaves and other wastes by sweepers on the road side. If open space is available then the leaves could be composted and used as organic manure for road side plantations. Use of appropriate tool play an important role in improving efficiency of work force. Using mechanical sweepers is ideal. Most mechanical sweepers are suction machines usually assisted by one or more revolving "Scarifying" brushes for dislodging adhering matter. Such type of sweeping machines are being introduced in a few cities in India. Special care is to be taken to clean the kerb side channel along the road kerb.

23

Suggested Reading Material

GOVT. / CPWD PUBLICATIONS

No.	Title	Price
1.	**Schedule of Rates (Electrical) Part-I Internal** 2007	150
2.	**Schedule of Rates (Electrical) Part- II External** 2007	150
3.	**General Specifications for Electrical Works Part-I Internal** 2005	350
4.	**General Specifications for Electrical Works (Part-III-Lift & Escalators)** 2003 (Bilingual)	275
5.	**General Specifications for Electrical Works Part IV Sub-Stations** 2007	200
6.	**General Specification for Electrical Works Part V (Wet Riser & Sprinkler Systems)** 2006	200
7.	**General Specification for Electrical Works Part VII (DG Sets)** 2006	150
8.	**General Specification for Heating, Ventilation & Air Conditioning (HVAC) Works** 2004	500
9.	**Commentary on CPWD Specifications** (Vol 1, 2 and 3) by Nabhi, 2008	1900
10.	**Delhi Schedule of Rates, 2007** with supplement	508
11.	**Nabhi's Compilation of CPWD Maintenance Mannual** 2008	250
12.	**CPWD Works Mannual** 2007	452
13.	**General Conditions of Contract C.P.W.D Works** 2008 (with supplement)	45
14.	**Manual of Water Supply and Treatment** Third -Revised and updated by CPHEEO, May 1999	250
15.	**Manual on Sewerage and Sewage Treatment** (Second Edition) by CPHEEO, 1993	210
16.	**Manual of Municipal Solid Waste,** May 2000	511
17.	**Rain Water Harvesting And Conservation-A Manual, 2002**	47
18.	**CPWD Telephone Directory, 2008**	75
19.	**Mesurement Book** (Big Size) Rs. 100 (Small Size)	50
20.	**Hindrance Register**	50
21.	**Materials-at-Site (M.A.S) Register**	50
22.	**Cement Register**	50
23.	**Site Order Book**	50
24.	**Inspection Register**	50
25.	Nabhi's **Practical Handbook on Energy Conservation in Buildings 2008** by Indian Building Congress	700
26.	**Energy Conservation Building Code** 2007 by Bureau of Energy Efficency	200
27.	**JMJs Compendium of Reference Circulars to CPWD Works Manual CON-Circulars,** Oct 2007	600
28.	**JMJs Compendium of Reference Circulars to CPWD Works Manual MAN-Circulars,** Oct 2007	300
29.	JMJs **Referencer** for **Civil Engineers, Architects, Valuers, Builders** (Based on CPWD Circlars), 2008	590

PUBLICATIONS AVAILABLE ON CPWD WEBSITE

1. **CPWD-Plinth Area Rates**
   http://cpwd.nic.in/newsitem/latestnewspdf/plintharearates.pdf
2. **CPWD-DELHI SCHEDULE OF RATES (DSR)-2007**
   http://cpwd.nic.in/final-DSR2007.pdf
3. **CPWD WORKS MANUAL-2007**
   http://cpwd.nic.in/nesitem/latestnewspdf/final-worksmanual.pdf

4. **CPWD-ANALYSIS OF RATES FOR DELHI-2007**
http://cpwd.nic.in/Deputation/AR-1-Final.pdf
http://cpwd.nic.in/Deputation/AR2-Final.pdf

BUREAU OF INDIAN STANDARDS

1. **National Building Code of India 2005 (containing All the groups 1 to 5)** 8700
2. **Handbook On Building Construction practices (Electrical Work)** Jan, 2001 3820
3. **Summaries of Indian Standards For Building Materials.** March, 1985 3820
4. **Handbook On Causes and Prevention of Cracks in Buildings,** March 1999 600
5. **Handbook On Concrete Mixes Based on Indian Standards,** Nov. 2001 1130
6. **Handbook For Structural Engineers 1, Structural Steel Sections** (Second Revision) Jan. 2002 1660
7. **Engineering Drawing Practice for Schools & Colleges SP 46:** Aug. 2004 1660
8. **Handbook On Masonry Design and Construction (First Revision)** SP20: 1991 Dec. 2003 1210
9. **Handbook On Structures with Reinforced Concrete portal Frames (Without Cranes)SP 43: 1987** 2620
10. **Handbook On Timber Engineering** Oct, 1988 1130
11. **Compendium of Indian Standards On Soils Engineering Part 1&2** 4060
12. **Hand-Operated Hand Tools Safety code for the Use, Care and Protection** June, 1995 980
13. **Location, Operation and Maintenance of Tube/Bore Wells-Guidelines** 980
14. **IS-456-Plain and Reinforced Concrete Code of Practice** 950
15. **IS-800-General Construction in Steel Code of Practice** 1130
16. **IS-1200-Methods of Measurement of Building and Civil Engineering Works** (set of 28Vol) 2820

MINISTRY OF SURFACE TRANSPORT : INDIAN ROAD CONGRESS

1. **Standard Data Book For Analysis of Rates For Rural Roads** (With Supplement) 800
2. **Standard Data Book For Analysis of Rates** (First Revision) 1200
3. **Guidelines On Quality Systems For Roads.** 400
4. **Hill Road Manual** 2006 600
5. **Standard Plans For Single, Double And Triple Cell Box Culverts With And Without Earth Cushion 2005 Indian** 500
6. **Standard Plans For 3.0 m To 10.0 m Span Reinforced Cement Concrete Solid Slab Superstructure with & Without Footpaths For Highways** 2005 400
7. **Bridge Inspectors Reference Manual, 1999** 140
8. **Guidelines For The Design of Small bridges And Culverts (IRC:SP.13-2004)** 400
9. **Guidelines On Quality Systems For Road Bridges, 2003 (IRC:SP.47-1998)** 280
10. **Pocket Book For Highway Engineer** (Second Revision) 2006 320
11. **Highway Safety Code** 2003 160
12. **Specifications for Rural Road** (Reprint 2006) 500
13. **Pocket Book For Bridge Engineers** 2006 200
14. **Standard Specifications and Code of Practice for Road Bridges** Sec-IX (Bearings Part I: Pot-cum-Ptfe, Pin and metallic Guide Bearings) 200

NABHI PUBLICATIONS

1. **Analysis of Rates for Civil Works** Vol. I & II Reprint 2009 1290
2. **Analysis of Rates for Electrical Works,** 2009 790
3. **Handbook of Modern Technology in Civil Engineering,** 2008 by Er. S.C. Basu Roy 290
4. **Practical Problems and Solutions in Civil Engineering Works** by Er. S. C. Basu Roy May, 2007 250
5. **Practical Handbook on Building Maintenance Vol. I (Civil Works including Horticulture)** by M K Gupta, 2007 240
6. **Practical Handbook on Building Maintenance Vol. II (Electrical & Mechanical Works),** by M. K. Gupta 2007 160
7. **Practical Handbook on Building Construction** by Er. M.K. Gupta, 2007 370
8. **Dictionary of Civil Engineering** by M K Gupta 2007 110
9. **Dictionary of Mechanical Engineering** 2007 90

10. **Compilation of Central Public Works Account Code** Oct, 2007 180
11. **Practical Handbook on Construction Management for Architects and Engineers** by R Ramaswamy, 2006 220
12. **Commentary on CPWD Specifications** Vol 1 and 2 & 3, 2008 1900
13. **Compilation of CPWD Maintance Manual,** 2008 250

OTHER CIVIL ENGINEERING PUBLICATIONS

1. **Quality Dimension In Civil Construction** by K.A.N Talpasai July, 2004 (J.M.Jaina) 120
2. **Rehabilitation of Structures Fundamental** 2006 by K.A.N Talpasai (J.M.Jaina) 150
3. **MPD - 2021 Delhi Master Plan, W. E. F. 7th February,** 2007 (J.M.Jaina) 330
4. **JMJ. Building Systems for Low Income Housing** by A.K. Jain 2008 600
5. **Ready Reckoner of R.C.C. Designs and Material Requirements** by D.K Bhargava 200
6. **Practical Handbook on Cracks and Leakages in Buildings** by D.K.Bhargava 200
7. **School Building Planning Design and Management** by A.K.Jain 400
8. **Handbook on Concrete Mixes with free mix design software** by Er. S.C Rathi, IDSE 995
9. **Ready Made Design for Common Items in Civil Works** by M.K.Gupta 350
10. **Delhi Master Plan, 1962** by Akalank's 500
11. **The Scientific Costing Guide Build Cost: Building Construction Cost Data** 2008 With CD 4100
12. **Encyclopedia of Civil Engineering** By R Kapoor 2007 795
13. **Implementing ISO 9001 : 2000** by Pradeep Kumar Mathur 300
14. **Specifications of Buildings And Methods of Measurement** by D K Bhargava, 2007 795
15 **International Building Code, 2008** by International Code Council 1900
16. **International Fire Code, 2008** by International Code Council 1500
17. **International Residental Code, 2008** by International Code Council 1900
18. **The Architects Handbook** - Quentin Pickard Riba 1195
19. **Site Engineers Manual** - David Doran 795
20. **Handbook on Seismic Retrofit of Buildings** 650
21. **Electrical and Mechanical Services in High Rise Building Design and Estimating Manual** A.K.Mittal 495
22. **Architecture - Form space and order** - Francis D.K.Ching 1500
23. **A Visual Dictionary of Architecture** - Francis D.K.Ching 1500
25. **Building Construction Illustrated** - Francis D.K.Ching 1600
26. **Landscape Architecture - A Manual of Enviormental Planning and Design** John Ormsbee Simonds Barry Starke $89.95
27. **A History of Architecture** : Sir Banister Flichers 1595
28. **Energy Efficient Buildings in India** edited by Mili Majumar, by Tata Energy Research 860
29. **Energy Efficient Buildings Systems** Dr. Lal Jayomaha $69.95
30. **Climate Responsive Architecture - A Design Handbook for Energy Efficient Buildings** - Arvind Krishan -Simosyannas Nick Baker - Svszokolay 695

TIME -SAVER STANDARDS

1. **Time-Saver Standards for Architectural Design, Technical Data for Professional Practice** by Donald Watson & Michael J. Crosbie Tat ($157.50)
2. **Time Saver Standards for Building Types** by Joseph De Chiara, Michael J. Crosbie ata McGraw Hill ($46.95)
3. **Time Saver Standards for Urban Design** by Donald Watson, Alam Plattus, Robert Shibley Tata McGraw Hill ($157.50)
4 **Time Saver Standards for Interior Design and Space Planning** by Joseph De Chiara, Julius Panero, and Martin Zelnik, ($75)
5. **Time Saver Standards for Landscape Architecture** ($157.50)
6. **Neufert Architects' Data** 2500
7. **Time Savers Standards Concise - Site Construction Manual** -Nichola1 T.Dines Kyle D.B. Rown $63

ESTIMATING AND COSTING

1. **Estimating, Costing Specification & Valuation in Civil Engineering** 2007 by M Chakraborti, 2008 310

2. **Civil Engineering Estimating & Costing** (including Quantity Survey, Valuation & Contracting) Oct. 2008 by V.N. Vazirani S.P Chandola, 2008 295
3. **Estimating and Costing in Civil Engineering** by B. N. Dutta, 2009 295
4. **Estimating Costing and Valuation** by Rangwala, 2006 250
5. **A Text Book of Estimating and Costing** by G.S.Birdie, 2005 295
6. Nabhi's **Compilation of Analysis of Rates for Civil Works,** Reprint 2009 1290
7. Nabhi's **Compilation of Analysis of Rates – Electrical** 2009 790

CIVIL ENGINEERING HANDBOOK

1. **Indian Practical Civil Engineers Handbook 2008** by P N Khanna 315
2. **Hand Book of Civil Engineering** Reprint 2007 by Gurcharan Singh 295
3. **Concise handbook of Civil Engineering** 2009 by V. N. Vazirani S. P. Chandola 500
4. **Handbook of Civil Engineering** by VN Vazirani, S.P.Chandola, 2009 550
5. **Handbook for Civil Engineering** by R.K.Gupta Ed.2008 295

BUILDING CONSTRUCTION

1. **Practical Handbook on Building Construction** by M. K. Gupta, 2007 370
2. **Structural Design of Multi-Storeyed Buildings 2002** by U H Varyani 295
3. **Handbook of Building Construction Vol. I and II** by M M Goyal 2006 2000
4. **Building Planning Designing and Scheduling** Reprint 2008 by Gurcharan Singh Jagdish Singh 150
5. **The A to Z of Practical Building Construction and its Management 2007-08** by Sandeep Mantri 750
6. **Building Construction** by Rangwala Ed. 2008 160
7. **Building construction** by Sushil Kumar Ed.2008 200

CONSTRUCTION & MAINTENANCE MANAGEMENT

1. **Construction & Foundation Engineering** by Dr. J. Jha, S. K. Sinha, 2008 180
2. **Construction Technology** 2006 (Volume I, II) by J T Grundy 200
3. **Heavy Construction** by V. N. Vazirani S. P. Chandola 40
4. **Construction Project Management** 2008 by Chitkara 525
5. **Construction Planning, Equipment, and Methods** by Peurifoy/Schexnayder, 2007 Tata Mcgraw Hill 325
6. **Construction Management and Machinery** 2008 by B. L. Gupta Amit Gupta 200
7. **Construction Products in India, The issues, The Potential and the way a head** 1250
8. **Handbook of Modern Technology in Civil Engineering, 2008** by Er. S.C. Basu Roy 290
9. **Practical Problems and Solutions in Civil Engineering Works** by Er. S. C. Basu Roy May, 2007 250
10. **Practical Handbook on Construction Management for Architects and Engineers** 2006 by R. Ramaswamy 220
11. **Construction Management Practice** 2005 by Dr. V. K. Raina 150
12. **Maintenance And Repairs of Buildings** 2006 by P. K. Guha 245
13. **Practical Handbook on Building Maintainance** Vol. I & II, by M.K.Gupta 2007 400

CONCRETE

1. **Concrete Technology Theory and Practice** By M. S. Shetty, 2009 390
2. **Concrete Technology** 2006 by A. M. Neville & J. J. Brooks, 2008 325
3. **Properties of Concrete** 2006 by A. M. Neville, 2009 550
4. **Concrete Technology** 2006 by M. L. Gambhir, 2008 275
5. **Concrete Materials Properties, Specification and Testing** 2002 by Sandor Popovics 550
6. **Concrete For Construction facts and Practice** 2005 by V. K. Raina 250
7. **Prestressed Concrete 2007** by N Krishna Raju, 2008 325
8. **Handbook of Concrete Engineering** edited by Mark Fintel 2004 1095
9. **Prestressed Concrete** by N. Rajagopalam ed. 2007 350
10. **Modern Prestressed Concrete** by James R. Libby 200

11. **Design of Concrete Structures** by J.N. Bandyopadhyan Ed. 2008 395
12. **Concrete Micro Structure Properties and Materials** by P.Kumar Mehta Ed. 2006 795
13. **Design of Concrete Structures** by Arthur H. Nilson Ed . 2008 615

ENGINEERING DRAWING

1. **Drawing and Design of Residential & Commercial Buildings** 2008 by S Kaleem A & Zaidi Suhail Siddiqui 250
2. **Engineering Drawing With a Primer on Auto CAD** 2005 by Arshad N.Siddiquee Zahid A. Khan Mukhtar Ahmad, 2006 225
3. **A Text of Engineering Drawing** 2006 (Geometrical Drawing) by R. K. Dhawan, 2009 340
4. **Building Drawing With an Integrated Approach to Built Environment** Fourth Edition by M G Shah C M Kale S Y Pathak 399

REINFORCED CONCRETE

1. **Advanced Reinforced Concrete Design** 2007 by P.C. Varghese 395
2. **Advanced Reinforced Concrete Design** 2005 by N. Krishna Raju, 2007 210
3. **Reinforced Concrete** 2006 by Dr. H. J. Shah Reinforced Concrete Vol.I & II 425
4. **Design of Reinforced Concrete Structures** by S. Ramamrutham, 2009 460
5. **Fundamental of Design of Reinforced Concrete Structures** 2006 by M. L. Gambhir 325
6. **Design of Reinforced concrete Structures** by M.L.Ghambhir Ed.2008 425
7. **Reinforced Concrete Design** by S Unnikrishna Pillai & Devdas Menon, 2007 325
8. **Limit State Theory and Design of Reinforced Concrete** by Dr. V. L. Shah, S. P. Karve, 2005 400
9 **Reinforced Concrete Constructions For 21st Century** by Er. Kollegal. K. Meghashyam, 2005 (JMJ) 990
10. **Reinforced Concrete Handbook for Building Design** by D.Krishnamurty, 2007 795
11. **Limit State Design of Reinforced Concrete** by P. C. Varghese 295
12. **Elements of Limit State Design of Concrete Structures** 2004 by Dr Ram Chandra Virendra Gehlot 240

RCC

1. **Treasure of R.C.C Designs** 2006 by Shushil Kumar 280
2. **RCC Theory and Design** Revised Edition by M G Shah C M Kale, 2007 215
3. **Reinforced Concrete Constructions for Multistoreyed Structures** ER Lollegal, K.Meghashyam, 2007(JMJ) 990
4. **Reinforced Concrete Structures** Vol.I&II by B.C.Punmia, 2005 555

SCIENCE MATERIAL

1. **Materials Science** by G K Narula K S Narula V K Gupta, 2007 215
2. **Materials Science and Engineering an Introduction** by William.D.Callister, JR 2007 499

STRENGTH OF MATERIALS

1. **Strength of Materials In S.I.Units 2007** by Dr. Sadhu Singh 275
2. **Strength of Materials (Mechanics of Solids) S.I. Units** by R.K. Rajput, 2009 625
3. **Mechanics of Materials** 2007 by Gere & Timushenko 310
4. **Handbook of Thermal Analysis of Construction Materials** 2004 by V S Ramachandran Ralph M Paroli James J Beaudoin, 2006 795
5. **Strength of Materials** 2007 by S Ramamrutham, 2008 375
6. **Mechanics of Materials** 2007 **by** Dr. B C Punmia Ashok Kumar Jain Arun Kumar Jain 395

ANALYSIS OF STRUCTURES

1. **Analysis of Structures Strength And Behavior** 2005 by Thandavamoorthy 295
2. **Analysis of Structures Vol I, II,** 2008 by V N Vazirani Dr. M M Ratwani Dr S K Duggal 515
3. **Basic Structural Analysis** 2007 by C S Reddy 285
4. **Theory of Structures** 2004 by B C Punmia Ashok Kumar Jain Arun Kumar Jain 250
5. **Dynamics of Structures** 2008 by Anil K Chopra 395
6. **Theory of Structures** 2008 by S Ramamrutham R Narayan 415

7. **Analysis and Design of Substructures** by Swami Saran 395
8. **Fundamental of Stuctural Analysis** : Kenneth M.Leet Chia Mine Vang anne M.Gilbert Ed. 2008 $33.95
9. **Elementary Structural Analysis** - Charles Head Norres John Benson Wilbur Senol Utru Ed. 2006 375
10. **Theory of Structures** R.S. Khurmi Ed. 2008 190
11. **Structures:** Daniel L. Schodek Fourth Ed. 325
12. **Structural Analysis** : Devdas Menon Ed. 2008 690

STEEL STRUCTURES

1. **Steel Handbook, A Guide to World Equivalent Steel Grades** 4100
2. **Design of Steel Structures** 2007 (Volume I, II) by Ram Chandra 640
3. **Design of Steel Structure** 2008 by S K Duggal 325
4. **Design of Steel Structure** 2008 by S Ramamrutham 310
5. **An Introduction to Modern Steel Making** 2007 by R H Tupkary V R Tupkary 175
6. **Westermann Tables For the Metal Trade** 2007 by Hermann Jutz Eduard Scharkus 125
7. **Steel Tables S I Units** by K L V Ramu 45
8. **Design of Steel Sturctures** : IC Sayal Satinders Singh Ed. 2007 225
9. **Design of Steel Structures** : A.S. Arya J.L.Ajhani Ed. 2007 250
10. **Design and Analysis of Steel Stuctures** : V.N. Vazirani M.M Ratwani Honey Mehra Ed. 2008 265
11. **Design of Steel Structures** : B.C.Punmia A.K. Jain Arun Kumar Jain Ed. 2008 475

SURVEYING

1. **A Text Book of Advanced Surveying** 2007 by R Agor 165
2. **A Text Book of Surveying** 2009 by C L Kochher 150
3. **Surveying And Levelling** Part I & II by Late T P Kanetkar S V Kulkarni May, 2007 380
4. **Surveying And Levelling** : N.N.Basak Ed. 2008 210
5. **Surveying Vol. 1, 2** by S.K. Duggal Ed. 2008 525
6. **Surveying and Levelling** R.Subrmaniam Ed. 2007 395
7. **Plane Surveying** Alak De Ed. 2007 300
8. **Surveying** Vol I&II by B.C. Punmia Ed.2008 535
9. **Higher Surveying** Vol III by B.C.Punmia Ed. 2008 135

INDUSTRIAL ENGINEERING

1. **Industrial Hygiene And Chemical Safety 2006** by M H Fulekar 195
2. **Industrial Engineering And Management** by Dr B Kumar, 2007 215
3. **Industrial Engineering And Production Management** Hartandtelsang Ed. 2008 295
4. **Industrial Engineering And Management Systems** S.Palela Hansoorali Ed. 2006 275
5. **Industrial Engineering And Management** Dr. Ravi Shankar Ed. 2008 190
6. **Industrial Safety Health And Enviorment Management Systems** R.K.Jain Sunil S.Rao Ed. 2008 275
7. **Industrial Safety Management** by L.M.Deshmukh Ed. 2008 575

METAL CUTTING

1. **Metal Cutting Principles 2005** by Milton C Shaw 375
2. **Metal Cutting 2006** by Edward by M Trent Paul K Wright 395
3. **Smithells Metals Reference Book 2000** by E A Brandes G B Brook 4000

HIGHWAY BRIDGE ENGINEERING, TUNNEL & HARBOUR

1. **Highway Engineering 2005** by S K Khanna C E G Justo 200
2. **Highway Material Testing** by S K Khanna C E G Justo Reprint 2006 80
3. **Roads, Railways Bridges Tunnels & Harbour Dock Engineering** by B L Gupta Amit Gupta, 2008 325
4. **Highway Engineering** 2004 by Paul H Wright Karen K Dixon 330
5. **Rainas Concrete Bridge Practice Aalysis Designs and Economics** by V.K.Raina, 2007 1000
6. **Concrete Bridge Practice: Construction Maintaince and Rehabilitation** by V.K.Raina, 2005 525

7. **Concrete Bridge: Inspection, Repair, Strengthing, Testing, Load, Capacity Evaluation** by V.K.Raina, 2005 675
8. **Bridge Engineering** by Ponnuswamy, 2008 750
9. **Essential of Bridge Engineering** by D.Johson. Victor, 2007 295
10. **Design of Bridge** by N.Krishna Raju 225
11. **Raina's Feild Manual For Highway And Bridge Engineers** by V.K.Raina 1000
12. **Tunnel Engineering Handbook** John O.Bickel Thomas R. Kuesel Elway H.King 795
13 **Highway Engineering Handbook** Roger L.Brocken Brough Kenneth J.Boedecker $131
14. **Bridge Superstructure** by N.Rajgopalan 295

FINITE ELEMENTS

1. **Introduction to Finite Element In Engineering** 2007 by Tirupati R Chandrupatla Ashok D Belegundu 295
2. **Finite Element Analysis Theory And Programming** 2006 by C S Krishnamoorthy 340

GEOTECHNICAL ENGINEERING

1. **Geotechnical Engineering 2008** by Shashi K Gulhati Manoj Datta 325
2. **Geotechnical Earthquake Engineering** by Steven.Kramer, 2008 425
3. **Geotechical Engineering Principles and Practice** by Donald.P.Coduto, 2007 350

DYNAMIC ANALYSIS AND EARTHQUAKE RESISTANT DESIGN

1. **Dynamic Analysis And Earthquake Resistant Design** Volume I, II, III, 2000 2725
2. **Masonry and Timber Structures Including Earthquake Resistant Design** by Anand S. Arya, 2006 175
3. **Earthquake Disaster Reduction Masnory Buildings Design and Construction** by Anan D.S.Arya, 2007 840
4. **Earhquake Resistant Design of Structutres** by Pankaj Agarwal 425
5. **Earthquake Resistant Design of Structure** by S.K.Duggal Ed.2007 285

VALUATION

1. **Principles & Practice of Valuation** 1998 by D N Banerjee 450
2. **Valuation of Plant And Machinery Theory and Practice** 2002 By P C Gupta 375
3. **Valuation of Real Property Principles & Practice** 2004 By Syamales Datta 585

INTERIOR DESIGN & ARCHITECTURE

1. **Creating Not so big House** by The Taunton Press $19.95
2. **The Magic of Small Spaces** by The Taunton Press $35.95
3. **The New Family Home** by the Taunton Press $24.95
4. **The Twentieth Century House in Britian From the Archieves of Country Life** by Aurum $65.95
5. **25 Houses under 3000 Square Feet** by Harper Coolins $39.95
6. **Cool Homes in Hot Places** by Harper Collins $39.95
7. **High Rise Living** by Gibbs Smith $39.95
8. **Design Ideas for Home Decorating** by Creative Homeowner $24.95
9. **Gnostic Architect** by The Monacelli Press $45.00
10. **Heikkinen + Komonen (Conversion Charges at the Rate Prevailing at the time of Purchase)** by The Monacelli Press $45.95
11. **New Country Houses** by Abbeville Press $55.95
12. **Plans of Row Houses, Duplex Houses and cottage Type Houses** 2001 by C H Gopinatha Rao 250
13. **Interior Design Principles and Practice** 2005 by M Pratap Rao 150
14. **Shopping Complex** Series C-2004 110
15. **Cinemas** Series C-2004 110
16. **Factories Series** C-2004 110
17. **Ashray** Series C-2005 140
18. **Dream Homes** Series A-2006 by M C Malhotra 250
19. **Interior Design 2005** by Ahmed A Kasu 1995
20. **The World of Contemporary Architecture** by Francisco Asensiocerver 1250

21. **Architecture, Engineering and Environment** by Dean Mawkes and Wayre Forster 995
22. **Architecture in India** by Marlia Albense 995
23. **The Guide to the Architecture of Indian Subcontinent** by Takeo Kamya Architect 995
24. **Houses of the World** by Francisco Asensio Cerver 1250
25. **Interior Design Atlas** by Francisco Asensio Cerver 1250
26. **Principles of Architecture** by G.Muthu Shoba Mohan 195

ENGINEERING MATHEMATICS

1. **Advanced Engineering Mathematics** by Dennis G.Zill 495
2. **Higher Engineering Mathematics** 2007 by Dr B S Grewal 345
3. **Advanced Engineering Mathematics** 2009 by H K Dass 425
4. **Advanced Engineering Mathematics** by Miachel D.Green Berg Ed.2008 525
5. **Advanced Engineering Mathematics** by N.Bali M.Goyal C.Walking Ed. 2007 795
6. **Text Book of Engineering** by N.P.Bali-Manish Goyal Ed. 2008 575
7. **Engennering Mathemetics** by Karaiya Ed. 2007 210

FOUNDATION ENGINEERING

1. **Foundation Analysis and Design** by Joseph.E.Bowles, 1997 2995
2. **Advanced Foundation Engineering** by V.N.S.Murthy, 2007 695
3. **Designs of Foundation System Principal and Practice** by Nainan.P.Kurain, 2008 540
4. **Foundation Engineering** by P.C.Vargheese, 2007 425
5. **Principles of Foundation Engineering** by Braja.M.Das, 2004 425

PLANNING COMMISSION GOVERNMENT OF INDIA

1. **Public Private Partnership in Ports (MCA)** 500
2. **Public Private Partnership in State Highway (MCA)** 500
3. **Public Private Partnership in Operation and Maintenance of Highways (MCA)** 500
4. **Public Private Partnership in National Highways (MCA)** 500
5. **Public Private Partnership pre qualification of Bidders** 500
6. **Two Lanning of Highways through Public Private Partnership** (Manual of specification & standards) 500
7. **Fourlaning of Highways through Public Private Partnership** (Manual of specification & standards) 300

SOFTWARE FOR CIVIL ENGINEERS

1. **Road Estimate Quantity Estimation and Project Control** - Y.Aagboatwala 1700
2. **HVAC Cost** - Yaagboatwala 2000
3. **Roads - Software for the Roads** -Y.A.Agbobatwala 1500
4. **Super Civil CD A Complete Material for Civil Engineers** - Y.Aagboatwalas 1350
5. **Super Rate Analysis - Analysis of Rates for Civil Engineers**
6. **2D Frame Analysis Discovers the Beauty of Structural Analysis** - Y.Aagboatwala 1500
7. **Learn Qty in 12 Easy Steps a Program for Quantity and Cost Estimation** - Y.Aagboatwala 1800
8. **Super Real Valuation Software for Immovable Poperties** - Y.Aagboatwala 2000
9. **Electric Cost** - Y. Aagboatwala 2000
10. **Interior Decoration** - A Complete Course 499
11. **Civil Engineering Technology** 499
12. **PC Achitecture** - A Complete Course 499
13. **Basic Electronics** 499

03